Introduction to Diffraction, Information Processing, and Holography

The MIT Press
Cambridge, Massachusetts, and London, England

**Introduction to Diffraction,
Information Processing, and Holography**

Francis T. S. Yu

This book was designed by the MIT Press Design Department.
It was set in Times New Roman, Series 327, 334, and 569 (Monophoto)
Printed on Mohawk Neotext Offset
by Halliday Lithograph Corp.
and bound in Columbia Milbank Vellum MBV–4321 (green)
by Halliday Lithograph Corp.
in the United States of America.

Library of Congress Cataloging in Publication Data

Yu, Francis T   S   1934 –
  Introduction to diffraction, information processing, and holography.

  Includes bibliographical references.
  1. Diffraction.   2. Optical data processing.   3. Holography.   I. Title.
QC415.Y8   535'.4   72–10113
ISBN 0-262-24015-7

In memory of my father

Contents

II   Information Processing

Preface

A new technique of image formation by means of wavefront reconstruction was initiated by D. Gabor in 1948, and was revived by E. N. Leith and J. Upatnieks in 1963. Since then it has attracted a great deal of interest. In recent years, there has occurred a great interaction between electrical engineering and coherent optics. This trend continues, not only because modern optical systems are capable of performing certain Fourier transform operations, signal filtering, and character recognition, but also because the underlying optical theory is very similar to the general system analog. Because of the urgent demand for engineering applications of coherent optics, about five years ago I developed a sequence of two courses in coherent optics in the electrical engineering department at Wayne State University.

This book is mostly based on the lecture notes that I wrote for my classes. The manuscript was originally written with electrical engineering students in mind; however, it may also serve as well interested physicists or members of research staffs. The book contains three main parts—diffraction, information processing, and holography. The first part is intended to serve as a foundation, particularly for readers who are not familiar with the basic concepts of diffraction. This part covers essentially the same materials as are found in some of the well-known texts cited at the ends of the chapters. The second and third parts of the book are the main body of this text. Much of the material therein was derived from recently published articles.

In writing this book, I have for the most part approached the analysis by means of an elementary point concept (impulse response) and linear system theory; this is particularly true when I discuss holographic processes. There are two major reasons for using this approach:
1. It simplifies the analysis, so that the solution of the problem may be directly calculated.
2. Electrical engineers are more familiar with the basic concepts of the impulse response and linear system theory.

The material of the book, together with a few additional seminars, was used in a series of two-quarter courses in coherent optics. The students were mostly in their first year of graduate studies. I have found that it is possible to teach the whole book without any significant omissions in a two-quarter period, with the addition of a few seminars. All the materials in the text may also be covered in a full semester course. It may be emphasized that the book is not intended to cover the vast area of coherent optics, but rather it is restricted to a region that I consider to be of practical importance.

In view of the great number of contributors in this area, I apologize for possible omissions of appropriate references in various parts of the book. In this connection, the excellent book *Principles of Optics* by Born and Wolf deserves a special mention.

I am indebted to Dr. H. K. Dunn, retired member of the technical staff of Bell Telephone Laboratories, for his encouragement, criticism, and enormous assistance during the preparation of my manuscript. I am grateful for the excellent research and publications of the members of the Radar and Optics Laboratory at the University of Michigan, particularly to E. N. Leith, J. Upatnieks, and A. B. Vander Lugt. Special mention must be given to Professor L. J. Cutrona, who provided encouragement at the time I needed it most.

I would also like to express my appreciation to the following: J. P. Morrison, who spent much of his valuable time making the photographs which are vitally important to the book; Mrs. Sylvia Wasserman, for her excellent typing of the manuscript; Mrs. Y. K. Ma, for her encouragement and proofreading of most parts of the book; and my students, for their enthusiasm and motivation. Finally, I wish to thank my wife, Lucy, without whose patience and encouragement this book could not have been brought to completion.

F. T. S. Yu
Detroit, Michigan
November, 1971

**Introduction to Diffraction,
Information Processing, and Holography**

Chapter 1
Linear System Theory and Fourier Transformation

It is well known that optical and quadrupolar systems exhibit similarities. For instance, any optical lens can be conveniently represented by a quadrupolar model. Thus the concepts of linear system theory are rather important in the analysis of modern coherent optical systems, for at least two reasons: (1) a great number of applications in modern optics may be assumed linear, at least within some specified ranges; and (2) an exact solution in the analysis of linear optical problems may be obtained by means of standard techniques.

Except for a very few special cases, there is no general procedure for analyzing nonlinear optical systems. Of course, there are practical ways of solving nonlinear optical problems which may involve some graphical or experimental approaches. Approximations are often necessary in solving nonlinear problems, and each situation may require special techniques. Fortunately, a great number of optical problems are linear, and these are generally solvable. However, we would like to emphasize that no optical system is strictly linear; linearization always involves the imposition of restrictions.

Since linearity leads to a simplification in mathematical analysis, it is our aim in this chapter to review some of the mathematical tools of linear analysis. However, it may be admitted that the development of the mathematics in this chapter is by no means rigorous, but it is rather along introductory lines. It may also be emphasized that optical information processing is generally two-dimensional in nature. Therefore in the development of the mathematical tools, we shall restrict our presentation to the two-dimensional case.

1.1 Linear Systems from a Physical Viewpoint

It is a common practice to describe the behavior of a physical system by relating the input excitation to the output response (fig. 1.1). Both excitation and response may be physically measurable quantities. Suppose an input excitation $f_1(x, y)$ produces an output response $g_1(x, y)$, and a

Fig. 1.1 A block diagram of a linear system.

second excitation $f_2(x, y)$ produces a second response $g_2(x, y)$. We may write

$$f_1(x, y) \rightarrow g_1(x, y) \tag{1.1}$$

and

$$f_2(x, y) \rightarrow g_2(x, y), \tag{1.2}$$

respectively. Then for a linear system we have

$$f_1(x, y) + f_2(x, y) \rightarrow g_1(x, y) + g_2(x, y). \tag{1.3}$$

Equation (1.3), in conjuction with eqs. (1.1) and (1.2), represents the *additivity* property of the linear system. Thus, a necessary condition for a system to be linear is that the *principle of superposition* hold. The principle of superposition implies that the presence of one excitation does not affect the response due to the other excitations.

Now, if the input excitation of a physical system is $Cf_1(x, y)$, where C is an arbitrary constant, then the output response is $Cg_1(x, y)$, that is

$$Cf_1(x, y) \rightarrow Cg_1(x, y). \tag{1.4}$$

Equation (1.4) represents the *homogeneity* characteristic of the linear system. Thus a property of a linear system is the preserving of the magnitude scale factor. Therefore, a physical system is linear if and only if eqs. (1.3) and (1.4) are satisfied. In other words, if a system possesses the additivity and homogeneity properties, then it is a linear system.

There is, however, another important physical aspect that characterizes a linear system with constant parameters. That is, if the input excitation $f(x, y)$ applied to such a system is an alternating function of x and y with spatial frequencies p and q, and results in an alternating output response $g(x, y)$ with the same spatial frequencies p and q, then the system is said to have a *spatial invariance*. In other words, a spatially invariant system will not generate new spatial frequencies. The qualification of spatial invariance implies that if

$$f(x, y) \rightarrow g(x, y),$$

then

$$f(x - x_1, y - y_1) \rightarrow g(x - x_1, y - y_1), \tag{1.5}$$

where x_1 and y_1 are arbitrary spatial constants. A linear system that possesses the spatial invariance property of eq. (1.5) is called a *linear spatially invariant system*.

The physical characteristics of linear systems will be much clearer from the viewpoint of Fourier analysis, the development of which will be given in the following few sections.

1.2 Fourier Transformation and the Spatial Frequency Spectrum

Fourier transforms are particularly important in the analysis of modern optical information processing and in wavefront reconstruction. In this section we will consider first a class of complex functions $f(x, y)$ which satisfy the following sufficient conditions:

1. $f(x, y)$ must be sectionally continuous in every finite region over the (x, y) plane, i.e., there must be only a finite number of discontinuities; and
2. $f(x, y)$ must be absolutely integrable over the (x, y) plane, i.e.,

$$\int_{-\infty}^{\infty}\int |f(x, y)| \, dx \, dy < \infty. \tag{1.6}$$

Then these functions can be represented by the equation

$$f(x, y) = \frac{1}{4\pi^2} \int_{-\infty}^{\infty}\int F(p, q) \exp[i(px+qy)] \, dp \, dq, \tag{1.7}$$

where

$$F(p, q) = \int_{-\infty}^{\infty}\int f(x, y) \exp[-i(px+qy)] \, dx \, dy, \tag{1.8}$$

and p and q are the corresponding spatial frequency variables.

Equations (1.7) and (1.8) are known as the two-dimensional *Fourier transform pair*. Equation (1.8) is often called the *Fourier transform*, and eq. (1.7) is known as the *inverse Fourier transform*. For brevity, eqs. (1.7) and (1.8) can be written as

$$f(x, y) = \mathscr{F}^{-1}[F(p, q)] \tag{1.9}$$

and

$$F(p, q) = \mathscr{F}[f(x, y)], \tag{1.10}$$

where $\mathscr{F}^{-1}$ and $\mathscr{F}$ denote the inverse and direct Fourier transformations, respectively.

It may be noted that $F(p, q)$ is, in general, a complex function:

$$F(p, q) = |F(p, q)| \exp[i\phi(p, q)], \tag{1.11}$$

where $|F(p, q)|$ and $\phi(p, q)$ are frequently referred to as the *amplitude* and *phase* spectra, and $F(p, q)$ is known as the *Fourier spectrum* or *spatial frequency spectrum*.

A complex function of two independent variables is called a *separable* function if and only if it can be written as a product of two functions, each of which is a single-variable function:

$$f(x, y) = f(x) f(y).\tag{1.12}$$

From eq. (1.12) we may conclude that the Fourier transform of a two-dimensional separable function is the product of the Fourier transform of each of the one-dimensional functions, i.e.,

$$F(p, q) = F(p) F(q).\tag{1.13}$$

By way of illustration, we will next consider the Fourier analyses of two elementary functions that are of considerable importance in optical information processes.

FOURIER TRANSFORMATION OF A CIRCULAR SYMMETRIC FUNCTION
Given the circular symmetric function

$$f(x, y) = f(r) = \begin{cases} 1, & r \leq 1 \\ 0, & \text{otherwise} \end{cases},\tag{1.14}$$

where $r^2 = x^2 + y^2$. To determine the Fourier transformation of eq. (1.13), we first apply the coordinate transformations:

$$\begin{array}{ll} x = r \cos\theta, & r^2 = x^2 + y^2, \\ y = r \sin\theta, & \theta = \tan^{-1}(y/x), \\ p = \rho \cos\phi, & \rho^2 = p^2 + q^2, \\ q = \rho \sin\phi, & \phi = \tan^{-1}(q/p), \\ dx\, dy = r\, dr\, d\theta. \end{array}\tag{1.15}$$

By proper substitution of eqs. (1.14) and (1.15) in (1.8), we have

$$F(\rho) = F(p, q) = \int_0^1 r\, dr \int_0^{2\pi} \exp[-ir\rho \cos(\theta - \phi)]\, d\theta.\tag{1.16}$$

By the use of the identity

$$J_0(z) = \frac{1}{2\pi} \int_0^{2\pi} \exp[-iz \cos(\theta - \phi)]\, d\theta,\tag{1.17}$$

where J_0 is the zero order Bessel function of the first kind, eq. (1.16) can be reduced to

$$F(\rho)=2\pi \int_0^1 J_0(r\rho)\, r\, dr. \tag{1.18}$$

By applying the well-known integral,

$$zJ_1(z)=\int_0^z \alpha J_0(\alpha)\, d\alpha, \tag{1.19}$$

eq. (1.18) becomes

$$F(\rho)=\frac{2\pi}{\rho}\, J_1(\rho), \tag{1.20}$$

where J_1 is the first order Bessel function of the first kind. Equation (1.20) is the required Fourier transform.

By the way, it can be shown that the Fourier transform of any circular symmetric function is

$$F(\rho)=2\pi \int_0^\infty f(r)\, J_0(\rho r)\, r\, dr, \tag{1.21}$$

and its inverse Fourier transform is

$$f(r)=2\pi \int_0^\infty F(\rho)\, J_0(\rho r)\, \rho\, d\rho. \tag{1.22}$$

Equation (1.21) is frequently referred to as the *Hankel transform* (of order zero) of $f(r)$. It may be emphasized that the Hankel transform is no more than a special case of the two-dimensional Fourier transform. Thus any property of the Fourier transform can equivalently be applied to the Hankel transform.

SPECTRUM OF THE TWO-DIMENSIONAL DIRAC DELTA FUNCTION

The two-dimensional Dirac delta function, $\delta(x-x_0,\, y-y_0)$, is defined on the real spatial domain of the xy plane, as existing at (x_0, y_0), but having zero value elsewhere on the xy spatial plane. Also by definition, it has the property

$$\int\limits_{-\infty}^{\infty}\!\!\int \delta(x-x_0, y-y_0) \, dx \, dy = 1. \tag{1.23}$$

Thus the delta function is a pulse of zero spatial duration but of infinite amplitude. The Fourier transform of the delta function is

$$F(p, q) = \mathscr{F}\left[\delta(x-x_0, y-y_0)\right] = \exp\left[-i(px_0+qy_0)\right]. \tag{1.24}$$

Therefore, the corresponding amplitude and phase components may be given respectively as

$$|F(p, q)| = 1 \tag{1.25}$$

and

$$\phi(p, q) = -(px_0+qy_0). \tag{1.26}$$

Equation (1.25) shows that the amplitude spectrum of the delta function is a continuous spatial frequency function of unit height, which extends over the whole spatial frequency domain.

If the delta function occurs at the origin $(0, 0)$, of the spatial domain, then the phase spectrum of eq. (1.26) vanishes, and eq. (1.24) becomes

$$F(p, q) = 1. \tag{1.27}$$

It should also be noted that, for any other value of (x_0, y_0), the unit spectral vector rotates continuously with the phase angle of $-(px_0+qy_0)$.

1.3 Properties of the Fourier Transform

We will now consider some of the basic properties of the Fourier transform that we will find useful in optical information processing and wavefront reconstruction. These basic properties will be presented here as mathematical theorems. Since the proofs are similar to those for the one-dimensional case, the proofs will simply be sketched in, with no pretense at rigor.

1. Linearity

Given that C_1 and C_2 are two arbitrary complex constants. If $f_1(x, y)$ and $f_2(x, y)$ are Fourier transformable, i.e., if

$$\mathscr{F}\left[f_1(x, y)\right] = F_1(p, q),$$

and

$$\mathscr{F}\left[f_2(x, y)\right] = F_2(p, q),$$

then

$$\mathscr{F}[C_1 f_1(x, y) + C_2 f_2(x, y)] = C_1 F_1(p, q) + C_2 F_2(p, q). \tag{1.28}$$

Proof: The proof of the property can be directly derived from the definition of the Fourier transform, eq. (1.8).

2. Translation Theorem

If $f(x, y)$ is Fourier transformable, so that $\mathscr{F}[f(x, y)] = F(p, q)$, then

$$\mathscr{F}[f(x - x_0, y - y_0)] = F(p, q) \exp[-i(px_0 + qy_0)], \tag{1.29}$$

where x_0 and y_0 are arbitrary real constants.

Proof: The proof of this theorem follows from the definition

$$\mathscr{F}[f(x - x_0, y - y_0)] = \int\!\!\int_{-\infty}^{\infty} f(x - x_0, y - y_0) \exp[-i(px + qy)] \, dx \, dy.$$

If we let the variables $x' = x - x_0$ and $y' = y - y_0$, then from the above equation it can readily be shown that

$$\mathscr{F}[f(x - x_0, y - y_0)] = \exp[-i(px_0 + qy_0)] \int\!\!\int_{-\infty}^{\infty} f(x', y')$$

$$\times \exp[i(px' + qy')] \, dx' \, dy' = \exp[-i(px_0 + qy_0)] F(p, q).$$

Essentially this theorem tells us that the translation of a function in the spatial domain causes a linear phase shift in the spatial frequency domain.

3. Reciprocal Translation Theorem

If $f(x, y)$ is Fourier transformable, so that $\mathscr{F}[f(x, y)] = F(p, q)$, then

$$\mathscr{F}\{f(x, y) \exp[-i(xp_0 + yq_0)]\} = F(p + p_0, q + q_0), \tag{1.30}$$

where p_0 and q_0 are arbitrary real constants.

Proof: The proof of this theorem can be obtained immediately by application of the Fourier transformation

$$\mathscr{F}\{f(x, y) \exp[-i(xp_0 + yq_0)]\}$$

$$= \int\!\!\int_{-\infty}^{\infty} f(x, y) \exp\{-i[(p + p_0) x + (q + q_0) y]\} \, dx \, dy = F(p + p_0, q + q_0).$$

4. Scale Change in a Fourier Transform

If $f(x, y)$ is Fourier transformable, so that $\mathscr{F}[f(x, y)] = F(p, q)$, then

$$\mathscr{F}[f(ax, by)] = \frac{1}{|ab|} F\left(\frac{p}{a}, \frac{q}{b}\right), \tag{1.31}$$

where a and b are arbitrary complex constants.

Proof:

$$\mathscr{F}[f(ax, by)] = \int\limits_{-\infty}^{\infty}\int f(ax, by) \exp[-i(px + qy)]\, dx\, dy$$

$$= \frac{1}{ab} \int\limits_{-\infty}^{\infty}\int f(ax, by) \exp\left[-i\left(\frac{p}{a}ax + \frac{q}{b}by\right)\right] d(ax)\, d(by)$$

$$= \frac{1}{ab} F\left(\frac{p}{a}, \frac{q}{b}\right).$$

This theorem shows that a magnification of the spatial domain results in demagnification in the spatial frequency domain, and an overall reduction in the amplitude spectrum.

5. Parseval's Theorem

If $f(x, y)$ is Fourier transformable, so that $\mathscr{F}[f(x, y)] = F(p, q)$, then

$$\int\limits_{-\infty}^{\infty}\int |f(x, y)|^2\, dx\, dy = \frac{1}{4\pi^2} \int\limits_{-\infty}^{\infty}\int |F(p, q)|^2\, dp\, dq. \tag{1.32}$$

Proof:

$$\int\limits_{-\infty}^{\infty}\int |f(x, y)|^2\, dx\, dy = \int\limits_{-\infty}^{\infty}\int f(x, y)\, f^*(x, y)\, dx\, dy,$$

where * denotes the complex conjugate. This equation can be written as

$$\int\limits_{-\infty}^{\infty}\int |f(x, y)|^2\, dx\, dy = \int\limits_{-\infty}^{\infty}\int dx\, dy \left\{\frac{1}{4\pi^2} \int\limits_{-\infty}^{\infty}\int F(p', q') \exp[i(xp' + yq')]\, dp'\, dq'\right\}$$

$$\times \left\{\frac{1}{4\pi^2} \int\limits_{-\infty}^{\infty}\int F^*(p'', q'') \exp[-i(xp'' + yq'')]\, dp''\, dq''\right\},$$

which can be reduced to

$$\int\int_{-\infty}^{\infty} |f(x, y)|^2 \, dx \, dy = \frac{1}{4\pi^2} \int\int_{-\infty}^{\infty} F(p', q') \, dp' \, dq' \int\int_{-\infty}^{\infty} F^*(p'', q'') \, dp'' \, dq''$$

$$\times \frac{1}{4\pi^2} \int\int_{-\infty}^{\infty} \exp\{i[x(p'-p'')+y(q'-q'')]\} \, dx \, dy$$

$$= \frac{1}{4\pi^2} \int\int_{-\infty}^{\infty} F(p', q') \, dp' \, dq' \int\int_{-\infty}^{\infty} F^*(p'', q'') \, dp'' \, dq'' \, \delta(p'-p'', q'-q'')$$

$$= \frac{1}{4\pi^2} \int\int_{-\infty}^{\infty} |F(p, q)|^2 \, dp \, dq.$$

It should be clear to the reader that Parseval's theorem implies the conservation of energy.

6. Convolution Theorem

If $f_1(x, y)$ and $f_2(x, y)$ are Fourier transformable, so that

$$\mathscr{F}[f_1(x, y)] = F_1(p, q)$$

and

$$\mathscr{F}[f_2(x, y)] = F_2(p, q),$$

then

$$\mathscr{F}\left[\int\int_{-\infty}^{\infty} f_1(x, y) \, f_2(\alpha-x, \beta-y) \, dx \, dy\right] = F_1(p, q) \, F_2(p, q). \tag{1.33}$$

(We may write this as

$$\mathscr{F}[f_1(x, y) * f_2(x, y)] = F_1(p, q) \, F_2(p, q), \tag{1.34}$$

where $*$ denotes the convolution operation).

   Proof:

$$\mathscr{F}\left[\int\int_{-\infty}^{\infty} f_1(x, y) \, f_2(\alpha-x, \beta-y) \, dx \, dy\right]$$

$$= \int\limits_{-\infty}^{\infty}\!\!\int f_1(x, y) \, \mathscr{F}\left[f_2(\alpha - x, \beta - y)\right] dx \, dy$$

$$= \left\{ \int\limits_{-\infty}^{\infty}\!\!\int f_1(x, y) \exp\left[-i(px + qy)\right] dx \, dy \right\} F_2(p, q) = F_1(p, q) \, F_2(p, q).$$

This convolution theorem will be found very useful in our study of linear spatially invariant systems.

7. Autocorrelation Theorem (Wiener-Khinchin Theorem)
If $f(x, y)$ is Fourier transformable, so that $\mathscr{F}\left[f(x, y)\right] = F(p, q)$, then

$$\mathscr{F}\left\{ \int\limits_{-\infty}^{\infty}\!\!\int f(x, y) \, f^*(x - \alpha, y - \beta) \, dx \, dy \right\} = |F(p, q)|^2. \tag{1.35}$$

We may write this as

$$\mathscr{F}\left[R(\alpha, \beta)\right] = \mathscr{F}\left[f(x, y) \circledast f^*(x, y)\right] = |F(p, q)|^2 \tag{1.36}$$

where

$$R(\alpha, \beta) = \int\limits_{-\infty}^{\infty}\!\!\int f(x, y) \, f^*(x - \alpha, y - \beta) \, dx \, dy$$

is known as the *autocorrelation function*, and $\circledast$ denotes the correlation operation.)

Conversely, we have

$$\mathscr{F}^{-1}\left[|F(p, q)|^2\right] = R(x, y). \tag{1.37}$$

Equations (1.36) and (1.37) constitute a Fourier transform pair. In other words, the theorem states that the autocorrelation function and power spectral density are Fourier transforms of each other.

Proof:

$$\mathscr{F}\left[\int\limits_{-\infty}^{\infty}\!\!\int f(x, y) \, f^*(x - \alpha, y - \beta) \, dx \, dy \right]$$

$$= \mathscr{F}\left[\int\limits_{-\infty}^{\infty}\!\!\int f(\xi + \alpha, \eta + \beta) \, f^*(\xi, \eta) \, d\xi \, d\eta \right]$$

$$= \int\limits_{-\infty}^{\infty}\int d\xi\, d\eta\, f^*(\xi, \eta)\, \mathscr{F}\left[f(\xi+\alpha, \eta+\beta)\right]$$

$$= \int\limits_{-\infty}^{\infty}\int d\xi\, d\eta\, f^*(\xi, \eta)\, \exp\left[i(p\xi+q\eta)\right] F(p, q) = F^*(p, q)\, F(p, q).$$

8. Crosscorrelation Theorem

If $f_1(x, y)$ and $f_2(x, y)$ are Fourier transformable, so that $\mathscr{F}\left[f_1(x, y)\right]$ $= F_1(p, q)$ and $\mathscr{F}\left[f_2(x, y)\right] = F_2(p, q)$, then

$$\mathscr{F}\left[R_{12}(\alpha, \beta)\right] = \mathscr{F}\left[f_1(x, y) \circledast f_2^*(x, y)\right] = F_1(p, q)\, F_2^*(p, q), \qquad (1.38)$$

and

$$\mathscr{F}\left[R_{21}(\alpha, \beta)\right] = \mathscr{F}\left[f_1^*(x, y) \circledast f_2(x, y)\right] = F_1^*(p, q)\, F_2(p, q) \qquad (1.39)$$

where

$$R_{12}(\alpha, \beta) = \int\limits_{-\infty}^{\infty}\int f_1(x+\alpha, y+\beta)\, f_2^*(x, y)\, dx\, dy,$$

$$R_{21}(\alpha, \beta) = \int\limits_{-\infty}^{\infty}\int f_1^*(x+\alpha, y+\beta)\, f_2(x, y)\, dx\, dy,$$

are the *crosscorrelation functions*.

   Proof: The proof of this theorem is similar to that of the autocorrelation theorem.

9. Some Symmetric Properties

(a) If $f(x, y)$ is real, i.e., if $f^*(x, y) = f(x, y)$, then

$$F^*(-p, -q) = F(p, q), \qquad (1.40)$$

and

$$F^*(p, q) = F(-p, -q). \qquad (1.41)$$

(b) If $f(x, y)$ is real and even, that is $f(x, y) = f^*(x, y) = f^*(-x, -y)$, then $F(p, q)$ is real and even,

$$F(p, q) = F^*(-p, -q) = F^*(p, q). \qquad (1.42)$$

(c) If $f(x, y)$ is real and odd, that is, $f(x, y) = f^*(x, y) = -f(-x, -y)$, then

$$F(p, q) = -F(-p, -q) = F^*(-p, -q).$$ (1.43)

Proof: The proof of these properties can be obtained by direct application of the definition of the Fourier transformation, eq. (1.8).

The Fourier transform properties that we have just mentioned may not be adequate for all the applications in modern coherent optics. However, we will use these properties frequently in the course of this book, and will thereby avoid much tedious calculation.

1.4 Response of a Linear Spatially Invariant System

In sec. 1.1 we discussed the basic properties of a linear system from a physical point of view. In this section we will develop the mathematical viewpoint more fully. It is well-known that a linear system may be represented by a linear operator L, as shown in the block diagram of fig. 1.1. Since the operator is linear, we have the following property:

$$L[C_1 f_1(x, y) + C_2 f_2(x, y)] = C_1 L[f_1(x, y)] + C_2 L[f_2(x, y)],$$ (1.44)

where C_1 and C_2 are arbitrary complex constants.

If the system has spatial invariance, then the linear operator must also have spatial invariance,

$$L[f(x - x_0, y - y_0)] = g(x - x_0, y - y_0),$$ (1.45)

where $g(x, y)$ is the output response.

It may be interesting to note that if the input excitation to a linear spatially invariant system is the Dirac delta function $\delta(x, y)$, then the *output impulse response* is

$$h(x, y) = L[\delta(x, y)].$$ (1.46)

Now if the input excitation is an arbitrary function of $f(x, y)$, which may be represented by the *convolution integral*

$$f(x, y) = \int\!\!\int_{-\infty}^{\infty} f(x', y') \, \delta(x' - x, y' - y) \, dx' \, dy',$$ (1.47)

then the output response may be written

$$g(x, y) = L\left[\int\!\!\int_{-\infty}^{\infty} f(x', y') \, \delta(x' - x, y' - y) \, dx' \, dy'\right]$$

$$= \int\!\!\int_{-\infty}^{\infty} f(x', y') \, L[\delta(x' - x, y' - y)] \, dx' \, dy'. \tag{1.48}$$

It follows that

$$g(x, y) = \int\!\!\int_{-\infty}^{\infty} f(x', y') \, h(x' - x, y' - y) \, dx' \, dy'. \tag{1.49}$$

Equation (1.49) shows the remarkable property of the output response of a linear spatially invariant system: the response is the convolution of the impulse response with the input excitation.

Furthermore, from the *Fourier convolution theorem* in the previous section, eq. (1.49) can be written in the Fourier transform form,

$$G(p, q) = F(p, q) \, H(p, q), \tag{1.50}$$

where $G(p, q) = \mathscr{F}[g(x, y)]$, $F(p, q) = \mathscr{F}[f(x, y)]$, and $H(p, q) = \mathscr{F}[h(x, y)]$. That is, the Fourier transform of the output response of a linear spatially invariant system is the product of the Fourier transform of the input excitation and the Fourier transform of the impulse response. We may emphasize the fact that, if the impulse response of a linear system is known, then the corresponding Fourier transform is the *system transfer function*.

1.5 Detection of Signal by Matched Filtering

A problem of considerable importance in coherent optical information processing is the detection of a signal corrupted by random noise. It is our aim in this section to discuss a special type of optimum linear filter, known as a *matched filter*, which is useful in complex spatial filtering and in optical pattern or character recognition. We will first derive an expression of the spatial filter transfer function on a somewhat general basis, namely, for a stationary additive noise. Then we will immediately move to a *white* (i.e., of uniform spectral density over the spatial frequency domain) stationary additive noise.

It is well-known in communication theory that the signal-to-noise ratio at the output end of a correlator can be improved to a large degree. Let us consider the input excitation to a linear filtering system to be an additive mixture of signal $s(x, y)$, and a stationary random noise $n(x, y)$: i.e.,

$$f(x, y) = s(x, y) + n(x, y). \tag{1.51}$$

Let the output response of the linear filter due to the signal $s(x, y)$ alone be $s_0(x, y)$, and that due to the random noise $n(x, y)$ alone be $n_0(x, y)$. The figure of merit, on which the filter design is based, is the output signal-to-noise ratio at $x = y = 0$,

$$\frac{S}{N} \triangleq \frac{|s_0(0, 0)|^2}{\sigma^2}, \tag{1.52}$$

where σ^2 is the mean-square value of the output noise.

In terms of the filter transfer function $H(p, q)$ and the Fourier transform $S(p, q)$ of the input signal $s(x, y)$, these quantities may be written

$$s_0(0, 0) = \frac{1}{4\pi^2} \int\limits_{-\infty}^{\infty}\!\!\int H(p, q)\, S(p, q)\, dp\, dq \tag{1.53}$$

and

$$\sigma^2 = \frac{1}{4\pi^2} \int\limits_{-\infty}^{\infty}\!\!\int |H(p, q)|^2\, N(p, q)\, dp\, dq, \tag{1.54}$$

where $|H(p, q)|^2\, N(p, q)$ is the power spectral density of the noise at the output end of the filter, and $N(p, q)$ is the power spectral density of the noise at the input end. Thus the output signal-to-noise ratio may be expressed explicitly in terms of the filter function $H(p, q)$,

$$\frac{S}{N} = \frac{1}{4\pi^2} \frac{\left| \int\limits_{-\infty}^{\infty}\!\!\int H(p, q)\, S(p, q)\, dp\, dq \right|^2}{\int\limits_{-\infty}^{\infty}\!\!\int |H(p, q)|^2\, N(p, q)\, dp\, dq}. \tag{1.55}$$

The objective of the filter designer is to specify a filter function such that the output signal-to-noise ratio is a maximum. To obtain such a filter transfer function, the designer may apply the Schwarz inequality, which states that

$$\frac{\left| \int\limits_{-\infty}^{\infty}\!\!\int u(x, y)\, v^*(x, y)\, dx\, dy \right|^2}{\int\limits_{-\infty}^{\infty}\!\!\int |u(x, y)|^2\, dx\, dy} \leq \int\limits_{-\infty}^{\infty}\!\!\int |v^*(x, y)|^2\, dx\, dy, \tag{1.56}$$

where $u(x, y)$ and $v(x, y)$ are arbitrary functions and * denotes the complex conjugate. The equality in eq. (1.56) holds if and only if $u(x, y)$ is proportional to $v(x, y)$.

To make the Schwarz inequality applicable to eq. (1.55), it may be expedient to express the output noise spectral density as the product of two conjugate factors:

$$N(p, q) = N_1(p, q) \, N_1^*(p, q). \tag{1.57}$$

Then eq. (1.55) may be written as

$$\frac{S}{N} = \frac{1}{4\pi^2} \frac{\left| \int\int_{-\infty}^{\infty} [H(p, q) \, N_1(p, q)] \left[\frac{S^*(p, q)}{N_1^*(p, q)} \right]^* dp \, dq \right|^2}{\int\int_{-\infty}^{\infty} |H(p, q) \, N_1(p, q)|^2 \, dp \, dq}. \tag{1.58}$$

If we identify the bracketed quantities of eq. (1.58) as $u(x, y)$ and $v(x, y)$, then, in view of the Schwarz inequality we have

$$\frac{S}{N} \le \frac{1}{4\pi^2} \int\int_{-\infty}^{\infty} \frac{|S(p, q)|^2}{N(p, q)} \, dp \, dq. \tag{1.59}$$

The equality in eq. (1.59) holds if and only if the filter function is

$$H(p, q) = K \frac{S^*(p, q)}{N(p, q)}, \tag{1.60}$$

where K is a proportionality constant. The corresponding value of the output signal-to-noise ratio is therefore

$$\frac{S}{N} = \frac{1}{4\pi^2} \int\int_{-\infty}^{\infty} \frac{|S(p, q)|^2}{N(p, q)} \, dp \, dq. \tag{1.61}$$

It may be interesting to note that if the stationary additive input noise is white, then the optimum filter function is

$$H(p, q) = KS^*(p, q), \tag{1.62}$$

which is proportional to the conjugate of the signal spectrum. This optimum filter is then said to be matched to the input signal $s(x, y)$. The output spectrum of the matched filter is therefore proportional to the power spectral density of the input signal, i.e.,

$$G(p, q) = K |S(p, q)|^2 . \tag{1.63}$$

Consequently, we see that the phase variation at the output end of the matched filter vanishes. In other words, the matched filter has the capability of eliminating all the phase variations of $S(p, q)$ across the spatial frequency domain.

In concluding this chapter, we would like to emphasize again that the Fourier treatment presented in this chapter makes no attempt at mathematical rigor and completeness, but it is rather introduced for the operational purpose of this book. For the reader interested in a more complete treatment, we would suggest the excellent texts by Papoulis (ref. 1.1), and by Bracewell (ref. 1.2). Between these two books, a complete treatment of Fourier transform theory and some of the applications of Fourier analysis can be found.

Problems

1.1 Determine the Fourier transforms of the following functions:

(a) $f(x) = \exp(ipx)$,

(b) $f(x) = \cos px$,

(c) $f(x) = \sum\limits_{n=-\infty}^{\infty} \delta(x - na)$,

(d) $f(x) = \begin{cases} 1, & |x| \leq a/2 \\ 0, & |x| > a/2 \end{cases}$,

(e) $f(x) = \exp(-ax^2)$,

where p is an angular spatial frequency, δ is the Dirac delta function, and α is an arbitrary positive constant.

1.2 Suppose the Fourier transforms of $f(x, y)$ and $g(x, y)$ exist, i.e., $\mathscr{F}[f(x, y)] = F(p, q)$ and $\mathscr{F}[g(x, y)] = G(p, q)$, respectively. Show that

$$\int\limits_{-\infty}^{\infty}\int f(x, y)\, g^*(x, y)\, dx\, dy = \frac{1}{4\pi^2} \int\limits_{-\infty}^{\infty}\int F(p, q)\, G^*(p, q)\, dp\, dq .$$

1.3 If $f(x, y)$ is Fourier transformable, i.e., $\mathscr{F}[f(x, y)] = F(p, q)$ exists, show that

$$\frac{\int\int_{-\infty}^{\infty} f(x, y)\, dx\, dy}{f(0, 0)} = \frac{4\pi^2 F(0, 0)}{\int\int_{-\infty}^{\infty} F(p, q)\, dp\, dq}.$$

1.4 With reference to the Hankel transform pair of eqs. (1.21) and (1.22), if $f(r)$ and $g(r)$ are Hankel transformable, i.e., $\mathscr{H}[f(r)] = F(\rho)$ and $\mathscr{H}[g(r)] = G(\rho)$ exist, derive the following properties:

(a) $\mathscr{H}[f(ar)] = \dfrac{1}{a^2} F\left(\dfrac{\rho}{a}\right)$, where a is an arbitrary constant.

(b) $\mathscr{H}[f(r) + g(r)] = F(\rho) + G(\rho)$.

(c) $\mathscr{H}\left[\displaystyle\int_0^\infty \int_0^{2\pi} f(r')\, g(\alpha)\, r'\, dr'\, d\theta\right] = F(\rho)\, G(\rho)$,

where $\alpha^2 = r^2 + r'^2 - 2rr' \cos\theta$.

(d) $\mathscr{H}\left[2\pi \displaystyle\int_0^\infty f(r)\, r\, dr\right] = F(0)$.

(e) $\mathscr{H}\left[2\pi \displaystyle\int_0^\infty r^2 f(r)\, r\, dr\right] = -\dfrac{1}{2\pi^2} \dfrac{d^2 F(\rho)}{d\rho^2}\bigg|_{\rho = 0}$.

1.5 Consider a one-dimensional band-limited function $f(x)$ sampled at Nyquist's rate (i.e., minimum sampling rate at p_0/π). Show that the function $f(x)$ may be written as

$$f(x) = \sum_{n=-\infty}^{\infty} f(nx_0) \frac{\sin p_0(x - nx_0)}{p_0(x - nx_0)},$$

where p_0 is the highest angular spatial frequency of the function $f(x)$ and $x_0 = \pi/p_0$, the Nyquist sampling interval. (This equation is also known as Shannon's sampling theorem.)

1.6 We shall extend the sampling theorem in the previous problem to a two-dimensional band-limited function $f(x, y)$. Show that the function $f(x, y)$ may be written as

$$f(x, y) = \sum_{n=-\infty}^{\infty} \sum_{m=-\infty}^{\infty} f(nx_0, my_0) \frac{\sin p_0(x-nx_0) \sin q_0(y-my_0)}{p_0(x-nx_0) q_0(y-my_0)},$$

where p_0 and q_0 are the highest angular spatial frequency with respect to the x and y coordinates of $f(x, y)$, and $x_0 = \pi/p_0$ and $y_0 = \pi/q_0$ are the respective Nyquist sampling intervals.

1.7 In a manner similar to that of the previous problem, the sampling theorem can also be applied to a circularly band-limited function, i.e., one where the spatial frequency is band limited within a circle of radius r_0 in the spatial frequency domain. Show that

$$f(x, y) = \frac{\pi}{2} \sum_{n=-\infty}^{\infty} \sum_{m=-\infty}^{\infty} f\left(\frac{n\pi}{r_0}, \frac{m\pi}{r_0}\right) \frac{J_1\left[\sqrt{(r_0 x - n\pi)^2 + (r_0 y - m\pi)^2}\right]}{\sqrt{(r_0 x - n\pi)^2 + (r_0 y - m\pi)^2}},$$

where J_1 is the first-order Bessel function of the first kind.

1.8 Given a spatial random noise $n(x, y)$, which is assumed to have a zero mean over the space Σ of (x, y) coordinates, i.e.,

$$\lim_{\Sigma \to \infty} \frac{1}{\Sigma} \int\int_{\Sigma} n(x, y) \, dx \, dy = 0.$$

Let us denote $s(x, y)$ as a useful spatial signal. If $s(x, y)$ and $n(x, y)$ are assumed uncorrelated, show that

$$\lim_{\Sigma \to \infty} \frac{1}{\Sigma} \int\int_{\Sigma} s(x', y') \, n(x'-x, y'-y) \, dx' \, dy' = 0$$

for every (x, y).

1.9 Show that the autocorrelation function of the sum of a given signal $s(x, y)$ and an additive random noise $n(x, y)$ is the sum of the individual autocorrelations and their crosscorrelations.

1.10 Suppose a useful signal is described by

$$s(x, y) = \exp[(-a+ib)(x^2 + y^2)],$$

where a and b are arbitrary positive constants. This signal is assumed to be embedded in random additive noise $n(x, y)$ of zero mean:

$f(x, y) = s(x, y) + n(x, y)$.

The received signal $f(x, y)$ is fed into an input-output linear spatially invariant system that has a spatial impulse response of

$h(x, y) = s^*(-x, -y)$,

Calculate the output response. Make a rough sketch of the functions $|s(x, y)|$ and

$$g(x, y) = \int\limits_{-\infty}^{\infty}\int s(x', y')\, s^*(-x' + x, -y' + y)\, dx'\, dy'.$$

Compare the envelopes of these two plots and give a brief discussion of their significance in regard to signal detection.

References
1.1 A. Papoulis, *The Fourier Integral and Its Applications*, McGraw-Hill, New York, 1962.

1.2 R. N. Bracewell, *The Fourier Transform and Its Applications*, McGraw-Hill, New York, 1965.

1.3 W. M. Brown, *Analysis of Linear Time-Invariant Systems*, McGraw-Hill, New York, 1963.

1.4 D. K. Cheng, *Analysis of Linear Systems*, Addison-Wesley, Reading, Mass., 1959.

1.5 W. B. Davenport, Jr. and W. L. Root, *An Introduction to the Theory of Random Signals and Noise*, McGraw-Hill, New York, 1958.

1.6 M. J. Lighthill, *Introduction to Fourier Analysis and Generalized Functions*, Cambridge University Press, New York, 1960.

Part I Diffraction

Chapter 2   Diffraction Theory

2.1 General Aspects

In fig. 2.1 a source of light is shown illuminating a screen, except for the shadow of an opaque object placed between source and screen. Let us examine the sharpness of the edge of the shadow. It is clear that if the source of light subtends an appreciable angle, then the shadow's edge will fade gradually from full illumination to total darkness, since there are points on the screen from which a part but not all of the source can be seen. To avoid this effect, let us make the source so small that we can consider it a single point of light. Under this condition, and with the usual assumption of geometrical optics that rays of light in a homogeneous medium travel in straight lines, we would expect the edge of the shadow to be quite sharp, so that points on the screen would either be fully illuminated or lie in total darkness.

If we set up such a system, the edge of the shadow at first glance appears to be sharp. If we examine it more closely, however, we find that the illumination on the screen fades gradually over a short distance. In addition, if the source of light is monochromatic, there will be seen narrow light and dark bands parallel to the edge of the geometrical shadow. These bands are called *fringes*. It is apparent that the light, in passing the edge of the obstacle deviates from a straight-line propagation. This effect is called *diffraction*.

Diffraction can be explained from the theory of wave motion. Historically, it was the observation of diffraction that led to general acceptance of the wave theory of light, as opposed to the corpuscular theory. The wave theory shows that the magnitude of the diffraction effect, i.e., the angle

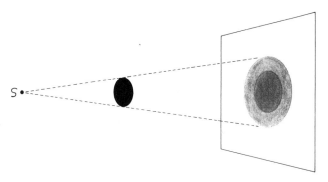

Fig. 2.1 Shadow formed by an opaque object and a point source *S*.

of deviation from straight-line propagation, is directly proportional to the wavelength. Thus it is that with a sound wave we are usually not conscious of a shadow at all, because of the large diffraction angle in the case of the long sound waves. Even with sound, however, shadows can be demonstrated if supersonic frequencies are used. Since the wavelengths of visible light are extremely small, the diffraction is also small. The straight-line propagation assumed in geometrical optics is only the limit approached as the wavelength is allowed to approach zero.

The theory of diffraction will be treated in the next chapter on the basis of *Huygens' principle*, in the form given by Kirchhoff. This predicts results very close to the effects actually observed and will at least give a qualitative understanding of the phenomenon. It is also a much less difficult approach than that which uses the wave equation in a field bounded by the shadowing object. The latter approach is generally satisfactory only in some special cases in which simplifying assumptions can be made. Nevertheless, some interesting conclusions about diffraction can be deduced from the complete wave theory, as will be elaborated in the remaining sections of this chapter.

2.2 Fraunhofer and Fresnel Diffraction

In the preceding section we have spoken of the diffraction of light as it passes the edge of an obstacle. Diffraction may be treated more simply if we consider the light as passing through one or more small apertures in a diffracting obstacle. The former case is then the limit as the dimensions of the aperture become infinite.

It is customary to divide diffraction into two cases, depending on the distances of light source and viewing screen from the diffracting screen, and these cases have been given the names of two early investigators of diffraction. If the source and viewing point are so far from the diffracting screen that lines drawn from the source or viewing point to all points of the apertures do not differ in length by more than a small fraction of a wavelength, the phenomenon is called *Fraunhofer diffraction*. If these conditions do not hold, it is called *Fresnel diffraction*. The boundary between these two cases is somewhat arbitrary, and depends upon the accuracy desired in the results. In most cases it is sufficient to use the Fraunhofer methods if the difference in distances does not exceed one-twentieth of a wavelength. Of course Fraunhofer diffraction may be achieved without a great physical distance of the source, if a collimating lens is used to make the rays of light from the source nearly parallel.

Figure 2.2 is drawn to illustrate the above considerations. A source of

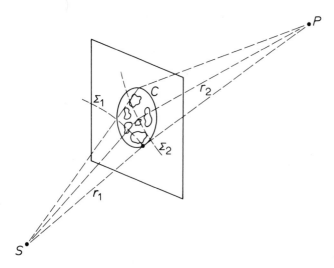

Fig. 2.2 Geometry for defining Fraunhofer and Fresnel diffraction.

monochromatic light is at S, and a viewing point at P, and between them is an opaque screen having a finite number of apertures. Let a circle C be drawn on the plane of the screen, and let it be as small as possible while still enclosing all the apertures. Let C be the base of cones with S and P as vertices. Also draw spherical surfaces Σ_1 and Σ_2, having S and P as centers, and as radii r_1 and r_2, the shortest distances from S and P to the circle C. If the largest distances from C to Σ_1 and to Σ_2 are not more than one-twentieth of the wavelength of the light used, then the diffraction is Fraunhofer, and the light falling on the observing screen at P forms a *Fraunhofer diffraction pattern*.

On the other hand, if by reason of the large size of C, or the shortness of the distances to S or P, the distances between C and Σ_1 or Σ_2 are greater than one-twentieth of the wavelength, then we must speak of Fresnel diffraction and a *Fresnel diffraction pattern* at P.

The radius of the circle C of fig. 2.3 is denoted by ρ, the shortest distance of S from the screen is l, and the greatest separation of sphere and screen is Δl. From the definition of Fraunhofer diffraction, Δl must be a small fraction of a wavelength. However, ρ may be many wavelengths long (as it can also be in the Fresnel case). In the right triangle of Fig. 2.3 we have

$$(l + \Delta l)^2 = l^2 + \rho^2 \tag{2.1}$$

and because of the small size of $(\Delta l)^2$ in comparison with the other quanti-

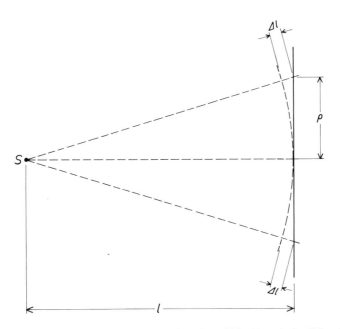

Fig. 2.3 Geometry for determining the value of l for Fraunhofer diffraction.

ties we may make the approximation

$$l \simeq \frac{\rho^2}{2\Delta l}. \tag{2.2}$$

As an example, suppose that ρ is 1 cm, and that the source has the longest wavelength in the visible red, i.e., about 8×10^{-5} cm. Let Δl be one-twentieth of this, or 0.4×10^{-5} cm. The distance l would then be approximately 1.25 km. If violet light of half the wavelength were used, l would be 2.5 km. The requirements for Fraunhofer diffraction can thus be rather stringent.

2.3 Fraunhofer Diffraction from Multiple Apertures

Let the diffracting screen be in the form of a plane, and let it have n apertures. Furthermore, let the apertures be identical in size, shape and orientation. Choose an origin O and axes of coordinates in the plane, by which the location of the apertures may be specified, as shown in fig. 2.4. Each aperture is identified by a point Q_i. The position of this point with respect to the ith aperture is the same for all i.

This diffracting screen is located between the source S and the viewing point P, as shown in fig. 2.5. (Only one of the many apertures is indicated

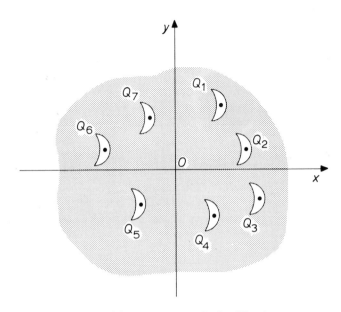

Fig. 2.4 Geometry of the open apertures in the diffracting screen.

in this figure.) Draw planes through S and P parallel to the plane of the diffracting screen, and draw a line through O perpendicular to this screen. Let the intersections of this line with the other planes, at O_1 and O_2, be the origins of coordinates in these planes, and let the ξ and η axes of the source plane and the α and β axes of the viewing plane be parallel to the x and y axes of the diffracting plane. The line from O to S makes angles ϕ_1 with the O_1Ox plane and θ_1 with the O_1Oy plane. Similarly, the line OP makes angles ϕ_2 and θ_2 with these planes. Arrows in the figure indicate the positive directions of these angles.

Consider for the moment the illumination of the viewing screen due to a single aperture in the diffracting screen. The geometrical spot of light is larger than the size of the aperture—on the order of twice as large if S and P are at equal distances from the aperture. In addition, however, there is a spreading due to diffraction at the aperture. Even though the diffraction angle is small, the large distance to P required in Fraunhofer diffraction permits the spread of light to be great, so that the diffraction pattern at P will be much larger than the geometrical spot. The angle of spread, in radians, is given approximately by the ratio of the wavelength of the light to the width of the aperture.

According to Huygens' principle (which will be explained more fully at the end of this chapter), the different points in the plane of the aperture may be treated as if they were secondary sources of light, each source radiating in all directions into the space beyond the aperture. The secondary sources all have the same amplitude, since their distances from the primary source S are all very nearly the same under the Fraunhofer conditions. Their slight differences in distance, however, amount to appreciable fractions of a wavelength of the light, hence these secondary sources differ in phase.

All points in the receiving plane are equally illuminated by the various secondary sources in the aperture, and, again because of the large distance, the amplitudes at the different points in the plane of P are substantially equal. If the aperture were reduced in size to a single point, the irradiance*

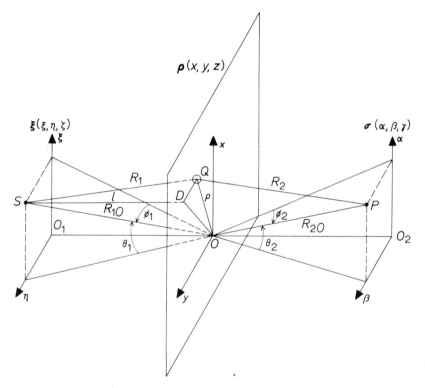

Fig. 2.5 The defined coordinate systems. The points (ξ, η) define the source plane; (x, y), the diffracting screen; and (α, β), the image or observing plane.

* The definition of *irradiance* follows eq. 2.13.

over the receiving plane would be uniform. However, because of the different phases in the secondary sources in an aperture of appreciable size, and the further phase changes brought about by the slight differences in distance from aperture to receiving plane, there are considerable phase differences at different points on this plane. At some points there will be increased irradiance because of favorable phase combination in the light from the secondary sources, while at other points there will be destructive interference. This is the reason for the light and dark fringes in the diffraction pattern.

Now let us suppose that a second aperture is opened up on the diffraction screen. This will be in a shifted position with respect to the first, but still lie within the circle C of fig. 2.2. Then all that we have said above about phase differences but lack of appreciable amplitude differences applies equally well to the pattern from the new aperture. The pattern from the new aperture alone would be like that from the first, but of course in a shifted position, as determined by drawing lines from S through the two apertures and continuing to the receiving plane. This shift will be small compared with the size of the diffraction pattern from either aperture. The combined pattern from both apertures will depend on the phase combinations from the two separate patterns.

The strength of the irradiance at P due to an aperture Q may be expressed in terms of that which would result if the aperture were at the origin O on the diffracting screen (see fig. 2.5). Thus we may suppose that the electric field at P, when the aperture is at O, is given by the complex number $Ae^{i\phi}$, where A is a constant and ϕ is a phase angle which varies with the position of the point P. The phase angle ϕ also changes if the aperture is moved to a point other than O. Now if the aperture at Q is considered instead of the one at O, the difference of paths from S to P may be reduced to a phase shift by dividing it by the wavelength λ, and multiplying by 2π. Thus in fig. 2.5, let $d_1 = R_1 - R_{10}$ and $d_2 = R_2 - R_{20}$. Then the complex electric field at P due to the aperture at Q is given by

$$Ae^{i\phi} \exp[ik(d_1 + d_2)] \tag{2.3}$$

where $k = 2\pi/\lambda$. The quantity $(d_1 + d_2)$ may be called the geometrical path difference between the apertures at Q and at O, and $k(d_1 + d_2)$ is the phase difference at P for the two positions.

The geometrical path difference may also be expressed in terms of the coordinates of S, Q, and P of fig. 2.5. Suppose a line is drawn from S perpendicular to the diffraction plane. This line is parallel to O_1O, and has the same length. Call this length l. Call the point of intersection with the diffraction plane D. It has the same coordinates as S, i.e., ξ and η. Draw lines from D to $Q(x, y)$ and to $O(0, 0)$. Two right triangles are formed, having R_1 and R_{1O} as hypotenuses and having l as a common leg. Also notice that

$$(DO)^2 = \xi^2 + \eta^2 \text{ and } (DQ)^2 = (\xi - x)^2 + (\eta - y)^2.$$

Then

$$l^2 = R_1^2 - (\xi - x)^2 - (\eta - y)^2 = R_{1O}^2 - \xi^2 - \eta^2, \tag{2.4}$$

from which

$$R_1^2 - R_{1O}^2 = x^2 + y^2 - 2(x\xi + y\eta). \tag{2.5}$$

Now $R_1 - R_{1O}$ has been defined as d_1, and $x^2 + y^2$, in fig. 2.5, is ρ^2. Also, R_1 and R_{1O} are so nearly equal that $R_1 + R_{1O}$ may be taken as $2R_{1O}$. Equation (2.5) then, becomes

$$d_1 \simeq \frac{\rho^2}{2R_{1O}} - \frac{x\xi + y\eta}{R_{1O}}. \tag{2.6}$$

The first term on the right side of eq. (2.6) represents the path difference when ξ and η are both zero, i.e., when S coincides with O_1. According to our requirements for Fraunhofer diffraction, as already stated, this difference might be as much as one-twentieth of the wavelength of the light used. In terms of phase shift, this would be an angle of $\pi/10$. We propose to neglect this first term of eq. (2.6), and if a phase shift of $\pi/10$ seems too large an angle to neglect, then we will make our conditions for Fraunhofer diffraction stricter. In other words, we can always arrange matters so that the wave front arriving at the diffraction screen is essentially plane.

As for the remaining term in eq. (2.6), it is clear that d_1 will not represent an appreciable phase shift unless either ξ or η, or both, are considerably larger in absolute value than the corresponding x and y when Q lies close to O, as is usually the case. Note also from fig. 2.5 that $\xi/R_{1O} = \sin\theta_1$, and $\eta/R_{1O} = \sin\phi_1$. Equation (2.6) then reduces to

$$d_1 \simeq -x \sin\theta_1 - y \sin\phi_1. \tag{2.7}$$

In an exactly similar manner, and with the same strict Fraunhofer conditions, we can treat the geometrical path difference d_2,

$$d_2 \simeq -x \sin\theta_2 - y \sin\phi_2. \tag{2.8}$$

The complete path difference for the apertures at Q and at O is the sum of eqs. (2.7) and (2.8). Note that it can be either positive or negative, depending on the signs of the angles. We would expect this ambiguity of sign from our definitions of $d_1 = R_1 - R_{10}$ and $d_2 = R_2 - R_{20}$. The angles $\phi_1, \theta_1, \phi_2, \theta_2$ are not restricted, up to values of $\pm \pi/2$.

Substituting eqs. (2.7) and (2.8) into eq. (2.3) gives the electric field at P due to the single aperture Q in terms of the coordinates of Q, S, and P. Let us go a step further, and assume now that there are N identical apertures (except for their coordinates) and add the respective fields. Note that each aperture is treated here as so small in itself that there is no appreciable phase shift involved when different points within the same aperture are considered. Then

$$E_p = Ae^{i\phi} \exp(-i\omega t) \sum_{n=1}^{N} \exp\{-ik[x_n(\sin\theta_1 + \sin\theta_2)$$
$$+ y_n(\sin\phi_1 + \sin\phi_2)]\}, \qquad (2.9)$$

where (x_n, y_n) are the coordinates of the defining point Q_n of the nth aperture. It is consistent with our previous assumption of long distances for S and P that we use the same angles for all the apertures. The factor $\exp(-i\omega t)$ is introduced to provide the time variation of the light signal; ω has the usual value, $\omega = 2\pi c/\lambda$, where c is the velocity of light.

The E_p of eq. (2.9) is the time-varying magnitude of the electric vector of the light wave at P. We may also write down the magnetic vector:

$$\mathbf{H}_p = \frac{\mathbf{k}}{k} \times \mathbf{E}_p. \qquad (2.10)$$

The $\mathbf{k}$ in eq. (2.10) is the wave vector, and $\times$ denotes the vector product.

We can now obtain the time average of the Poynting vector of the wave, given by

$$\langle \mathbf{S} \rangle = \mathrm{Re}[\mathbf{S}] = \tfrac{1}{2}\,\mathrm{Re}(\mathbf{E} \times \mathbf{H}^*) \qquad (2.11)$$

In this equation $\mathbf{S}$ is the Poynting vector, and $\langle \ \rangle$ represents the time average; Re is the real part. Combining (2.11) with (2.10) and (2.9),

$$\langle \mathbf{S}_p \rangle = \langle \mathbf{S}_p \rangle_1 \left| \sum_{n=1}^{N} \exp\{-ik[x_n(\sin\theta_1 + \sin\theta_2) + y_n(\sin\phi_1 + \sin\phi_2)]\} \right|^2, \qquad (2.12)$$

where $\langle \mathbf{S}_p \rangle_1 = (\mathbf{k}/2k)\,A^2$ is the time average of the Poynting vector at P when there is a single aperture at the origin. For convenience, let us abbreviate eq. (2.12) as

$$I = I_1 |G|^2, \qquad (2.13)$$

where I and I_1 represent the time averages of the Poynting vectors, and

$$G = \sum_{n=1}^{N} \exp\{-ik[x_n(\sin\theta_1 + \sin\theta_2) + y_n(\sin\phi_1 + \sin\phi_2)]\} \tag{2.14}$$

The I in equation (2.13) represents the *irradiance* at P, that is, the average time rate of the energy falling on a unit area at P, perpendicular to the direction of wave propagation. This I combines light from all apertures of the diffraction screen. The irradiance I_1 is that due to light that has passed through the central aperture only.

It may be noted that the x and y axes of fig. 2.5 were freely chosen, the only restriction being that the origin O should be close to the system of apertures. A different set of axes would change the x's and y's of equation (2.12), and along with them the θ's and ϕ's, but the value of I would be unchanged. It is therefore expedient to choose the axes in such a way as to make the calculation of $|G|^2$ simple.

2.4 Special Case of Two Apertures

In 1801, Thomas Young performed an experiment which is credited with bringing about general acceptance of the wave theory of light. This experiment was a demonstration of the diffraction of light by two close apertures. He provided a very small source by using the sunlight passing through a pinhole in an otherwise opaque screen, then letting the light from this source fall on a screen containing two pinholes. Of course his light was not monochromatic, and the fringes produced by the different wavelengths fell in different places on the viewing screen.

Let us assume a single wavelength λ from a point source, and let the light fall on two similar apertures separated by a distance d. Let one of these be at the origin O, and the other on the x axis, as shown in fig. 2.6. We can now apply the theory of the preceding section. The summation in equation (2.14) will have only two terms, with x and y coordinates of $(0, 0)$ and $(d, 0)$. It therefore reduces to

$$G = 1 + \exp[-ikd(\sin\theta_1 + \sin\theta_2)]. \tag{2.15}$$

The angles θ_1 and θ_2 are unknown, since in our choice of coordinate axes we have made no assumption with regard to the positions of S and P.

In eq. (2.15), $k = 2\pi/\lambda$, so that $kd = 2\pi d/\lambda$, and $d(\sin\theta_1 + \sin\theta_2)$ is the geometrical path difference $SQ_1P - SQ_2P$, found by the sum of eqs. (2.7) and (2.8) (with the signs reversed). Let us define the *optical path difference*, δ, as the geometrical path difference (without regard to sign) divided by the wavelength. In other words, δ is the path difference from S to P, for light passing through Q_1 or through Q_2, as measured in wavelengths. In this case it is given by

$$\delta = \frac{d}{\lambda}(\sin\theta_1 + \sin\theta_2), \tag{2.16}$$

and (2.15) reduces to

$$G = 1 + e^{-i2\pi\delta} = 2e^{-i\pi\delta}\cos(\pi\delta). \tag{2.17}$$

Then,

$$|G|^2 = 4\cos^2(\pi\delta). \tag{2.18}$$

A vector representation of eq. (2.17) is given in fig. 2.7. Equation (2.18) may be substituted into eq. (2.13), thus,

$$I = 4I_1\cos^2(\pi\delta) \tag{2.19}$$

As the point P is moved about in the observing plane, the angle θ_2 changes if the motion of P is parallel with the x axis, that is, parallel to the line be-

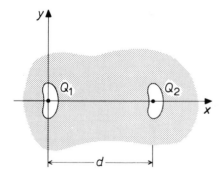

Fig. 2.6 Two identical open apertures.

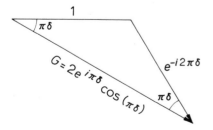

Fig. 2.7 Vector representation of eq. 2.17.

tween the two apertures. By eq. (2.16) this changes the value of δ. We see then from eq. (2.19) that the irradiance I goes through a series of maxima and minima, having a maximum of $4I_1$ when δ is zero or any integer (positive or negative), and a minimum of zero when δ is an odd multiple of one-half. In other words, when the path difference is an exact number of wavelengths there is constructive interference between the waves from the two apertures; when the difference is an odd number of half-wavelengths, the interference is destructive.

The variation of irradiance for two apertures is illustrated in fig. 2.8. The solid curve along the α axis is a plot of eq. (2.19), while the dashed envelope represents the variation of $4I_1$.

For a photograph of the actual pattern produced on an observing screen by a single circular aperture of appreciable size, refer to fig. 4.11 (p. 62). The dark ring surrounding the central bright region has an angular diameter (with the aperture as center) of approximately λ/d radians, where d is the diameter of the aperture. The center of the circular spot is at the point on the observing plane which is directly opposite the aperture in the diffraction screen. If l is the distance between these planes, the linear diameter of the dark ring is approximately $\lambda l/d$.

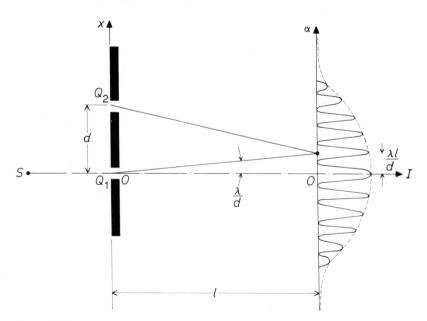

Fig. 2.8 Diffraction due to two identical apertures.

Finally, with the two circular apertures discussed in this section, the diffraction pattern looks like that of fig. 2.9. The variation of irradiance vertically across this figure corresponds to the curve drawn on the α axis in fig. 2.8. In the other direction the $|G|^2$ of equation (2.18) does not apply.

2.5 The Reciprocity Theorem

The reciprocity theorem states that if the source is placed at an image point P, the same irradiance will be observed at the source point S as appeared at P when the source was at S. Thus there is a symmetry between a motion of P and a motion of S. The theorem is equivalent to saying that the diffraction system is both *linear* and *bilateral*. The proof of the theorem lies in showing that there is no change in the I in eq. (2.13) if ϕ_1 and ϕ_2 are interchanged, along with θ_1 and θ_2. That this is true for the factor $|G|^2$ of eq. (2.13) is at once apparent from an examination of eq. (2.14). The determination of I_1 has not yet been explained in detail. It will be shown in the next chapter that I_1 is also unchanged by the exchange of source point and image point.

Fig. 2.9 Diffraction pattern of two circular apertures.

The reciprocity theorem suggests a method of designating the irradiance at P as the source is moved about over a surface Σ near S. Let a source of unit intensity be placed at P, and let the resulting irradiance be observed at each point of the pattern on Σ. Call this the *sensitivity*, a function of the surface, $\Lambda(\Sigma)$. Then if the source is placed at any point of Σ, the resulting irradiance at P can be calculated by multiplying the Λ of the point where the source is placed by the intensity of the source.

2.6 Huygens' Principle

By means of Huygens' principle it is possible to obtain by graphical methods the shape of a wave front at any instant if the wave front at an earlier instant is known. The principle may be stated as follows: Every point of a wave front may be considered as the source of a small secondary wavelet, which spreads in all directions from the point at the wave propagation velocity. A new wave front is found by constructing a surface tangent to all the secondary wavelets. If the velocity of propagation is not constant for all parts of the wave front, then each wavelet must be given its appropriate velocity.

An illustration of the use of Huygens' principle is given in fig. 2.10. The known wave front is shown by the arc $\Sigma\Sigma$, and the directions of propagation are indicated by small arrows. To determine the wave front after an interval Δt, with a wave velocity v, simply construct a series of spheres of radius $r = v\Delta t$ from each point of the original front $\Sigma\Sigma$. These spheres represent the secondary wavelets. The envelope enclosing the surfaces of the spheres represents the new wave front. This is the surface marked $\Sigma'\Sigma'$ in the figure. In this example the magnitude of the wave velocity is considered to be the same at all points.

It may be noted that Huygens' principle predicts the existence of a back-

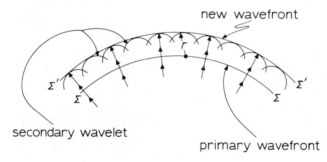

Fig. 2.10 Huygens' principle.

ward wave, which, however, is never observed. The explanation for this discrepancy can be found by examining the interference of the secondary wavelets throughout the space surrounding Σ. The main use of Huygens' principle is that it enables us to predict diffraction patterns. This will be elaborated in later chapters. The principle when first stated was a useful method for finding the shape of a new wave front, and little physical reality was attached to the secondary wavelets at that time. Later, as the wave nature of light came to be more fully understood, Huygens' principle took on a deeper significance than was at first supposed.

Problems

2.1 Given a diffraction screen which consists of a linear array of five identical apertures, which are located on the diffraction screen at $x=0$, $x=d$, $x=2d$, $x=4d$, and $x=8d$, respectively, then (a) draw a vector diagram to find the function G, and (b) make a rough sketch of $|G|^2$ as a function of the path difference δ, as described in sec. 2.4.

2.2 Consider a rectangular array of $m \times n$ identical apertures, as shown in fig. 2.11. (a) Determine the function G for the Fraunhofer diffraction pattern. (b) Write out all the terms of the function $|G|^2$, for the case $m=n=2$.

2.3 A monochromatic plane wave is normally incident on an array of identical apertures. The diffraction pattern from any one of the apertures is confined to small values of θ_2 and ϕ_2 (see fig. 2.5). The diffraction screen is

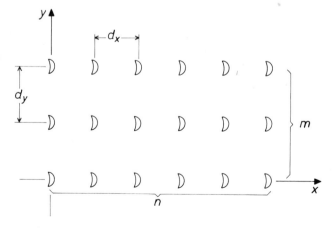

Fig. 2.11

then stretched in the x direction such that an aperture originally situated at (x, y) is shifted to (mx, y). Show that the function $|G|^2$ changes after such a deformation of the diffraction screen. Also discuss briefly the changes in the fringe pattern.

2.4 Show that when a rectangular diffracting array is deformed to a square array, or vice versa, the change in shape of the principal maximum of the diffraction pattern is in agreement with the result of the preceding problem.

2.5 Given a large number of identical apertures placed at random in a diffraction screen. Assume that no two apertures are so far apart that the condition of Fraunhofer diffraction is not satisfied or so close that they do not contribute independently to the diffraction. Assume that a monochromatic plane wave is incident normally on the diffraction screen. Determine the probabilistic distribution of $|G|$.

References

2.1 M. Born and E. Wolf, *Principles of Optics*, 2nd rev. ed., Pergamon Press, New York, 1964.

2.2 B. Rossi, *Optics*, Addison-Wesley, Reading, Mass., 1957.

2.3 J. M. Stone, *Radiation and Optics*, McGraw-Hill, New York, 1963.

2.4 F. W. Sears, *Optics*, Addison-Wesley, Reading, Mass., 1949.

2.5 A. Sommerfeld, *Optics* (Lectures on Theoretical Physics, vol. 4), Academic Press, New York, 1954.

2.6 M. Kline and I. W. Kay, *Electromagnetic Theory of Geometrical Optics*, Interscience, New York, 1965.

2.7 M. Françon, *Diffraction Coherence in Optics*, Pergamon Press, New York, 1966.

Chapter 3 A Scalar Treatment of Light

In Chap. 2 we treated the diffraction of light from a representation of Maxwellian electromagnetic theory. In studying diffraction in this way we were able to show how the interaction between the light from different apertures affects the complete diffraction pattern. This, however, was not a complete solution of the diffraction problem, since it did not explain in detail the role played by the sizes and shapes of the apertures. Thus the factor $|G|^2$ of eq. (2.13) was written out explicitly in eq. (2.14). The factor I_1, however, was only introduced as being the irradiance due to a single aperture, without explaining how this factor is affected by the characteristics of the aperture. In most cases it is extremely difficult to reach such a complete solution by the Maxwell technique. One problem that has been solved in this way is that of monochromatic light passing the straight edge of a thin conducting sheet. The solution was obtained by Sommerfeld (ref. 3.4, p. 247). However, most cases remain unsolved by the vector method.

Fortunately, however, there are scalar methods which can be applied, and which in most cases give diffraction results of high accuracy. It is not to be expected that these scalar methods can be effectively employed where the *polarization* of the light is an important factor. The early theories of Huygens and Young on the propagation of light waves lead readily to scalar methods of treatment. However, it was not until after the introduction of the Maxwell theory, in 1856, that Kirchhoff brought out, in 1882, a generally satisfactory scalar method. This is the method we propose to follow here. Readers interested in further details of the historical development are referred to *Principles of Optics* by Born and Wolf (ref. 3.1).

3.1 The Scalar Development

In Maxwell's theory a *wave equation* is derived in the form

$$\nabla^2 u - \frac{1}{c^2}\frac{\partial^2 u}{\partial t^2} = 0, \tag{3.1}$$

where c is the velocity of the wave and ∇^2 is the Laplacian differential operator

$$\nabla^2 = \frac{\partial^2}{\partial x^2} + \frac{\partial^2}{\partial y^2} + \frac{\partial^2}{\partial z^2}.$$

The u in equation (3.1) may represent either the electric vector **E** or the magnetic vector **H** of the field in free space. In the scalar treatment, we shall assume that the same eq. (3.1) applies, but with u in this case representing a purely scalar quantity, the *amplitude* of the wave.

In the following, we shall attempt to find solutions of eq. (3.1) which may be applied to diffraction phenomena. The results will not be exact, but in many cases will be very close to the facts disclosed by experiment. Note also that the treatment is not confined to light alone, but may also be applied to other forms of wave propagation (such as sound) in which the wavelength is small compared with aperture dimensions and in which polarization is unimportant.

Given a point source which radiates uniformly in all directions, (i.e., an *isotropic* source), the amplitude in the free field at a distance r from the source may be expressed as

$$u(r, t) = \frac{1}{r} f(r - ct). \tag{3.2}$$

It may be verified that the solution (3.2) satisfies eq. (3.1), except at $r = 0$. This singularity is unimportant, since in fact a physical source cannot have a zero radius. The f in eq. (3.2) represents some function of the quantity in parentheses. The form of f depends upon the nature of the source. If we are dealing with what we call a monochromatic wave, then f is a sine or cosine function, or in general

$$f(r - ct) = a \cos[k(r - ct) + \theta], \tag{3.3}$$

where a is a positive constant, k is the wave number $2\pi/\lambda$, and θ is a phase angle. If now eq. (3.3) is written in complex form and then substituted into eq. (3.2), but with $ae^{i\theta}$ replaced by the complex constant A,

$$u(r, t) = \frac{A}{r} e^{i(kr - \omega t)}. \tag{3.4}$$

The ω in eq. (3.4) is the angular frequency of the wave; $\omega = kc$. To recapitulate, eq. (3.4) gives the scalar amplitude of the wave from a point source of light which is monochromatic and isotropic, at a distance r from the source, in a free (i.e., unbounded) field.

It may also be said for the scalar eq. (3.1) that if two different forms of u are solutions then their sum is also a solution. This is called the *principle of superposition*, which holds for the scalar treatment of light, as it is in the vector theory.

3.2 Kirchhoff's Integral

Suppose that we wish to know the scalar amplitude at a point P that is in the combined field of a number of monochromatic sources, all of the same wavelength. Suppose also that none of the sources coincides with P. Then if we have complete knowledge of all the sources—their positions, amplitudes, and phases—the amplitude at P can be found from the principle of superposition. Kirchoff, however, proposed that such a knowledge of the sources is unnecessary, provided that at each point of an arbitrary surface Σ, surrounding P but not enclosing any source, the amplitude and its normal spatial derivatives are known.

The above idea was developed by Kirchhoff into what is known as Kirchhoff's integral, which is very useful in the scalar theory of diffraction. The development makes use of Green's second identity, which may be described as follows: Let U and V be any two complex scalar fields, functions of position only, and let a closed surface Σ enclose a region R in the space where U and V exist. Now if U and V are continuous and have continuous second derivatives both on and within the surface Σ, then

$$\iint_{\Sigma} \left(V \frac{dU}{dn} - U \frac{dV}{dn} \right) d\Sigma = \iiint_{\sigma} (V\nabla^2 U - U\nabla^2 V)\, d\sigma, \qquad (3.5)$$

where d/dn is a derivative normal to Σ in the outward direction, $d\Sigma$ is an element of surface, and $d\sigma$ an element of volume.

The space σ has the outer boundary Σ. Let us also give it an inner boundary Σ', which we set arbitrarily as a small sphere with its center at P, the point where the field amplitude is to be found. Figure 3.1 illustrates this situation. Here σ is the space between Σ and Σ' only, and the positive normals on the bounding surfaces are oriented away from σ, as indicated by the small arrows.

We have assumed that the entire scalar light wave is monochromatic. We can therefore express its amplitude u in complex form

$$u = u_0 e^{-i\omega t}, \qquad (3.6)$$

where u_0 is a function of position only. It follows from eq. (3.1) that u_0 satisfies the equation

$$\nabla^2 u_0 = -k^2 u_0, \qquad (3.7)$$

which is known as the Helmholtz equation. Here $k = \omega/c$, just as in eq. (3.4).

It was said in connection with eq. (3.5) that V and U could be any two scalar fields. Let us take them as

$$V = u_0, \qquad (3.8)$$

$$U = \frac{1}{r} e^{ikr}, \qquad (3.9)$$

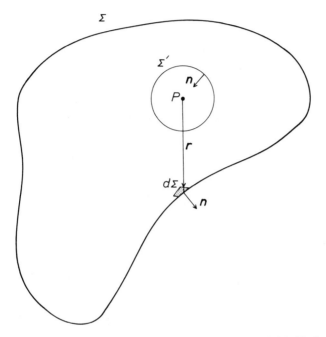

Fig. 3.1 The enclosed surfaces used in determining Kirchhoff's integral from Green's second identity.

where r is the radial distance from P. Note that the V and U given in (3.8) and (3.9) satisfy the continuity conditions on solutions of (3.5).

Just as with u_0 in eq. (3.7), U satisfies the wave equation

$$\nabla^2 U = -k^2 \frac{1}{r} e^{ikr}, \tag{3.10}$$

except at $r=0$. But since the point has been excluded from the space σ, eq. (3.10) is true for all of this space.

We can now substitute (3.8), (3.9), (3.10) and (3.7) into eq. (3.5), remembering that the surface integral must now be taken over both Σ and Σ'. We also note that U has been defined in such a way that the volume integral in eq. (3.5) becomes zero. Then

$$\int\!\!\int_{\Sigma}\left[u_0 \frac{d}{dn}\left(\frac{1}{r} e^{ikr} \right) - \frac{1}{r} e^{ikr} \frac{du_0}{dn} \right] d\Sigma$$
$$+ \int\!\!\int_{\Sigma'}\left[-u_0 \frac{d}{dr}\left(\frac{1}{r} e^{ikr} \right) + \frac{1}{r} e^{ikr} \frac{\partial u_0}{\partial r} \right] d\Sigma' = 0. \tag{3.11}$$

In the last term of the Σ' integral in eq. (3.11), d/dn has been replaced by $-\partial/\partial r$, since Σ' is a spherical surface. If the derivative of the first term of the Σ' integral in eq. (3.11) is carried out, this integral becomes

$$\int\int_{\Sigma'}\left(\frac{u_0}{r^2}-\frac{iku_0}{r}+\frac{1}{r}\frac{\partial u_0}{\partial r}\right)e^{ikr}\,d\Sigma'.$$

Let us now make the secondary spherical surface Σ' very small, but still with the point P at its center. The value of u_0 at every point of Σ' will then be practically the same as at P itself, so that u_0 may be called $u_0(P)$ everywhere in the Σ' integral. We also note that with r very small, e^{ikr} is approximately 1. Now the second and third terms of the Σ' integral, as expressed above, approach zero as r does, and the entire integral depends on the first term. The integral then reduces to $4\pi u_0(P)$. Putting this value into eq. (3.11) gives

$$u_0(P)=-\frac{1}{4\pi}\int\int_{\Sigma}\left[u_0\frac{d}{dn}\left(\frac{1}{r}e^{ikr}\right)-\frac{1}{r}e^{ikr}\frac{du_0}{dn}\right]\,d\Sigma.\qquad(3.12)$$

This is Kirchhoff's integral, and it says that the scalar field at a point P can be found, provided the field and its outward derivative is known at every point on a surface which surrounds P but does not contain any sources. In this equation, u_0 is the known field at each elementary area $d\Sigma$, and r is the distance of $d\Sigma$ from P.

Let us apply the Kirchhoff integral to the case of a single monochromatic point source, but for the moment without a diffraction screen. The situation is pictured in Fig. 3.2a, where r_1 and r_2 represent the distances of any point on the surface Σ from the source S and from the point P, respectively. Fig. 3.2b shows how the angle between $\mathbf{r}_1$ and $\mathbf{n}$, the normal to Σ, and the angle between $\mathbf{r}_2$ and $\mathbf{n}$ are designated.

The complex amplitude u_0 at the element $d\Sigma$ of the surface may be expressed as

$$u_0=\frac{A}{r_1}\exp(ikr_1),\qquad(3.13)$$

where A is a complex constant, and $k=\omega/c$. We can also express the normal derivative at $d\Sigma$ in terms of the angles designated in Fig. 3.2b:

$$\frac{du_0}{dn}=\frac{du_0}{dr_1}\cos(\mathbf{n},\mathbf{r}_1)=A\cos(\mathbf{n},\mathbf{r}_1)\left(-\frac{1}{r_1^2}+\frac{ik}{r_1}\right)\exp(ikr_1).\qquad(3.14)$$

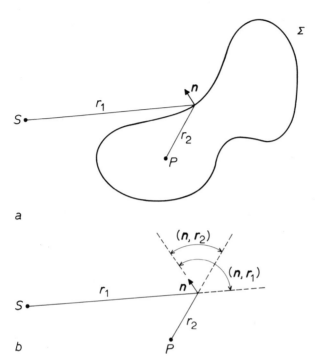

Fig. 3.2 Application of Kirchhoff's theorem to a single point source, without diffraction screen.

The r of eq. (3.12) is now r_2, and in a manner similar to the above,

$$\frac{d}{dn}\left[\frac{1}{r_2}\exp(ikr_2)\right]=\cos(\mathbf{n},\mathbf{r}_2)\left(-\frac{1}{r_2^2}+\frac{ik}{r_2}\right)\exp(ikr_2). \tag{3.15}$$

In eqs. (3.13), (3.14), and (3.15) we have evaluated the parts of eq. (3.12). Before the substitution, however, let us note that in most situations r_1 and r_2 are many wavelengths long. Suppose $r_1=N\lambda$, where N is a large number. Then $1/r_1^2=1/N^2\lambda^2$. The constant k can be expressed as $2\pi/\lambda$, so that $k/r_1=2\pi/N\lambda^2$. Since N is large, $1/r_1^2$ may be neglected in comparison with k/r_1 and the same is true of $1/r_2^2$. The Kirchhoff integral (3.12) in this case may then be written

$$u_0(P)=-\frac{ikA}{4\pi}\int\!\!\int_\Sigma\frac{1}{r_1r_2}\left[\cos(\mathbf{n},\mathbf{r}_2)-\cos(\mathbf{n},\mathbf{r}_1)\right]\exp\left[ik(r_1+r_2)\right]d\Sigma. \tag{3.16}$$

The quantity $\left[\cos(\mathbf{n},\mathbf{r}_2)-\cos(\mathbf{n},\mathbf{r}_1)\right]$ is known as the *obliquity factor*.

The actual shape of the enclosure around P is at our disposal, and in order to lead up to the use of a plane diffraction screen let us give Σ the form shown in fig. 3.3. The figure shows a cross section of a spherical surface of radius R, intersected by a plane which we later will make the plane of the diffraction screen. P is at the center of the spherical part of Σ. Let us call this spherical section Σ_0, and that part of the integral in eq. (3.16) contributed by Σ_0 will be called W. As shown in Fig. 3.3, the normal to Σ_0 at any point is called $\mathbf{n}'$, and the distance from S is called r_1'. Note also that the angle $(\mathbf{n}', \mathbf{R})$ is zero. Then the Σ_0 part of the integral in eq. (3.16) is

$$W = \iint_{\Sigma_0} \frac{1}{r_1' R} \left[1 - \cos(\mathbf{n}', \mathbf{r}_1') \right] \exp\left[ik(r_1' + R) \right] d\Sigma_0. \tag{3.17}$$

The obliquity factor in this equation can be expressed in terms of the distances involved, including the distance between S and P, which we will call l. For, by the law of cosines,

$$l^2 = r_1'^2 + R^2 - 2r_1' R \cos(\mathbf{n}', \mathbf{r}_1')$$
$$= (r_1' - R)^2 + 2r_1' R \left[1 - \cos(\mathbf{n}', \mathbf{r}_1') \right].$$

Then,

$$1 - \cos(\mathbf{n}', \mathbf{r}_1') = \frac{l^2 - (r_1' - R)^2}{2r_1' R}. \tag{3.18}$$

Substituting this in eq. (3.17), and considering only the absolute value

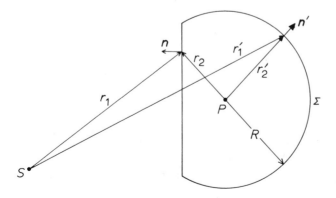

Fig. 3.3 The surface Σ in the form of a sphere intersected by a plane.

(to reduce the exponential to unity), we obtain

$$|W| = \int\int_{\Sigma_0} \frac{l^2 - (r_1' - R)^2}{2 r_1'^2 R^2} \, d\Sigma_0 \, .$$

It can be seen from eq. (3.18) that $l^2 - (r_1' - R)^2$ is always a positive number, hence l^2 must be the greater term, and $l^2 - (r_1' - R)^2 < l^2$. We can therefore put

$$|W| < \int\int_{\Sigma_0} \frac{l^2}{2 r_1'^2 R^2} \, d\Sigma_0 \, .$$

We will now let R become infinitely large. This can be done without including the source S inside the surface Σ, even though l is a fixed finite distance, provided the plane part of Σ lies between S and P. Note also that as R increases, r_1' also increases and in the limit can be taken as equal to R. Then

$$|W| < \frac{l^2}{2 R^4} \int\int_{\Sigma_0} d\Sigma_0 < \frac{2\pi l^2}{R^2} \, . \tag{3.19}$$

The last inequality comes from the fact that the integral over Σ_0 is less than that over a whole sphere, and the latter is given by $4\pi R^2$. Now as R becomes infinite, $|W|$, and therefore the whole contribution of Σ_0 to the integral of eq. (3.16), becomes zero. The scalar amplitude at P can then be found from the amplitudes and their normal derivatives at all points of a plane of infinite extent, lying between S and P.

It may now be seen how the scalar amplitude due to a plane diffracting screen may be evaluated. Let the plane of the screen coincide with the plane part of the surface Σ of fig. 3.3, and let the screen be opaque everywhere except for a finite region in which there are apertures. So far as the point P is concerned, u_0 is zero on all the opaque parts of the screen, and the integration of eq. (3.16) need be taken only over the areas where the apertures are located.

Figure 3.4 illustrates the situation for a single source and a diffraction screen with a single aperture. The surface Σ is shown as lying behind the screen, to emphasize the fact that the amplitude of the light reaching P from the opaque part of the screen is zero. In the case of Fraunhofer diffraction, both r_1 and r_2, and also the shortest distances of S and P from the screen, are to be taken as large compared with the wavelength of light.

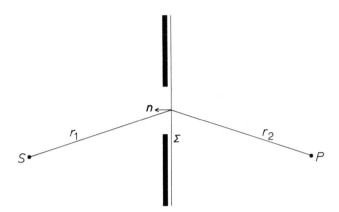

Fig. 3.4 Application of Kirchhoff's integral to a plane diffracting screen.

Over the unshadowed parts of Σ we assume that the complex wave amplitude is given by eq. (3.13). Then by the Kirchhoff theorem the amplitude over an element of surface $d\Sigma$ will be given by

$$Kd\Sigma = -\frac{ikA}{4\pi r_1} \left[\cos(\mathbf{n}, \mathbf{r}_2) - \cos(\mathbf{n}, \mathbf{r}_1)\right] \exp(ikr_1)\, d\Sigma, \tag{3.20}$$

and its contribution to the amplitude at P is

$$du_0(P) = \frac{Kd\Sigma}{r_2} \exp(ikr_2). \tag{3.21}$$

This is to be integrated only over the unshadowed parts of Σ. Application of this method to diffraction will be made in the next chapter.

It should be mentioned that the choice of a plane surface Σ is arbitrary, and that other surfaces may be chosen. For example, in fig. 3.5, with a circular aperture and S on the axis of the aperture, it might be convenient to take Σ as a segment of a sphere with S at its center. In this case r_1 is constant, and the angle $(\mathbf{n}, \mathbf{r}_1)$ is always π. Also, the angle θ of the figure is the same as angle $(\mathbf{n}, \mathbf{r}_2)$. Hence eq. (3.16) simplifies to

$$u_0(P) = -\frac{ikA}{4\pi r_1} \exp(ikr_1) \int\int_\Sigma \frac{1}{r_2} \left[1 + \cos\theta\right] \exp(ikr_2)\, d\Sigma. \tag{3.22}$$

However, in most cases to be treated, the plane form of Σ will be more convenient.

It will be recognized that the Kirchhoff scalar method, using the wave

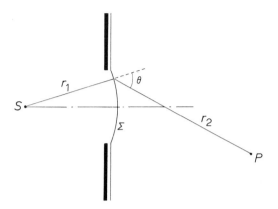

Fig. 3.5 Integration surface in the form of a spherical segment.

amplitude at each element of area on an intermediate surface, agrees with
Huygens' principle, which regards each point of a wave front as a new
source. However, if we ask whether the Kirchhoff method agrees with
the electromagnetic theory, the answer must be "not quite." We have
assumed that the amplitude at all points of an aperture is the same as
it would be without the presence of the screen. This is true for all parts
of the aperture which are more than a few wavelengths of light distant
from an edge of the screen. For points very close to an edge, however,
the field is affected, and it is also different from zero close to the edge on
the shadowed side. Now an aperture does not have to be very large for
most of its area to be far from the edges, and in these cases the result
of using the scalar method is accurate, and agrees with experiment. If
the aperture is a very small hole or very narrow slit, however, the scalar
calculation does not give the real conditions. The scalar treatment of
diffraction remains an approximate method.

It may be seen that for the scalar method embodied in eq. (3.16), the.
Reciprocity theorem mentioned in chap. 2 still holds. An interchange
of S and P, the source and the point of observation, involves the interchange of r_1 and r_2 in eq. (3.16) and a reversal in the direction of the
normal **n**. These changes leave the complete equation exactly as it was,
and the reciprocity is confirmed.

3.3 Babinet's Principle

Let us assume two diffraction screens such that when one is superimposed
on the other, the open region of one of the screens is the opaque region of

the other. Such two diffraction screens are said to be *complimentary* to each other. Let one of the diffraction screens be illuminated by a mono-chromatic point source of light; then the complex amplitude $U_{01}(P)$ of the diffraction pattern may be found. Now if the diffraction screen is replaced by its complimentary screen, then the complex light field $U_{02}(P)$ may be obtained without direct calculation by the following subtraction:

$$U_{02}(P) = U_{00}(P) - U_{01}(P), \qquad\qquad (3.23)$$

where $U_{02}(P)$ and $U_{00}(P)$ are the respective complex light amplitudes at P due to the complimentary screen and due to no screen. The equation is known as *Babinet's principle*. If any two of the complex amplitudes in eq. (3.23) were obtained, say, from the evaluation of Kirchhoff's integral, then the third could be found by simple addition or subtraction of the complex quantities. It may be emphasized that the generalization of Babinet's principle will be extremely useful in a number of cases, par-ticularly in the Fraunhofer diffractions.

References

3.1 M. Born and E. Wolf, *Principles of Optics*, 2nd rev. ed., Pergamon Press, New York, 1964.

3.2 J. M. Stone, *Radiation and Optics*, McGraw-Hill, New York, 1963.

3.3 B. Rossi, *Optics*, Addison-Wesley, Reading, Mass., 1957.

3.4 A. Sommerfeld, *Optics* (Lectures on Theoretical Physics, vol. 4), Academic Press, New York, 1954.

3.5 J. W. Goodman, *Introduction to Fourier Optics*, McGraw-Hill, New York, 1968.

3.6 W. K. H. Panofsky and M. Phillips, *Classical Electricity and Mag-netism*, 2nd ed., Addison-Wesley, Reading, Mass., 1962.

3.7 J. A. Stratton, *Electromagnetic Theory*, McGraw-Hill, New York, 1941.

3.8 H. Margenau and G. M. Murphy, *The Mathematics of Physics and Chemistry*, vol. 1, 2nd ed., Van Nostrand, New York, 1956.

Chapter 4 Fraunhofer and Fresnel Diffraction

Our understanding of both Fraunhofer and Fresnel diffraction (defined in chap. 2) may be improved by considering the application of Kirchhoff's integral. This application will be made in the present chapter. In both cases it will be assumed that aperture dimensions are large compared with the wavelengths of the incident light, and the results will apply only where this assumption is valid. Rectangular and circular apertures will be treated in both cases. In the Fraunhofer case it will be found that a simplified form of the Kirchhoff integral may be used. Finally, the resolving power for a general imaging system will be discussed.

4.1. Fraunhofer Diffraction

Assume a situation like that of fig. 4.1, where S is a point source of monochromatic light, P is the point of observation, and between them is a plane diffraction screen having an open aperture. A strict Fraunhofer case is assumed; i.e., the distances of both S and P from the screen are large compared with the aperture dimensions, and no lenses or other devices are used to increase the effective distances. The surface for Kirchhoff integration, Σ, is assumed to lie very closely behind the diffracting screen. That is to say, the plane of Σ is the same as the plane of the screen, but Kirchhoff integration takes place only where there are open apertures. The coordinate axes are indicated in the figure.

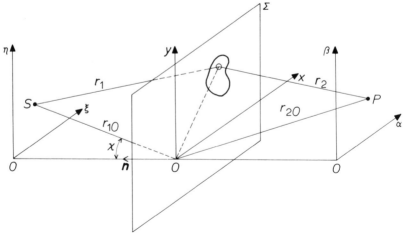

Fig. 4.1 Relation of source, screen, and observation point for the application of Kirchhoff's integral to Fraunhofer diffraction.

Under the Fraunhofer conditions, some simplifying approximations can be made. Thus $\mathbf{n}$, the normal to the screen at O, makes very nearly the same angle with r_1 as it does with r_{1O}, and the same is true on the P side of the screen.

That is,

$$\cos(\mathbf{n}, \mathbf{r}_1) \simeq \cos(\mathbf{n}, \mathbf{r}_{1O})$$

and

$$\cos(\mathbf{n}, \mathbf{r}_2) \simeq \cos(\mathbf{n}, \mathbf{r}_{2O}).$$

These approximations may be substituted in the Kirchhoff integral, eq. (3.16). Also, in the same equation we can put

$$\frac{1}{r_1 r_2} \simeq \frac{1}{r_{1O} r_{2O}}. \tag{4.1}$$

The $r_1 + r_2$ in the exponent of eq. (3.16) may be written

$$r_1 + r_2 \equiv r_{1O} + r_{2O} + (r_1 - r_{1O}) + (r_2 - r_{2O}).$$

But by eqs. (2.7) and (2.8), the difference of path $r_1 - r_{1O}$ may be replaced by $-(x \sin\theta_1 + y \sin\phi_1)$, and $r_2 - r_{2O}$ by $-(x \sin\theta_2 + y \sin\phi_2)$, where

$$\sin\theta_1 = \frac{\xi}{r_{1O}}, \quad \sin\phi_1 = \frac{\eta}{r_{1O}}, \quad \sin\theta_2 = \frac{\alpha}{r_{2O}}, \quad \text{and} \quad \sin\phi_2 = \frac{\beta}{r_{2O}}.$$

Then

$$r_1 + r_2 = r_{1O} + r_{2O} - (x \sin\theta_1 + y \sin\phi_1) - (x \sin\theta_2 + y \sin\phi_2). \tag{4.2}$$

We can also take advantage of the fact that the deviation of the diffracted light is small. Thus if we take a point (α_0, β_0) in the observing plane for which $\theta_2 = -\theta_1$ and $\phi_2 = -\phi_1$, only points close to (α_0, β_0) in angular distance as seen from the aperture will receive appreciable irradiance. Then all points which lie within the appreciable diffraction pattern will be *almost* in line with S and the aperture. Thus if we compare the angle χ, as shown in fig. 4.1, with the angles of fig. 3.2 under the condition of r_2 opposite to r_1, we see that $\chi \simeq (\mathbf{n}, \mathbf{r}_{2O})$ and that $(\mathbf{n}, \mathbf{r}_{1O}) \simeq \pi - \chi$. We then have

$$\cos(\mathbf{n}, \mathbf{r}_{2O}) - \cos(\mathbf{n}, \mathbf{r}_{1O}) \simeq 2\cos\chi. \tag{4.3}$$

The angle χ may be called the *angle of incidence*.

Now by substituting the approximations (4.1), (4.2) and (4.3) in Kirchhoff's integral, eq. (3.16), we have

$$U_0(P) = -\frac{ikA}{4\pi r_{10} r_{20}} \, 2 \cos\chi \, \exp\left[ik(r_{10}+r_{20})\right]$$

$$\times \iint\limits_{\substack{\text{over the}\\ \text{apertures}}} \exp\left\{-ik\left[x(\sin\theta_1+\sin\theta_2)+y(\sin\phi_1+\sin\phi_2)\right]\right\} dx\,dy.$$

$$(4.4)$$

It will be convenient to make the integral in (4.4) apply over the entire Kirchhoff surface Σ by introducing a complex transmission function

$$T(x, y) = |T(x, y)| \, e^{i\phi(x, y)} \tag{4.5}$$

which would be defined in the present case by

$$T(x, y) = \begin{cases} 1, & \text{over the apertures} \\ 0, & \text{otherwise} \end{cases}$$

Equation (4.4) is then written

$$U_0(P) = K \iint\limits_{\Sigma} T(x, y) \exp\left\{-ik\left[x(\sin\theta_1+\sin\theta_2)\right.\right.$$

$$\left.\left. +y(\sin\phi_1+\sin\phi_2)\right]\right\} dx\,dy, \tag{4.6}$$

where

$$K = -\frac{ikA}{4\pi r_{10} r_{20}} \, 2\cos\chi \, \exp\left[ik(r_{10}+r_{20})\right]$$

is a complex constant.

Now at the point (α_0, β_0) of the observing plane, where $\theta_2 = -\theta_1$ and $\phi_2 = -\phi_1$, the exponential in the integral of (4.6) would be unity, and

$$U_0(\alpha_0, \beta_0) = K \iint\limits_{\Sigma} T(x, y) \, dx\,dy.$$

If we now define another function $Q(P)$, such that

$$Q(P) = \frac{\displaystyle\iint\limits_{\Sigma} T(x, y) \exp\left\{-ik\left[x(\sin\theta_1+\sin\theta_2)+y(\sin\phi_1+\sin\phi_2)\right]\right\} dx\,dy}{\displaystyle\iint\limits_{\Sigma} T(x, y) \, dx\,dy},$$

$$(4.7)$$

then eq. (4.6) may be rewritten as

$$U_0(P) = U_0(\alpha_0, \beta_0) \, Q(P). \tag{4.8}$$

Note that $Q(\alpha_0, \beta_0) = 1$. Since eq. (4.8) deals in amplitudes, the average rate of energy per unit area (irradiance) may be obtained by squaring absolutes,

$$I(P) = I_0 |Q(P)|^2, \tag{4.9}$$

where $I_0 = |U_0(\alpha_0, \beta_0)|^2$, the irradiance in the undeviated direction of light from the source. When this is known, eq. (4.9) tells us that irradiance elsewhere in the diffraction pattern can be found by multiplying I_0 by the factor $|Q(P)|^2$, which, from (4.7), depends on the overall disposition of the apertures. Equation (4.9) is very similar to eq. (2.13). In fact, for identical apertures the two equations are identical.

4.2 The Fourier Transform in Fraunhofer Diffraction

Let us return to the simplified form of Kirchhoff's integral for Fraunhofer diffraction, eq. (4.6).

If we recall the following relations in the defined coordinate systems:

$$\sin\theta_1 = \frac{\xi}{r_{10}}, \quad \sin\phi_1 = \frac{\eta}{r_{10}}$$

and

$$\sin\theta_2 = \frac{\alpha}{r_{20}}, \quad \sin\phi_2 = \frac{\beta}{r_{20}},$$

it is clear that eq. (4.6) can be thought of as a two-dimensional Fourier integral. For example if we let the source be a monochromatic point source, located at (ξ_0, η_0) of the source plane, then eq. (4.6) may be written as

$$U_0(P) = K \iint_\Sigma T(x, y) \exp[-i(\mu x + v y)] \, dx \, dy, \tag{4.10}$$

where

$$\mu = k\left(\frac{\xi_0}{r_{10}} + \frac{\alpha}{r_{20}}\right), \quad v = k\left(\frac{\eta_0}{r_{10}} + \frac{\beta}{r_{20}}\right).$$

Equation (4.10) is the Fourier transform of $T(x, y)$ which occurs in the observation plane. This occurrence of the Fourier transformation in Fraunhofer diffraction can be seen from the examples of rectangular and circular apertures, which will be given in the next section. More specifically, we can define two new variables

$$\mu' = k\left(\frac{\xi}{r_{10}} + \frac{\alpha}{r_{20}}\right), \quad v' = k\left(\frac{\eta}{r_{10}} + \frac{\beta}{r_{20}}\right).$$

Then eq. (4.6) can be written

$$U_0(P) = K \int\int_\Sigma T(x, y) \exp\left[-i(\mu'x + v'y)\right] dx\, dy. \tag{4.11}$$

4.3 Examples of Fraunhofer Diffraction

RECTANGULAR APERTURE

In fig. 4.2 we show a rectangular aperture with dimensions a and b. In this case the transmission function for the diffraction screen will be

$$T(x, y) = \begin{cases} 1 & \left(-\frac{a}{2} \leq x \leq \frac{a}{2}, \quad -\frac{b}{2} \leq y \leq \frac{b}{2}\right). \\ 0 & \text{otherwise} \end{cases} \tag{4.12}$$

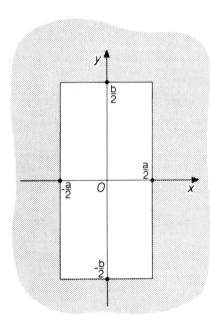

Fig. 4.2 Geometry of a rectangular aperture.

For convenience in notation, let us define two new variables,

$$\mu = \frac{ka}{2\pi}(\sin\theta_1 + \sin\theta_2), \tag{4.13}$$

$$v = \frac{kb}{2\pi}(\sin\phi_1 + \sin\phi_2). \tag{4.14}$$

By substituting (4.12), (4.13), and (4.14) in eq. (4.7), we have for the normalized complex amplitude

$$Q(P) = \frac{1}{ab}\int_{-a/2}^{a/2}\exp\left(-i\frac{2\pi\mu x}{a}\right)dx\int_{-b/2}^{b/2}\exp\left(-i\frac{2\pi v y}{b}\right)dy$$

$$= \frac{\sin\pi\mu}{\pi\mu}\cdot\frac{\sin\pi v}{\pi v}. \tag{4.15}$$

This normalized amplitude is plotted in fig. 4.3 as a function of μ, for a fixed value of v. It has a maximum when $\mu = 0$. Although $Q(P)$ is in general complex, by proper choice of the coordinate system in this special problem $Q(P)$ can be made real for all values of μ and v. The manner in which $Q(P)$ alternates between positive and negative values is shown in fig. 4.4. On the lines separating the regions in this figure, $Q(P)$ is zero.

From the amplitude (4.15), the irradiance can be written as

$$I(P) = I_0\frac{\sin^2\pi\mu}{(\pi\mu)^2}\frac{\sin^2\pi v}{(\pi v)^2}, \tag{4.16}$$

where I_0 is the irradiance when both μ and v are zero. $I(P)$ is plotted in fig. 4.5, again for a fixed value of v. The actual Fraunhofer diffraction pattern for a rectangular aperture is shown in fig. 4.6.

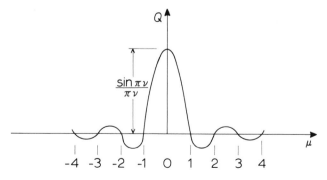

Fig. 4.3 The diffraction amplitude $Q(p)$ as a function of μ for a given value of v.

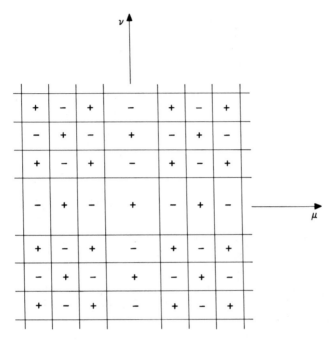

Fig. 4.4 Positive and negative regions of $Q(p)$ in the (μ, v) plane.

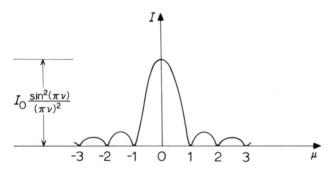

Fig. 4.5 Irradiance $I(p)$ as a function of μ for a given value of v.

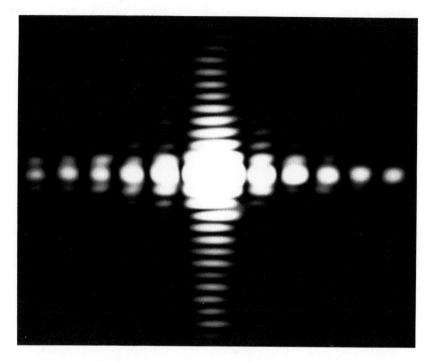

Fig. 4.6 Fraunhofer diffraction pattern of a rectangular aperture.

The variables μ and v, defined in eqs. (4.13) and (4.14) and used in (4.15) and (4.16), are dimensionless. Actual positions of the pattern on the observing plane, with reference to source and aperture, can be found from the θ's and ϕ's of (4.13) and (4.14). The coordinates α and β on the observing screen, as used in fig. 4.1, may be found from

$$\alpha = \frac{\lambda l}{a}\,\mu, \qquad \beta = \frac{\lambda l}{b}\,v, \tag{4.17}$$

where l is the separation between the diffraction aperture and the observing screen.

The variation of irradiance along the α axis of the diffraction pattern is shown in fig. 4.7. It shows a strong central maximum, having an angular half-width of λ/a radians, and a series of much smaller secondary maxima on each side. The photograph of fig. 4.6 shows the same effect. The same fig. 4.7 also holds for the β axis, if α and a in the figure are replaced by β and b.

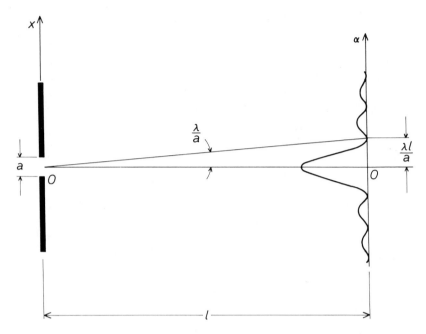

Fig. 4.7 Irradiance along one axis for a rectangular aperture. Fraunhofer diffraction.

CIRCULAR APERTURE

In examining the Fraunhofer diffraction pattern from a circular aperture, it will be assumed that the incident light is normal to the plane of the aperture. This is not the most general case, but it simplifies the calculation. Let a circular aperture of diameter D be centered at the origin of the diffraction screen [the (x, y) plane] of fig. 4.1. To satisfy the normal incidence assumption, the source S in fig. 4.1 would lie at the origin of the (ξ, η) coordinates. This results in making both θ_1 and ϕ_1, as defined in eq. (4.2), equal to zero, and these values can be substituted in eq. (4.7).

It is apparent that, with the circular aperture and the normal incidence of light, there must be complete symmetry of diffraction about the normal line. That is, if we find the variation of pattern along one radius from O in the observing plane, it will be the same along every other radius from O. Let us then find this pattern along the α axis, for which we can put $\phi_2 = 0$, and need only consider positive values of θ_2.

Let us remind ourselves that the double integrals over Σ in eq. (4.7), together with the factor $T(x, y)$ in the integrand, means that integration occurs only over the area of the aperture. In this case both x and y vary from $-D/2$ to $+D/2$. The denominator of eq. (4.7) means simply the area of the aperture, in this case $\pi D^2/4$. In order to obtain the variation along the α axis, we write eq. (4.7) as

$$Q(P) = \frac{4}{\pi D^2} \int\int_{\Sigma} \exp[-ikx \sin\theta_2]\, dx\, dy.$$

Considering for the moment a constant x, the corresponding limits of y are

$$\pm\sqrt{\frac{D^2}{4} - x^2},$$

and the entire integration with respect to y gives the factor

$$2\sqrt{\frac{D^2}{4} - x^2}.$$

If we make the further substitution

$$\mu = \frac{kD}{2\pi} \sin\theta_2, \tag{4.18}$$

then

$$Q(P) = \frac{4}{\pi D^2} \int_{-D/2}^{D/2} 2\left(\frac{D^2}{4} - x^2\right)^{1/2} \exp\left[-i\frac{2\pi}{D}\mu x\right] dx.$$

Now let

$\tau = 2x/D$, so that $x = D\tau/2$, $dx = (D/2)\,d\tau$, and $\tau = 1$ when $x = D/2$.

Then

$$Q(P) = \frac{2}{\pi} \int_{-1}^{1} (1-\tau^2)^{1/2} \exp[-i\pi\mu\tau]\, d\tau. \tag{4.19}$$

The variable of integration here is τ, which is directly related to x, which is one coordinate in the circular aperture. The quantity μ is related by eq. (4.18) to θ_2, which gives angular position in the diffracted beam, and is not related explicitly to x.

If we divide the integral (4.19) into two, from -1 to 0, and from 0 to 1, we notice that $(1-\tau^2)$ remains the same for each $\pm$ value of τ, and that the sign is changed in the exponential. Now the sum of $e^{i\pi\mu\tau}$ and $e^{-i\pi\mu\tau}$ is $2\cos(\pi\mu\tau)$, which is a real quantity. Equation (4.19) therefore becomes

$$Q(P) = \frac{4}{\pi} \int_0^1 (1-\tau^2)^{1/2} \cos(\pi\mu\tau)\, d\tau. \tag{4.20}$$

Integrating, we obtain the first order Bessel function

$$Q(P) = \frac{2J_1(\pi\mu)}{\pi\mu}, \quad \mu \geq 0. \tag{4.21}$$

The amplitude Q in the observing plane is plotted as a function of μ in fig. 4.8, with the variation shown along a complete radius of the pattern. The pattern will be the same along every other radius, so that the complete pattern is a series of concentric rings about O in the observing plane. The amplitude is a maximum where μ (and therefore θ_2) is zero. Figure 4.9 indicates how the amplitudes in the pattern vary between positive and negative values. Note that μ is directly proportional to the radius of the pattern. For, if we call such a radius ρ, then $\rho = l\tan\theta_2$, where l is the separation between the aperture plane and the observing plane. But under Fraunhofer conditions, θ_2 is so small that the tangent may be put equal to the sine, and by eq. (4.18) we have

$$\mu = \frac{kD}{2\pi l}\rho = \frac{D}{\lambda l}\rho.$$

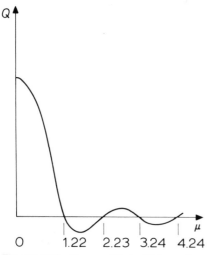

Fig. 4.8 Diffraction amplitude $Q(p)$ as a function of μ for a circular aperture.

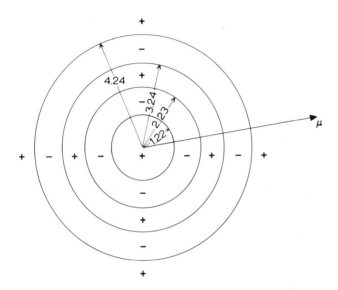

Fig. 4.9 Positive and negative regions of $Q(p)$.

Corresponding to the amplitude eq. (4.21), the irradiance in the diffraction pattern from a circular aperture is given by

$$I(P) = I_0 \left[\frac{2J_1(\pi\mu)}{\pi\mu} \right]^2, \quad \mu \geq 0. \tag{4.22}$$

The variation of $I(P)$ across a radius of the pattern is plotted in fig. 4.10.

Equation (4.22) was first derived by G. B. Airy, and the pattern formed by diffraction from a circular aperture is known as the *Airy pattern*. A photograph of the Airy pattern is given in fig. 4.11. The *Airy disk*, as the central circle of irradiance in the pattern is called, extends outward from the center to $\mu = 1.22$, or to a radius $\rho = 1.22\lambda l/D$. Eighty-four percent of the total power in the pattern is within this central disk. The first null in the pattern comes at a somewhat larger radius than with the rectangular aperture, where (by fig. 4.7) it is given as $\lambda l/a$, the a being comparable with D in the circular case.

4.4 Fresnel Diffraction

As explained in chap. 2, we apply the term "Fresnel" when the distances from source to diffraction screen, or from the latter to the observing plane, or both, are not great compared with the dimensions of the diffracting apertures. We are forced to abandon the assumption that the rays from

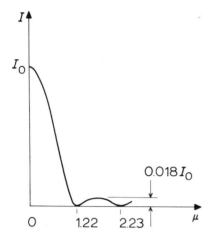

Fig. 4.10 Irradiance $I(p)$ as a function of μ for a circular aperture.

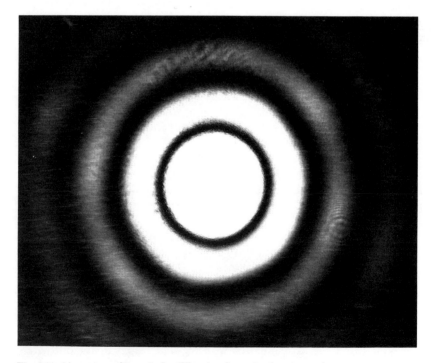

Fig. 4.11 Airy pattern (Fraunhofer diffraction for a circular aperture).

a point source to the different points of an aperture are essentially parallel, or that the geometrical rays from these points to the point of observation are parallel. Thus the simplified equations which we have found for Fraunhofer diffraction can no longer be used, and the problem of finding the diffraction pattern becomes much more difficult. However, the patterns from apertures of certain shapes may be treated, at least in part. In particular, we shall take up the case of the rectangular aperture, and then the more difficult case of the circular aperture.

RECTANGULAR APERTURE

In fig. 4.12 a rectangular aperture with dimensions a and b is shown. A monochromatic point source is at S, and let SO be the perpendicular from S to the plane of the aperture. Let this line SO be projected to the observing plane at P, and let us observe the irradiance at P. To explore a pattern, we should of course examine other points about P. Instead of this, let us find the changes in irradiance at P as the aperture is moved to different positions in the diffraction screen, with S, O, and P remaining fixed. The pattern found in this way will not differ greatly from that with a fixed aperture and P moved, provided the displacements given the aperture are small compared with the distances SO and OP (r_{10} and r_{20}). There are three cases which can be treated, for different assumptions with regard to the sizes of a and b.

For the first case, let a and b both be small compared with r_{10} and r_{20}. This would seem to be a return to the conditions assumed for the

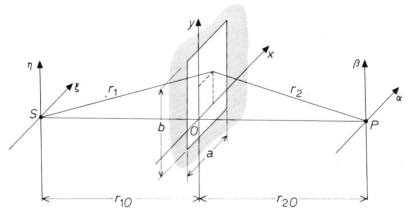

Fig. 4.12 Geometry for the application of Kirchhoff's integral to Fresnel diffraction at a rectangular aperture.

Fraunhofer diffraction. However, the treatment here will be somewhat different. For one thing, it will be assumed that the incidence of light on the plane of the aperture is nearly normal. If we compare fig. 4.12 with fig. 4.1, we see that the angle χ in the latter has been made zero in the present case. Thus we now have a further simplification of eq. (4.3),

$$\cos(\mathbf{n}, \mathbf{r}_{20}) - \cos(\mathbf{n}, \mathbf{r}_{10}) \simeq 2.$$

We can, however, make the same assumption as we did in eq. (4.1), that

$$\frac{1}{r_1 r_2} \simeq \frac{1}{r_{10} r_{20}}.$$

If these conditions are substituted in the Kirchhoff equation (3.16), we can then write down the complex amplitude at the observation point:

$$U_0(P) = -\frac{iA}{\lambda r_{10} r_{20}} \int_{y_1}^{y_2} \int_{x_1}^{x_2} \exp\left[ik(r_1 + r_2)\right] dx\, dy, \tag{4.23}$$

where λ in the constant part has been substituted for its equivalent $2\pi/k$, and the surface integral over Σ has been split into the two integrals over x and y.

Our estimate of $r_1 + r_2$ must be a little different than in the Fraunhofer case, eq. (4.2). If we compare fig. 4.12 with fig. 2.9, which defines the θ's and ϕ's, we see that now all these angles are zero, and eq. (4.2) could be written $r_1 + r_2 = r_{10} + r_{20}$. This comes about by the neglect of certain quantities in the formulation of eq. (4.2), and is a more extreme simplification than we wish to make in the present case. Instead, we will point out that $r_1^2 = r_{10}^2 + x^2 + y^2$ (see fig. 4.12), and it is a sufficiently accurate approximation to set

$$r_1 \simeq r_{10} + \frac{x^2 + y^2}{2r_{10}}. \tag{4.24}$$

Similarly, on the other side of the aperture,

$$r_2 \simeq r_{20} + \frac{x^2 + y^2}{2r_{20}}. \tag{4.25}$$

Adding these,

$$r_1 + r_2 \simeq r_{10} + r_{20} + (x^2 + y^2)\frac{r_{10} + r_{20}}{2r_{10} r_{20}}. \tag{4.26}$$

Before substituting (4.26) in (4.23), it will be convenient to introduce two new dimensionless variables,

$$\mu = x \left[\frac{2(r_{10} + r_{20})}{\lambda r_{10} r_{20}} \right]^{1/2}, \qquad v = y \left[\frac{2(r_{10} + r_{20})}{\lambda r_{10} r_{20}} \right]^{1/2}. \tag{4.27}$$

Equation (4.23) now becomes

$$U_0(P) = -\frac{iA}{2(r_{10} + r_{20})} \exp\left[ik(r_{10} + r_{20})\right] \int_{\mu_1}^{\mu_2} \exp\left(\frac{i\pi\mu^2}{2}\right) d\mu$$

$$\times \int_{v_1}^{v_2} \exp\left(\frac{i\pi v^2}{2}\right) dv. \tag{4.28}$$

If we designate by $U_{00}(P)$ the amplitude of the irradiance at P when no screen is interposed between S and P, we would have

$$U_{00}(P) = \frac{A}{r_{10} + r_{20}} \exp\left[ik(r_{10} + r_{20})\right],$$

and the constant factor in eq. (4.28) is thus $-(i/2)\, U_{00}(P)$.

The exponentials under the integral signs in eq. (4.28) may be separated into real and imaginary parts by means of Euler's equation:

$$\exp\left(\frac{i\pi t^2}{2}\right) = \cos\left(\frac{\pi t^2}{2}\right) + i \sin\left(\frac{\pi t^2}{2}\right),$$

where t is used for either μ or v. Let us also use the variable v to represent any of the limits of integration, and define the quantities

$$C(v) = \int_0^v \cos\left(\frac{\pi t^2}{2}\right) dt, \qquad S(v) = \int_0^v \sin\left(\frac{\pi t^2}{2}\right) dt. \tag{4.29}$$

These are called *Fresnel integrals*. Tables of the Fresnel integrals may be found in *Tables of Higher Functions* by Jahnke, Emde, and Lösch (ref. 4.7).

Using the notation of (4.29), eq. (4.28) can be written

$$U_0(P) = -\frac{i}{2} U_{00}(P) \{C(\mu_2) - C(\mu_1) + i[S(\mu_2) - S(\mu_1)]\}$$

$$\times \{C(v_2) - C(v_1) + i[S(v_2) - S(v_1)]\}. \tag{4.30}$$

Equation (4.30) can be evaluated by a graphical method. Let v be given different values, both positive and negative, and let the corresponding values of $C(v)$ and $S(v)$, from (4.29), be plotted against each other in the complex plane, with $S(v)$ on the imaginary axis. The resulting curve is called *Cornu's spiral* (fig. 4.13). The values of v used in calculating the C and S coordinates are plotted along the curve. When v is zero both $C(v)$ and $S(v)$ are zero, and as v is increased either positively or negatively the curve spirals inwardly to the points F and E, which are given by $C(\infty)=S(\infty)=1/2$, and $C(-\infty)=S(-\infty)=-1/2$. The curve is symmetrical about the origin. It may be pointed out that equal increments in v, anywhere along the curve, add equal lengths to the curve. The turns of the curve become more and more tightly packed together as points E and F are approached.

In making use of Cornu's spiral for finding the irradiance at P, one must first find the limits of μ and v. Let the center of the rectangular aperture have the coordinates x_0 and y_0. The limits of x are then $x_0-(a/2)$ and $x_0+(a/2)$, and those of y are $y_0-(b/2)$ and $y_0+(b/2)$. These four limits

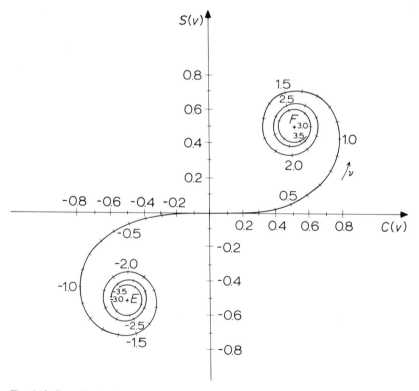

Fig. 4.13 Cornu's spiral.

are then substituted in the appropriate parts of eqs. (4.27) to get μ_1, μ_2, v_1, and v_2. Each of these, used as the value of v on the curve, gives a point from which the C and S values are obtained. Putting these into eq. (4.30) gives $U_0(P)$ in terms of the unrestricted amplitude $U_{00}(P)$. Changing the position of the aperture just means the choice of a different x_0 and y_0.

The foregoing procedure may be simplified as follows. Let the two points on the curve corresponding to $v=\mu_1$ and $v=\mu_2$ be joined by a line, as shown in fig. 4.14. This line represents a vector of magnitude A (length of the line) and phase angle ϕ (angle line makes with the real axis). A and ϕ are found graphically, by ruler and protractor, and they can be put into the equation

$$C(\mu_2)-C(\mu_1)+i[S(\mu_2)-S(\mu_1)]=Ae^{i\phi}. \tag{4.31}$$

The same process can be used for v_1 and v_2, finding the amplitude B and the phase ψ. Then

$$C(v_2)-C(v_1)+i[S(v_2)-S(v_1)]=Be^{i\psi}. \tag{4.32}$$

Putting (4.31) and (4.32) into eq. (4.30) gives a new expression for the amplitude and phase of the irradiance at P. Going from this to the

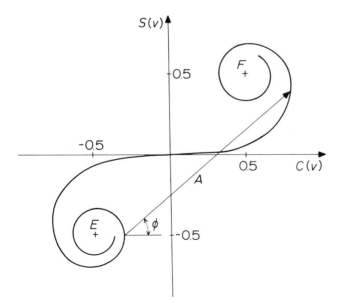

Fig. 4.14 Using Cornu's spiral to find $Ae^{i\phi}$.

absolute value of the energy (eliminating the phase) we have, for the irradiance at P,

$$I(P) = \tfrac{1}{4} A^2 B^2 I_0(P), \tag{4.33}$$

where $I_0(P)$ is the unobstructed irradiance at P.

We can now examine the variation which takes place in the irradiance at P as the aperture is moved to different positions in the diffraction screen relative to the point O (fig. 4.12). Let this motion first be only in the x direction, affecting the limits of μ but not those of v, and therefore causing changes in A but not in B [eqs. (4.31) and (4.32)]. It will be convenient to define two new quantities,

$$\Delta v = \mu_2 - \mu_1 = a \left[\frac{2(r_{10} + r_{20})}{\lambda r_{10} r_{20}} \right]^{1/2} \tag{4.34}$$

and

$$\mu_0 = \tfrac{1}{2}(\mu_2 + \mu_1) = x_0 \left[\frac{2(r_{10} + r_{20})}{\lambda r_{10} r_{20}} \right]^{1/2}. \tag{4.35}$$

Equation (4.34) follows at once from the method of finding μ_1 and μ_2 of eq. (4.27). For a given width of aperture a, Δv is a constant independent of how the aperture is moved about. In fact, Δv is the arc length along Cornu's spiral between the points on the curve corresponding to the limits $v = \mu_1$ and $v = \mu_2$.

In eq. (4.35), x_0 is the x coordinate of the center of the aperture, and μ_0 is the value of μ corresponding to this value of x. Actually, μ_0 is the midpoint of the section of Cornu's spiral lying between the points for μ_1 and μ_2. When x_0 is zero it means that the origin O (which lies on the direct line between S and P) is midway between two sides of the aperture. The quantity μ_0 is zero at this value of x_0. When $x_0 = -a/2$, the origin is at one edge of the aperture, and when $x_0 = a/2$, it is at the opposite edge. The corresponding values of μ_0 are $\pm \Delta v/2$. Between these limits there is a clear path between S and P, but μ_0 must be extended beyond the limits to find the extent of bending of light by diffraction into the geometrical shadow of the screen.

From the distances r_{10} and r_{20}, the width a of the aperture, and the wavelength λ of the light, the value of Δv is found by eq. (4.34). Keeping this constant, let x_0 be varied, the corresponding μ_0 being given by eq. (4.35). Also let the points on Cornu's spiral corresponding to μ_1 and μ_2 be found, and the straight-line distance A between them be measured as shown in fig. 4.14 [this measurement must be in the same units used for plotting $C(v)$ and $S(v)$]. If a given motion of x_0 causes the point for μ_1 to move a certain distance along Cornu's spiral, the points for μ_0 and for μ_2 move the same distance along the curve. The length A, however, will in general be different. Now let the square of the A found be plotted against μ_0 for the different positions of the center line (x_0) of the aperture. The result is a curve like one of those in fig. 4.15. The curves of fig. 4.15 have been drawn for six different values of Δv. The heavy lines along the μ_0 axis show the extension of the clear path from S to P, within the region $-\Delta v/2 \leq \mu_0 \leq \Delta v/2$. Beyond this is geometrical shadow.

If the motion of the aperture is in the y direction instead of x, we have only to replace a, μ, and x_0 in (4.34) and (4.35) by b, v, and y_0. Curves exactly like those of fig. 4.15 would be found, with the coordinates now v_0 and B^2.

The cross section of the diffraction pattern in the x direction, for an aperture of width a, is given by finding the Δv corresponding to a and then using the correct curve like those of fig. 4.15. Similarly, the cross section in the y direction is found from the height b of the aperture and the curve for the corresponding Δv. If patterns not on the axes are desired, the product of A^2 and B^2 must be used.

Most of the curves of fig. 4.15 show a considerable diffraction effect in the region of the geometrical shadow (beyond the heavy lines). When Δv is 10 or more, however, the light diffracted into the shadow is very small, and even the fringes outside of the geometrical shadow are small in their fluctuation except near the edge of the shadow. No matter how large Δv is made, the fringes near the shadow edge will always appear. When Δv is small, say less than unity, the extension of the pattern into the shadow is very marked. As Δv is decreased, the shape of the Fresnel pattern approaches that of the Fraunhofer pattern.

To obtain some notion of the relation of dimensions required for a small Δv, let us take a particular case. Suppose r_{10} and r_{20} are each 100 cm, and that λ is 0.8×10^{-4} cm, which is the longest wavelength of the visible spectrum. By eq. (4.34) we would have $\Delta v = 22.4a$, with a in centimeters. For $\Delta v = 1$, the distance a would be less than half a millimeter. To obtain an a of half a centimeter, we would have to have a Δv of approximately 11.

RECTANGULAR APERTURE, a OR b LARGE.

In the last section, the dimensions of the aperture were assumed small enough to permit some approximations to be used in evaluating the Kirchhoff integral. Let us now suppose that a or b is too large for these

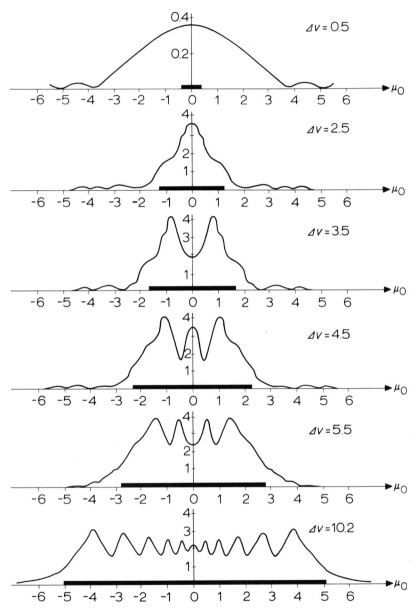

Fig. 4.15 The variation of A^2 along one axis of a rectangular aperture for six different values of Δv. Heavy lines extend over those values of μ_0 for which there is a clear geometric path through the aperture from S to P.

approximations to be used. At the upper limit (a and b infinite), there would be no screen at all between light source and observation point, so that the irradiance $I(P)$ would simply be the unobstructed value $I_0(P)$. Rather surprisingly, eq. (4.33) takes this form if we substitute infinite values for the μ and v limits in (4.31) and (4.32). For,

$$Ae^{i\phi} = C(\infty) - C(-\infty) + i[S(\infty) - S(-\infty)] = 1 + i = \sqrt{2}\,e^{i\pi/4}. \tag{4.36}$$

It will be noted that $\sqrt{2}$ and $\pi/4$ are, respectively, the length of the straight line from E to F in fig. 4.13, and the angle which this line makes with the real axis. Similarly, with the v limits infinite,

$$Be^{i\psi} = \sqrt{2}\,e^{i\pi/4}. \tag{4.37}$$

Using the absolute values of A and B from (4.36) and (4.37) and substituting in eq. (4.33), $I(P) = I_0(P)$, which is the exact value already stated. We can, in fact, use eqs. (4.36) and (4.37) in eq. (4.30), and obtain

$$U_0(P) = U_{00}(P). \tag{4.38}$$

We have thus shown that eq. (4.30), derived for the case of small aperture dimensions, also holds true when these dimensions are infinite.

From this fact alone it cannot be concluded that (4.30) is also exact for intermediate values of a and b. However, if we limit λ to wavelengths of visible light, while r_{10} and r_{20} have convenient practical values, we find that the limits of μ and v [see eqs. (4.27)] and the value of Δv [eq. (4.34)] can become quite large while x, y, a, and b are still small enough for the approximations used for (4.30) to remain valid. Thus we can say that, as a and b increase, they are still small enough for eq. (4.30) to be accurate when the limits of μ and v become near enough infinity for (4.30) to be correct for that reason. Then for small wavelengths eq. (4.30) can be expected to hold, approximately, for any size apertures.

Let us now consider an aperture in the shape of a long, narrow slit. In this case we take b large, but a small. For any motion of the aperture in the y direction, the B in eq. (4.33) would be constant, following eq. (4.37). The variation in $I(P)$ therefore depends only on A, which varies as the aperture is moved in the x direction, as shown by the curves of fig. 4.15. The amplitude equation is

$$U_0(P) = \frac{-i}{\sqrt{2}}\,U_{00}(P)\,A\,\exp\left\{i\left(\phi + \frac{\pi}{4}\right)\right\},$$

and the corresponding irradiance is

$$I(P) = \frac{A^2}{2} I_0(P).$$

The factor A^2 can be obtained from fig. 4.14, using the curve for the correct Δv. The pattern consists of long fringes parallel to the slit, and unvarying in this direction except where the ends of the slit are approached.

A third case arising from the rectangular aperture may also be treated: that of the diffraction caused by a straight edge. Let the edge run parallel to the y axis, so that in the y direction the limits are $-\infty$ and $+\infty$. In the x direction the space is semi-infinite, with the limits $x = -\infty$ and $x = x_2$, where x_2 is the position of the screen edge, and $x > x_2$ corresponds to opaque parts of the screen. Let the point O (fig. 4.12) be so close to the screen edge that $|x_2|$ is always small compared with r_{10} and r_{20}. The point P will then always be close to the geometrical shadow. By eqs. (4.27), there will be a finite μ_2 corresponding to x_2, but $\mu_1 = -\infty$, $v_1 = -\infty$, and $v_2 = +\infty$. In eq. (4.32), the coefficient B will have the constant value given by eq. (4.37), while from (4.31) we have

$$A e^{i\phi} = C(\mu_2) - C(-\infty) + i[S(\mu_2) - S(-\infty)].$$

One end of the line A in fig. 4.14 will always be at the point E (where $C = S = -1/2$), while the other end varies according to the position chosen for x_2, which determines μ_2. It is now possible to draw a curve for A^2 versus μ_2, from graphical measurements on fig. 4.14; this has been done in fig. 4.16. This curve is similar to those of fig. 4.15, but does not have the symmetrical quality of the latter. The pattern is one of fringes parallel to the straight edge, mostly on the clear side of the geometrical shadow but decreasing in amplitude as distance from the edge increases, and with some irradiance extending into the shadow. An example of the

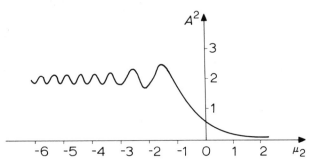

Fig. 4.16 The quantity A^2 as a function of μ_2 for diffraction at a straight edge. The value $\mu_2 = 0$ corresponds to the edge of the geometric shadow.

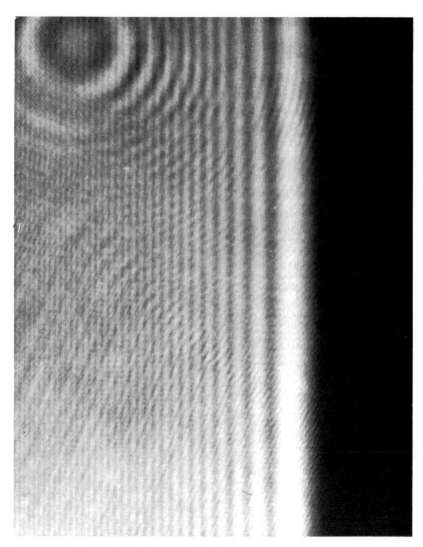

Fig. 4.17 Fresnel diffraction from a straight edge.

Fresnel diffraction pattern from a straight edge is photographed in fig. 4.17.

CIRCULAR APERTURE

In discussing Fresnel diffraction by a circular aperture, we will deal only with the situation where the line from the light source to the point of observation is perpendicular to the plane of the aperture, and passes through the center of the circle. Figure 4.18 illustrates this situation. Let c be the radius of the aperture, and let ρ be the radial distance to any point in the aperture.

Take first the case where c is much smaller than r_{10} or r_{20}. We can then make the same assumptions that we used in arriving at eq. (4.23). Also, with the circular form of the surface Σ, we can take $d\Sigma$ to be the ring-shaped element $2\pi\rho\,d\rho$, and carry the integration from $\rho=0$ to $\rho=c$. Thus the Kirchhoff integral is

$$U_0(P)=-\frac{iA}{\lambda r_{10}r_{20}}\int_0^c \exp\left[ik(r_1+r_2)\right]2\pi\rho\,d\rho. \tag{4.39}$$

From fig. 4.18 it can be seen that $r_1^2=r_{10}^2+\rho^2$ and $r_2^2=r_{20}^2+\rho^2$. Since r_{10} and r_{20} are constants, differentiation of these equations and division by 2 gives

$$\rho\,d\rho=r_1\,dr_1=r_2\,dr_2.$$

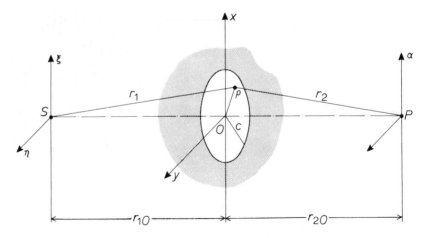

Fig. 4.18 Geometry for the application of Kirchhoff's integral to Fresnel diffraction at a circular aperture.

Thus $dr_1 = (1/r_1)\, \rho\, d\rho$ and $dr_2 = (1/r_2)\, \rho\, d\rho$, and, adding, we obtain

$$d(r_1 + r_2) = \left(\frac{1}{r_1} + \frac{1}{r_2}\right) \rho\, d\rho.$$

Then

$$\rho\, d\rho = \frac{r_1 r_2}{r_1 + r_2}\, d(r_1 + r_2).$$

Since c has been assumed small, it will not be far wrong to make the approximation

$$\rho\, d\rho \simeq \frac{r_{10} r_{20}}{r_{10} + r_{20}}\, d(r_1 + r_2).$$

Making the substitution $l = r_1 + r_2$, we can now write eq. (4.39) in the form

$$U_0(P) = -i\, \frac{2\pi A}{\lambda\,(r_{10} + r_{20})} \int_{l(0)}^{l(c)} e^{ikl} dl, \tag{4.40}$$

where $l(0) = r_{10} + r_{20}$.

The path l, from S to P, is always slightly longer than the direct distance $r_{10} + r_{20}$. Let us express the difference in the form first introduced by Fresnel,

$$l = r_{10} + r_{20} + \Delta\, \frac{\lambda}{2}. \tag{4.41}$$

Since λ here is the wavelength of light, the path difference is expressed in half-wavelengths. The variable Δ is dimensionless, and $\Delta\lambda/2$ expresses in half-wavelengths the increase of path from S to P as ρ increased from zero. This notation will be found useful later, when we discuss the *Fresnel zone plate*.

In applying eq. (4.41) to (4.40) let us note that $dl = (\lambda/2)\, d\Delta$, and that $\exp(ikl) = \exp[ik(r_{10} + r_{20})] \cdot \exp(i\pi\Delta)$, since $k\lambda = 2\pi$. The first of these exponentials is constant and may be removed from the integral, and it combines with other constants to form $U_{00}(P)$, the unobstructed amplitude of the irradiance at P [see the equation after eq. (4.28)]. Removing $\lambda/2$ from the integral, and noting that $\Delta = 0$ at $l(0)$, eq. (4.40) becomes

$$U_0(P) = -i\pi U_{00}(P) \int_0^{\Delta(c)} \exp(i\pi\Delta)\, d\Delta, \tag{4.42}$$

where $\Delta(c)$is the value of Δ where $\rho = c$.

By integrating eq. (4.42), we have

$$U_0(P) = U_{00}(P)\left\{1 - \exp[i\pi\Delta(c)]\right\}, \tag{4.43}$$

and the corresponding irradiance is

$$I(P) = |U_0(P)|^2 = 2I_0(P)\left[1 - \cos\pi\Delta(c)\right] = 4I_0(P)\sin^2\tfrac{1}{2}\pi\Delta(c). \tag{4.44}$$

By eq. (4.41), the quantity $\Delta(c)$ expresses in half-wavelengths the excess of the path from S to P by way of the aperture perimeter, rather than straight through the center. As c varies from zero to moderate values, $\sin^2\tfrac{1}{2}\pi\Delta(c)$ and therefore $I(P)$ passes through a series of values which fluctuate between zero and a maximum. For $I(P)$, the maximum is four times the irradiance when no diffraction screen is present. Thus, depending on the radius of the aperture, the center of the diffraction pattern may be either dark or bright. Figure 4.19 show the patterns for different values of $\Delta(c)$.

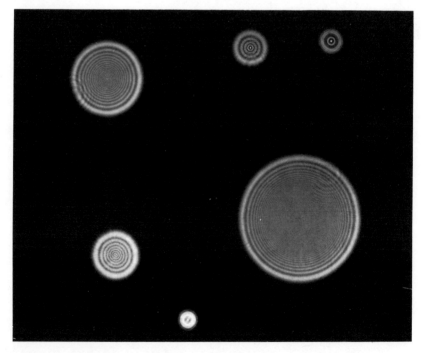

Fig. 4.19 Fresnel diffraction patterns of circular apertures for various values of $\Delta(c)$.

The value of $\Delta(c)$, in terms of c and the other dimensions shown in fig. 4.18, plus the wavelength λ, may be deduced from eq. (4.41). When the radius is c, the l in this equation becomes

$$l=(r_{10}^2+c^2)^{1/2}+(r_{20}^2+c^2)^{1/2}.$$

Or, approximately, since c is small,

$$l\simeq r_{10}+\frac{c^2}{2r_{10}}+r_{20}+\frac{c^2}{2r_{20}}.$$

Using this value for l in eq. (4.41),

$$\Delta(c)=\frac{c^2}{\lambda}\left(\frac{1}{r_{10}}+\frac{1}{r_{20}}\right)=\frac{c^2(r_{10}+r_{20})}{\lambda r_{10}r_{20}}. \tag{4.45}$$

Our derivation of the irradiance from a circular aperture, expressed in eq. (4.44), has dealt only with this irradiance at the center of the pattern, with light from the source striking perpendicularly at the center of the aperture. Nothing is disclosed, in this equation alone, regarding the diffraction pattern away from the central point P. Experimentally, the pictures of fig. 4.19 show what the pattern is like. We would certainly expect the fringes to be ring-shaped, about P as center. We would also expect that as $\Delta(c)$ is decreased toward unity, the fringes would become less extensive in total area, but more widely spaced, as they do in fig. 4.15 for the rectangular aperture as Δv decreases. If we compare $\Delta(c)$ in eq. (4.45) with Δv in eq. (4.34), we find that $\Delta(c)$ is comparable to $(\Delta v)^2$, provided that c^2 is comparable to $2a^2$. For Δv and $\Delta(c)$ both to be one, we would have $c=\sqrt{2}\,a$.

It is to be noticed from eq. (4.44) that $\Delta(c)=1$ is the lowest value of $\Delta(c)$ which will give the maximum irradiance at the center of the observed pattern. As $\Delta(c)$ is decreased below one, we would expect the diffraction to spread further into the geometrical shadow (as in fig. 4.15), but with decreased irradiance. The pattern would approach that of the Fraunhofer case [eq. (4.22)]. Lord Rayleigh concluded that $\Delta(c)=0.9$ gave the most distinct image from a pinhole camera: i.e., the least spreading of the image by diffraction.

LARGE CIRCULAR APERTURE

It is interesting to compare the irradiance at the center of the pattern from a circular aperture, as the radius of the aperture increases, with that from a square aperture of comparable dimensions. We can equate the width of the square with the diameter of the circle: $a = 2c$. Put the values $x_1 = y_1 = -a/2 = -c$ and $x_2 = y_2 = a/2 = c$ into eq. (4.27); thus,

$$\mu_2 = c \left[\frac{2(r_{10} + r_{20})}{\lambda r_{10} r_{20}} \right]^{1/2} = v_2 \tag{4.46}$$

and $\mu_1 = v_1 = -\mu_2$. We also note that for the Fresnel integrals, eq. (4.29), $C(-\mu_2) = -C(\mu_2)$ and $S(-\mu_2) = -S(\mu_2)$. The two main factors of eq. (4.30) are alike, and each becomes $2[C(\mu_2) + iS(\mu_2)]$. Thus (4.30) is now

$$U_0(P) = -2iU_{00}(P) [C(\mu_2) + iS(\mu_2)]^2 .$$

The irradiance is the square of the absolute value of this, or

$$I_s(P) = 4I_0(P) [C^2(\mu_2) + S^2(\mu_2)]^2 , \tag{4.47}$$

where the subscript s is used to denote that the equation applies to the square.

If we compare eqs. (4.45) and (4.46) we see that $\Delta(c) = \frac{1}{2}\mu_2^2$. Substituting this into eq. (4.44), and using a subscript c to refer to the circle,

$$I_c(P) = 4I_0(P) \sin^2 \tfrac{1}{4}\pi\mu_2^2 . \tag{4.48}$$

Equations (4.47) and (4.48) are both stated in terms of μ_2, which by (4.46) is proportional to the radius c of the circular aperture. The irradiances $I_s(P)$ and $I_c(P)$ are plotted against μ_2 in fig. 4.20, where it is seen that while the irradiance in the center of the pattern from the square converges toward the unobstructed value I_0 as size increases, that from the

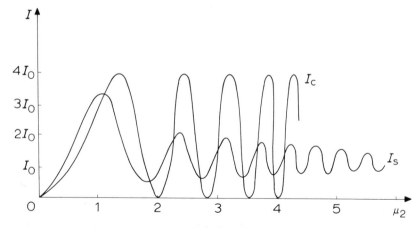

Fig. 4.20 Irradiance at the center of the diffraction pattern as a function of μ_2: I_c, irradiance for a circular aperture of radius c; I_s, irradiance for a square aperture of edge $2c$.

circle continues to oscillate between zero and its maximum value. This comparison is valid for moderate values of μ_2, and can be observed. Of course, we realize that as μ_2 approaches infinity the two cases must be the same. This only means that for extreme size of the apertures, the approximations which have been used in both cases are no longer valid, and a more exact way to treat the Kirchhoff integral in this region must be found.

4.5 Fresnel Zone Plate

In eq. (4.42), instead of taking the integral over the whole open circle from $\Delta = 0$ to $\Delta = \Delta(c)$, let us consider the contribution to $U_0(P)$ of a ring-shaped zone from $\Delta = \Delta_1$ to $\Delta = \Delta_2$:

$$U_0(P) = -i\pi U_{00}(P) \int_{\Delta_1}^{\Delta_2} \exp(i\pi\Delta)\, d\Delta = U_{00}(P)\left[\exp(i\pi\Delta_1) - \exp(i\pi\Delta_2)\right]$$

$$= U_{00}(P)\left[\cos\pi\Delta_1 - \cos\pi\Delta_2 + i(\sin\pi\Delta_1 - \sin\pi\Delta_2)\right].$$

Now let us suppose that Δ_1 is an even integer and that $\Delta_2 = \Delta_1 + 1$, and is therefore odd. We see that in this case $\mu_0(P) = 2U_{00}(P)$. Alternatively, when Δ_1 is odd and Δ_2 the next even number, $U_0(P) = -2U_{00}(P)$.

Suppose the complete circular aperture is divided into such zones, the boundaries of which have Δ equal to successive integers. The corresponding radii from the center of the aperture are found by solving eq. (4.45) for c, with $\Delta(c)$ being replaced by n:

$$c_n = \left[\frac{n\lambda r_{1O}r_{2O}}{r_{1O} + r_{2O}}\right]^{1/2}, \qquad n = 1, 2, 3, \ldots \tag{4.49}$$

Remembering the significance of Δ, we see that each integral increase in n in eq. (4.49) means that the light path from S to P has been increased by one-half wavelength. Such zones are called *Fresnel half-period zones*, or simply *Fresnel zones*. Each successive zone, in the open circular aperture, cancels the preceding zone in its effect on the amplitude at P. The net $U_0(P)$ can be considered as coming from the part of the circle remaining after the greatest even value of Δ.

With the circular aperture divided into Fresnel zones, as above, let alternate zones be covered by opaque material. We have then what is called the *Fresnel zone plate*. This is illustrated in fig. 4.21, where the central zone is shown open. The effect would be the same if we started with a closed zone in the center. Suppose there are N of these zones, with the screen

Fig. 4.21 Fresnel zone plate with center zone open.

entirely opaque for all parts outside the last open zone. The total number of open zones is then $N/2$ if the central zone is blocked, or $(N+1)/2$ if the central zone is open. Each open zone contributes $2\,U_{00}(P)$ to $U_0(P)$. (We need pay no attention to the negative sign if the central zone is opaque.) The total effect is then, with the central zone open,

$$U_0(P) = (N+1)\,U_{00}(P), \tag{4.50}$$

and the irradiance is

$$I(P) = (N+1)^2\,I_0(P). \tag{4.51}$$

With a large number of zones, the brightness of the spot at P is quite high. There is, in effect, a focusing of the source S at the point P. A more exact treatment for the complex light amplitude distribution can be obtained by means of the Fresnel-Kirchhoff integral, as will be shown in the chapter on wavefront reconstructions. It will also be shown later that the focusing effect of the Fresnel zone plate is the foundation of modern interference photography, i.e., holography.

We can compare the focusing effect of the Fresnel zone plate, which is due to diffraction, with that produced by refraction in the ordinary lens. If we make r_{1o} the object distance and r_{2o} the image distance, the focal length f of the lens is given by

$$\frac{1}{f} = \frac{1}{r_{1o}} + \frac{1}{r_{2o}} \qquad \text{or} \qquad f = \frac{r_{1o}r_{2o}}{r_{1o}+r_{2o}}.$$

The physical meaning of the focal length of the lens, defined by this equation, is that f is the value taken by the image distance r_{2o} when the object distance r_{1o} is made infinite. It can be given exactly the same meaning for the zone plate, and the same equation substituted into eq. (4.49), which on squaring becomes

$$c_n^2 = n\lambda f \qquad \text{or} \qquad f = \frac{c_n^2}{n\lambda}.$$

Since c_n^2 and n are proportional, let us make $n=1$ and $c_n = c_1$. The focal length of the zone plate is then

$$f = \frac{c_1^2}{\lambda}. \tag{4.52}$$

The focal length is thus dependent on λ, and if white light is used there will be a great deal of chromatic aberration. For this reason, monochromatic light is much to be preferred for use with the Fresnel zone plate.

No restriction has been placed on the value of λ, and thus zone plates can be designed to focus ultraviolet light or even X rays. This is not possible with lenses because of the lack of a material which has the refractive properties needed for a lens at these wavelengths.

4.6 Rayleigh's Criterion and Abbe's Sine Condition

Thus far we have been considering diffraction simply as a phenomenon in the propagation of light. It must also be considered as a source of imperfection in the performance of optical systems.

For example, in fig. 4.22 a simple lens is shown, projecting the images P_1 and P_2 of two sources, S_1 and S_2. The optical system could just as well be more complicated, with multiple lenses and mirrors, but forming,

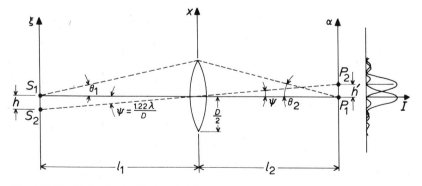

Fig. 4.22 Rayleigh criterion for two point sources.

as here, two images of two sources. We will assume, also, that it is a "stigmatic" system, by which is meant that any optical ray coming from a source S and striking the system at any point, must eventually also pass through the same image point P, and that the optical lengths (lengths measured in wavelengths of light) of all such paths from S to P are the same. Obviously, the geometrical length of the ray passing through the center of the lens in fig. 4.22 is less than the lengths of other rays from S to P. However, the lens is thicker in the center, and since the velocity of light is less inside the lens, the number of wavelengths in this section of the lens is increased.

Now in addition to the geometrical-ray properties of the system, diffraction also takes place. Thus the space occupied by the lens in fig. 4.22 may be regarded as a circular aperture in a diffracting screen, and because of the stigmatic character of the system, the diffraction may be taken as Fraunhofer. The result is that the images at P_1 and P_2 are not sharp, but each has a diffraction pattern around it. This tends to confuse the images of the points when they are close together, and the question arises as to how close they can be and still be *resolved*.

Rayleigh's criterion is that the images are resolved if the central maximum in the pattern of one point coincides with the first dark fringe in the pattern of the second point. Irradiance curves for P_1 and P_2 are drawn at the right of fig. 4.22, with the spacing just corresponding to Rayleigh's criterion. These curves are given by eq. (4.22), for the Fraunhofer diffraction from a circular aperture. As was explained in connection with this equation, the first dark fringe comes at $\mu = 1.22$, and this can be put into eq. (4.18). Let θ_2 of that equation be called here the angle ψ, and with it quite small we have

$$\psi_{min} \simeq \frac{1.22\lambda}{D}. \tag{4.53}$$

This ψ_{min} is the angle at the center of the lens subtended by the distance between images just resolvable under Rayleigh's criterion. The diameter of the lens is D, and λ is the wavelength of light. For an angle less then ψ_{min}, the images would not be resolved.

It can be seen from fig. 4.22 that the angles between the central rays are equal on the two sides of the lens. If the lens is the objective lens of an astronomical telescope, the distance l will be infinitely great, and the least angular separation given by eq. (4.53) will be satisfactory to rate the resolving power. For much closer sources, as with a compound microscope, it may be more significant to replace ψ by h/l, and express the least h as

$$h_{\min} = \frac{1.22\lambda l}{D}. \tag{4.54}$$

In applying eq. (4.54) to a particular optical instrument, it is customary to use another relation, known as *Abbe's sine condition*. This is developed as follows. In fig. 4.23, again let a focusing optical system be represented by a single convex lens, although the system may be more complicated than this. Let a source S_1 be located on the axis of the lens, with its image at P_1. Let a second source S_2 be located at a very small distance h off the axis, and let its image at P_2 be at a small distance h' from P_1. The system is assumed to be stigmatic, but let it be assumed that there is a different medium on the image side of the lens than that on the source side. Although the frequency of the light remains constant, it will have a different velocity and therefore a different wavelength on the two sides. Call the wavelengths λ and λ' on the source and image sides, respectively.

Now draw a set of near-axial rays from S_1 to P_1, and from S_2 to P_2. Let the optical lengths of these rays be b_{10} and b_{20}. Draw other rays of optical length b_1 and b_2 from S_1 and S_2 at very nearly the same angle as the first

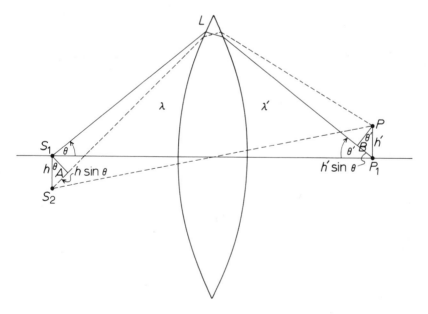

Fig. 4.23 Diagram for Abbe's sine condition.

set, but intersecting those rays inside the lens. The emission angles for the two sets of rays, being very nearly the same, may be assigned the same value θ. Both sets of rays may then be said to arrive at P_1 and P_2 at the same angle θ'. Since the system is stigmatic, we have at once that

$$b_1 = b_{10} \quad \text{and} \quad b_2 = b_{20}.$$

But since h and h' are very small, we also have $b_{10} \simeq b_{20}$, and therefore,

$$b_1 \simeq b_2. \tag{4.55}$$

These are the total optical distances from S_1 and S_2 to P_1 and P_2.

Let us drop a perpendicular line from S_1 to A on the line $S_2 L$. The geometrical distances $S_1 L$ and AL are very nearly the same, so that the line $S_2 L$ exceeds $S_1 L$ in length by the quantity $h \sin \theta$. In the same way on the image side, LP_1 exceeds LP_2 by $h' \sin \theta'$. The optical excess lengths are found from these by dividing by their respective wavelengths, and in view of eq. (4.55) we must have

$$\frac{h \sin \theta}{\lambda} \simeq \frac{h' \sin \theta'}{\lambda'}. \tag{4.56}$$

Equation (4.56) is known as *Abbe's sine condition*. It is very nearly true for all values of θ if h and h' are very small and the optical system is stigmatic.

Let us use this sine condition to find the resolving power of a microscopic objective. In the first case, let there be air on both sides of the lens, so that λ and λ' are equal. Notice also, from fig. 4.23, that $h/h' = l/l'$. Putting these conditions into eq. (4.56),

$$l \sin \theta = l' \sin \theta'. \tag{4.57}$$

In the microscope the angle θ' is quite small, so that we can put

$$\sin \theta' \simeq \theta' = \frac{D}{2l'}. \tag{4.58}$$

Putting this into (4.57) we have

$$\frac{D}{l} = 2 \sin \theta,$$

and this can be applied to eq. (4.54), giving

$$h_{\min} = \frac{1.22\lambda}{2 \sin \theta} = \frac{0.61\lambda}{\text{N.A.}}, \tag{4.59}$$

where the initials N.A. stand for "Numerical Aperture", which is equal to $\sin\theta$. It is the value of the N.A. which is usually given by manufacturers for the rating of the microscopic objective. It represents the angular radius of the pencil of rays from an object in focus, to the objective lens. The least separation of points to be resolved is found from eq. (4.59).

There is an instrument, called the oil immersion microscope, for which λ and λ' are not equal. Although the medium on the image side of the objective lens is air, the object and the space between it and the lens is immersed in oil with an index of refraction η. The index of refraction of air may be taken as very close to one. Now the velocity of light in a medium, and therefore also the wavelength of light of a given frequency, is inversely proportional to the index of refraction of the medium. In the above case, then, we have

$$\frac{\lambda'}{\lambda} = \frac{\eta}{1} = \eta.$$

The sine condition (4.56) then becomes

$$\eta h \sin\theta \simeq h' \sin\theta'. \tag{4.60}$$

We must modify the conditions for resolution, based on fig. 4.20. The angles ψ and ψ' are now not equal, but with an index η on the object side we must put $\eta\psi = \psi'$. Equation (4.53) still holds, if we put

$$\psi'_{\min} \simeq \frac{1.22\lambda'}{D},$$

but in terms of ψ, it is

$$\psi_{\min} = \frac{1.22\lambda'}{\eta D}. \tag{4.61}$$

Since we may replace ψ by h/l, the minimum separation of objects is

$$h_{\min} \simeq \frac{1.22\lambda' l}{\eta D}. \tag{4.62}$$

We continue to use λ', the wavelength in air, to designate the character of the light. Since $\eta\psi = \psi'$, we also have

$$\frac{\eta h}{l} = \frac{h'}{l'},$$

and if we put this into eq. (4.60), we get for the sine condition,

$$l \sin\theta \simeq l' \sin\theta', \tag{4.63}$$

which is the same as eq. (4.57). We can also use eq. (4.58) with this, and when the result is substituted in eq. (4.62),

$$h_{\min} \simeq \frac{0.61\lambda'}{\eta \sin\theta}. \tag{4.64}$$

This may be expressed as in eq. (4.59), provided we set, in this case,

$$\text{N.A.} = \eta \sin\theta. \tag{4.65}$$

From (4.65), it may be seen that for the oil immersion microscope, it is possible to have a numerical aperture greater than unity.

All of the equations which have been derived for resolving power, such as (4.53), (4.54), and (4.59), are significant and useful relations. There are conditions, however, such as an inhomogeneous or turbulent medium, under which they would not serve accurately. It must also be said that our assumption has been the separation of two close but independent sources. Although these may emit light of equal wavelength, they are assumed to be incoherent, in the sense that there is no fixed phase link between them. Our conclusions are not strictly applicable to separate objects that are illuminated by a common source of light, since in that case there is some degree of coherence in the light from the objects. Equation (4.59) would then require some modification.

Problems

4.1 Given a diffraction screen containing a pair of rectangular apertures, as shown in fig. 4.24. Let a monochromatic plane wave be normally

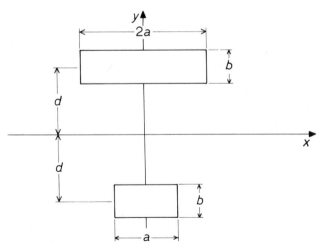

Fig. 4.24

incident on this diffraction screen. Determine the irradiance of the corresponding Fraunhofer diffraction pattern. Make a rough plot of the corresponding irradiance along the vertical axis of the observation screen.

4.2 The expression of the normalized amplitude diffraction pattern $Q(\rho)$ of a circular aperture in the Fraunhofer case is given as

$$Q(\rho) = \frac{2J_1(\pi\delta)}{\pi\delta}, \qquad \delta \geq 0,$$

where J_1 is the Bessel function of order one, $\delta = (kD/2\pi)\sin\theta_2$, $\sin\theta_2 = \alpha/r_{20}$, and D is the radius of the circular aperture. Given a screen having an annular aperture as shown in fig. 4.25, determine the normalized irradiance of the Fraunhofer diffraction pattern.

4.3 A diffraction screen contains an $n \times m$ rectangular array of identical circular apertures of radius c, arranged as shown in fig. 4.26. Determine the irradiance of the corresponding Fraunhofer diffraction pattern.

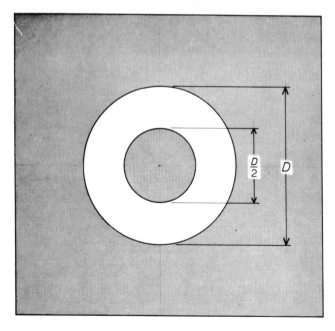

Fig. 4.25

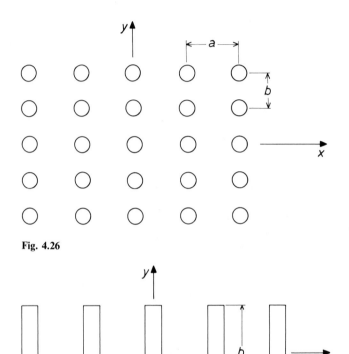

Fig. 4.26

Fig. 4.27

4.4 If a diffraction screen contains a linear array of five rectangular apertures, as shown in fig. 4.27, determine the complex light field of the Fraunhofer diffraction pattern. Make a rough sketch of the corresponding irradiance.

4.5 Let a plane wave of wavelength $\lambda = 5000$ Å and irradiance I be normally incident on a diffraction screen which has the open aperture shown in fig. 4.28. Determine the complex amplitude and the irradiance at a point P on the axis of the circles about 2 meters behind the diffraction screen.

4.6 By interchange of the opaque and transparent regions of the diffraction screen in the previous problem, show that the sum of the disturbances observed with the two complimentary screens satisfies Babinet's principle.

4.7 A plane wave of wavelength $\lambda = 5000$ Å is normally incident upon a semi-infinite straight-edge diffraction screen. By the application of Cornu's spiral, determine the locations of the maxima and minima of the diffraction pattern at a distance of 2 meters behind the screen.

4.8 It is assumed that a plane wave of wavelength λ is normally incident on

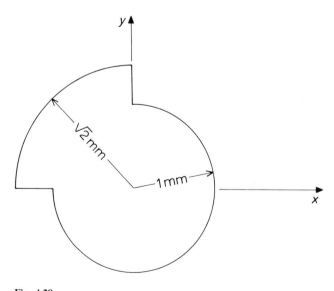

Fig. 4.28

a diffraction screen that has a transmission function

$$T(x, y) = \tfrac{1}{2} + m \sin p_0 x,$$

where $T(x, y) \leq 1$, and p_0 is the angular spatial frequency.

(a) Determine the Fresnel pattern behind the diffraction screen.
(b) If $m \ll 1/2$, determine the distances behind the diffraction screen at which the amplitude and phase modulations take place.

4.9 In fig. 4.18 let r_{10} be infinitely large. Show that the diffraction pattern goes from Fresnel to Fraunhofer as the radius of the aperture becomes very small.

4.10 (a) Calculate the minimum angular separation between two distant stars that may be barely resolved by a 75-cm diameter telescope, when a filter which selects the violet light of wavelength $\lambda = 4000$ Å is used.
(b) If the focal length of the objective is assumed to be 8 meters long, compute the minimum resolvable separation between the two images of the distant stars.

References
4.1 J. M. Stone, *Radiation and Optics*, McGraw-Hill, New York, 1963.
4.2 M. Born and E. Wolf, *Principles of Optics*, 2nd rev. ed., Pergamon Press, New York, 1964.
4.3 A. Sommerfeld, *Optics* (Lectures on Theoretical Physics, vol. IV) Academic Press, New York, 1954.
4.4 F. W. Sears, *Optics*, Addison-Wesley, Reading, Mass., 1949.
4.5 B. Rossi, *Optics*, Addison-Wesley, Reading, Mass., 1957.
4.6 J. W. Goodman, *Introduction to Fourier Optics*, McGraw-Hill, New York, 1968.
4.7 E. Jahnke, F. Emde, and F. Losch, *Table of Higher Functions*, 6th edition, McGraw-Hill, New York, 1960.

Chapter 5 Introduction to Coherent Theory

5.1 General Aspects of Mutual Coherence

The development and widespread application of the laser has made a discussion of the principles of coherence in radiation more important.

If the radiations from two point sources maintain a fixed phase relation between them, they are said to be mutually coherent. An extended source is coherent if all points of the source have fixed phase differences between them. In this chapter we will discuss some elementary concepts of coherence theory, as it applies to optical information processing and wave-front reconstruction. In the process, we will find it necessary to modify and extend the foregoing definitions.

In a classical discussion of electromagnetic radiation, as in the development of Maxwell's equations, it is usually assumed that the electric and magnetic fields at any position are at all times measurable. In this case no account need be taken of coherence or incoherence. There are problems, however, in which this assumption of known fields cannot be made; in these cases it is often helpful to apply coherence theory. For example, if a diffraction pattern as a result of radiation from several sources is to be worked out, an exact result cannot be obtained unless the degree of coherence of the separate sources is taken into account. It may be desirable, in such a case, to obtain an average that would represent the statistically most likely result from any such combination of sources. It may be more useful to provide a statistical description than to follow the dynamical behavior of a system in detail.

Our treatment of coherence will be on such an averaging basis. In particular, following Born and Wolf (ref. 5.1, pp. 499–503) we shall choose the second order moment as the quantity to be averaged. Thus, what we will call the *mutual coherence function* will be set down as

$$\Gamma_{12}(\tau) = \langle u_1(t+\tau)\, u_2^*(t)\rangle, \tag{5.1}$$

where $u_1(t)$ and $u_2(t)$ are the respective complex fields at points P_1 and P_2, and $\Gamma_{12}(\tau)$ is the mutual coherence function between these points for a time delay τ; the symbols $\langle\ \rangle$ indicate a time average. From eq. (5.1) we can define a *normalized mutual coherence function*,

$$\gamma_{12}(\tau) = \frac{\Gamma_{12}(\tau)}{[\Gamma_{11}(0)\, \Gamma_{22}(0)]^{1/2}}; \tag{5.2}$$

$\gamma_{12}(\tau)$ may also be called the complex degree of coherence or the degree of correlation.

A clearer idea of $\Gamma_{12}(\tau)$ and a demonstration of how it can be measured may be obtained from a consideration of Young's experiment on interference (see sec. 2.4). In fig. 5.1, Σ is an extended source of light, which is assumed to be incoherent, but nearly monochromatic; that is, its spectrum is of finite width, but narrow. The light from this source falls upon a screen at a distance r_{10} from the source, and upon two small apertures (pinholes) in this screen, Q_1 and Q_2, separated by a distance d. On an observing screen at a distance r_{20} from the diffracting screen, an interference pattern is formed by the light passing through Q_1 and Q_2. Now let us suppose that the changing characteristics of the interference fringes are observed as the parameters of fig. 5.1 are changed. As a measurable quantity, let us adopt Michelson's (ref. 5.1, p. 267) *visibility* $\mathcal{V}$ of the fringes, which he defines as

$$\mathcal{V} = \frac{I_{\max} - I_{\min}}{I_{\max} + I_{\min}}, \tag{5.3}$$

where $I_{\max}$ and $I_{\min}$ are the maximum and minimum irradiances of the fringes.

For the visibility to be measurable, the conditions of the experiment, such as narrowness of spectrum and closeness of optical path lengths, must be such as to permit $I_{\max}$ and $I_{\min}$ to be clearly defined. Let us assume that these ideal conditions exist.

As we begin our parameter changes, we find, first, that the average visibility of the fringes increases as the size of the source Σ is made smaller. Next, as the distance apart of the pinholes Q_1 and Q_2 is changed with Σ constant (and circular in form), the visibility changes in the manner

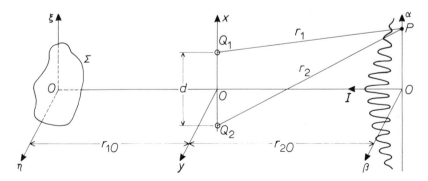

Fig. 5.1 Young's experiment. Here Σ is an extended but nearly monochromatic source.

shown in fig. 5.2. When Q_1 and Q_2 are very close together the irradiance between the fringes falls to zero, and the visibility is unity. As d is increased, the visibility falls rapidly and reaches zero as I_{max} and I_{min} become equal. Further increase in d causes a reappearance of fringes, although they are shifted on the screen by half a fringe; that is, the previously light areas are now dark, and vice versa. Still further increase in d causes the repeated fluctuations in visibility shown in the figure. A curve similar to that of fig. 5.2 is obtained if the hole spacing is kept constant while the size of Σ is changed. These effects can be predicted from the theorem of Van Cittert-Zernike (ref. 5.1, p. 507). The visibility versus pinhole separation curve is sometimes used as a measure of spatial coherence, as will be discussed in the next few sections. The screen separations r_{10} and r_{20} are both assumed to be large compared with the aperture spacing d and with the dimensions of the source. Beyond this limitation, changes in r_{10} or r_{20} change the scale of effects such as are shown in fig. 5.2, without changing their general character.

As the point of observation P (see fig. 5.1) is moved away from the center of the observing screen, the visibility decreases as the path difference

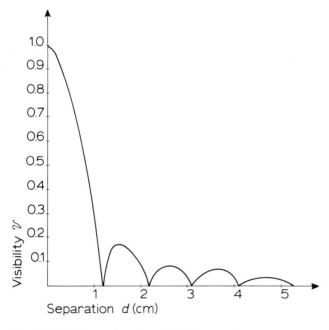

Fig. 5.2 Visibility as a function of pinhole separation.

$\Delta r = r_2 - r_1$ increases, eventually becoming zero. The effect also depends upon how nearly monochromatic the source is. It is found that the visibility of the fringes is appreciable only for a path difference

$$\Delta r \ll \frac{2\pi c}{\Delta\omega}, \tag{5.4}$$

where c is the velocity of light and $\Delta\omega$ is the width of the spectrum of the source of light. The inequality (5.4) is often used to define the *coherence length* of the source.

The foregoing example shows that it is not necessary to have completely coherent light to produce an interference pattern, but that under the right conditions such a pattern may be obtained from an incoherent source. This effect may properly be called one of partial coherence, and we need a method for defining and measuring it.

A further development of the preceding equations will be helpful. Thus, $u_1(t)$ and $u_2(t)$ of eq. (5.1), the complex fields at the points Q_1 and Q_2, are subject to the scalar wave equation in free space

$$\nabla^2 u = \frac{1}{c^2} \frac{\partial^2 u}{\partial t^2}. \tag{5.5}$$

This is a linear equation, and the field at the point P on the observing screen is a sum of those from Q_1 and Q_2:

$$u_p(t) = c_1 u_1\left(t - \frac{r_1}{c}\right) + c_2 u_2\left(t - \frac{r_2}{c}\right), \tag{5.6}$$

where c_1 and c_2 are appropriate complex constants. The corresponding irradiance at P may be written

$$\begin{aligned} I_p &= \langle u_p(t)\, u_p^*(t) \rangle \\ &= I_1 + I_2 + 2\,\mathrm{Re}\left\langle c_1 u_1\left(t - \frac{r_1}{c}\right) c_2^* u_2^*\left(t - \frac{r_2}{c}\right) \right\rangle, \end{aligned} \tag{5.7}$$

where I_1 and I_2 are proportional to the squares of the magnitudes of $u_1(t)$ and $u_2(t)$. In eq. (5.7) let us put

$$t_1 = \frac{r_1}{c} \quad \text{and} \quad t_2 = \frac{r_2}{c}, \tag{5.8}$$

and rewrite the irradiance at P,

$$I_p = I_1 + I_2 + 2c_1 c_2^*\, \mathrm{Re}\langle u_1(t - t_1)\, u_2^*(t - t_2)\rangle, \tag{5.9}$$

since c_1 and c_2 are not time variable. The quantity averaged in eq. (5.9) is the crosscorrelation of the two complex fields.

If we put $t_2 - t_1 = \tau$, eq. (5.9) can be written

$$I_p = I_1 + I_2 + 2c_1 c_2^* \operatorname{Re} \langle u_1(t+\tau) u_2^*(t) \rangle,$$

and combining this with eq. (5.1), we obtain

$$I_p = I_1 + I_2 + 2c_1 c_2^* \operatorname{Re}[\Gamma_{12}(\tau)]. \tag{5.10}$$

The autocorrelations (i.e., the self-coherence functions) of the radiations from the two pinholes are

$$\Gamma_{11}(0) = \langle u_1(t) u_1^*(t) \rangle \quad \text{and} \quad \Gamma_{22}(0) = \langle u_2(t) u_2^*(t) \rangle. \tag{5.11}$$

We can also put

$$|c_1|^2 \Gamma_{11}(0) = I_1 \quad \text{and} \quad |c_2|^2 \Gamma_{22}(0) = I_2.$$

The irradiance at P, eq. (5.10), can then be put in terms of the degree of complex coherence, eq. (5.2),

$$I_p = I_1 + I_2 + 2(I_1 I_2)^{1/2} \operatorname{Re}[\gamma_{12}(\tau)]. \tag{5.12}$$

Let us put $\gamma_{12}(\tau)$ in the form

$$\gamma_{12}(\tau) = |\gamma_{12}(\tau)| \exp[i\phi_{12}(\tau)], \tag{5.13}$$

and assume also that $I_1 = I_2 = I$, which may be called the best condition. Then eq. (5.12) becomes

$$I_p = 2I[1 + |\gamma_{12}(\tau)| \cos\phi_{12}(\tau)]. \tag{5.14}$$

The maximum of I_p is $2I[1 + |\gamma_{12}(\tau)|]$, while the minimum is $2I[1 - |\gamma_{12}(\tau)|]$. Putting these into the visibility equation, eq. (5.3), we find that

$$\mathcal{V} = |\gamma_{12}(\tau)|. \tag{5.15}$$

That is, under the best conditions the visibility of the fringes is a measure of the absolute value of degree of coherence.

5.2 Mutual Coherence Function

The complex crosscorrelation is usually defined as the second-moment time average. In these terms,

$$\Phi_{12}(\tau) \triangleq \langle f_1^*(t) f_2(t+\tau) \rangle = \lim_{T \to \infty} \frac{1}{2T} \int_{-T}^{T} f_1^*(t) f_2(t+\tau) \, dt. \tag{5.16}$$

Accordingly, we shall define the mutual coherence function as

$$\Gamma_{12}(\tau) \triangleq \langle u_1(t+\tau)\, u_2^*(t) \rangle = \lim_{T\to\infty} \frac{1}{2T} \int_{-T}^{T} u_1(T, t+\tau)\, u_2^*(T, t)\, dt. \qquad (5.17)$$

It is now possible, however, to give a more general definition, fundamental to the theory of coherence. This can be done by working from two different aspects of the phenomenon: first, from an ensemble viewpoint; and later, from time-averaging.

To obtain an ensemble average, we define the mutual coherence function as

$$\Gamma_{12}(\boldsymbol{\rho}_1, t_1; \boldsymbol{\rho}_2, t_2) = \lim_{N\to\infty} \frac{1}{N} \sum_{n=1}^{N} u_{1n}(\boldsymbol{\rho}_1, t_1)\, u_{2n}^*(\boldsymbol{\rho}_2, t_2). \qquad (5.18)$$

The summation is over an ensemble of N systems, and $\boldsymbol{\rho}_1$ and $\boldsymbol{\rho}_2$ are the respective position vectors.

If the statistics of the ensemble of systems are stationary, then

$$\Gamma_{12}(\boldsymbol{\rho}_1, t_1; \boldsymbol{\rho}_2, t_2) = \Gamma_{12}(\boldsymbol{\rho}_1, \boldsymbol{\rho}_2, \tau), \qquad (5.19)$$

where $\tau = t_1 - t_2$. Equation (5.18) can then be rewritten,

$$\Gamma_{12}(\boldsymbol{\rho}_1, \boldsymbol{\rho}_2, \tau) = \lim_{N\to\infty} \frac{1}{N} \sum_{n=1}^{N} u_{1n}(\boldsymbol{\rho}_1, t_2+\tau)\, u_{2n}^*(\boldsymbol{\rho}_2, t_2). \qquad (5.20)$$

It is well-known from the theory of random processes (ref. 5.6, p. 15) that when the statistics are stationary the time and ensemble averages may be equated. Thus we have

$$\Gamma_{12}(\boldsymbol{\rho}_1, \boldsymbol{\rho}_2, \tau) = \Gamma_{12}(\tau) = \langle u_1(\boldsymbol{\rho}_1, t+\tau)\, u_2^*(\boldsymbol{\rho}_2, t) \rangle. \qquad (5.21)$$

This definition of the mutual coherence function as a time average has been made dependent on having stationary statistics. It therefore is valid where the source of radiation is periodic. In general, however, the time average definition may be written as

$$\Gamma_{12}(\tau) = \langle u(\boldsymbol{\rho}_1, t+\tau)\, u^*(\boldsymbol{\rho}_2, t) \rangle. \qquad (5.22)$$

The *complex degree of coherence* will be defined as in eq. (5.2),

$$\gamma_{12}(\tau) = \frac{\Gamma_{12}(\tau)}{[\Gamma_{11}(0)\, \Gamma_{22}(0)]^{1/2}}, \qquad (5.23)$$

with the limits $0 \le |\gamma_{12}(\tau)| \le 1$. The lower limit represents complete incoherence, the upper limit complete coherence between the radiations at

$\boldsymbol{\rho}_1$ and $\boldsymbol{\rho}_2$. Note that $|\gamma_{12}(\tau)|$ is a function of τ. It is therefore possible that the radiations at two points may be coherent at one value of τ, but incoherent at another value.

The *self-coherence*, or autocorrelation, function is given by

$$\Gamma_{11}(\tau)=\langle u_1(\boldsymbol{\rho}_1, t+\tau)\, u_1^*(\boldsymbol{\rho}_1, t)\rangle. \tag{5.24}$$

The self-coherence function is very important in the analysis of the operation of Michelson's interferometer. Its zero value, $\Gamma_{11}(0)$, is the highest irradiance at the point $\boldsymbol{\rho}_1$, i.e.,

$$\Gamma_{11}(0)\geq\Gamma_{11}(\tau). \tag{5.25}$$

The zero value, $\Gamma_{12}(0)$, of the mutual coherence function is called the *mutual irradiance function*. It is essential in an examination of the "stellar" form of the interferometer.

It will also be found useful to take the Fourier transforms of both the mutual and self-coherence functions. That of the mutual function is called the *mutual power spectrum*, and is given by

$$L_{12}(\omega)=\begin{cases}\displaystyle\int_{-\infty}^{\infty}\Gamma_{12}(\tau)\,e^{-i\omega\tau}\,d\tau, & \omega>0 \\ 0, & \omega<0\end{cases}. \tag{5.26}$$

That the second of these conditions holds can be seen from the fact that $L_{12}(\omega)$ is analytic. The Fourier transform of a self-coherence function is the power spectrum of that particular radiation,

$$L_{11}(\omega)=\begin{cases}\displaystyle\int_{-\infty}^{\infty}\Gamma_{11}(\tau)\,e^{-i\omega\tau}\,d\tau, & \omega>0 \\ 0, & \omega<0\end{cases}, \tag{5.27}$$

and of course a similar equation for $L_{22}(\omega)$ in terms of $\Gamma_{22}(\tau)$.

A further remark must be made concerning coherence. The phrase *spatial coherence* is applied to those effects that are due to the size, in space, of the source of radiation. If we consider a point source, and look at two points at equal light-path distances from the source, the radiations reaching these points will be exactly the same. The mutual coherence will be equal to the self-coherence at either point. That is, if the points are Q_1 and Q_2,

$$\Gamma_{12}(Q_1, Q_2, \tau)=\langle u(Q_1, t+\tau)\, u^*(Q_2, t)\rangle=\Gamma_{11}(\tau). \tag{5.28}$$

As the source is made larger we can no longer claim an equality of mutual coherence and self-coherence. The lack of complete coherence is a *spatial* effect. *Temporal coherence* is an effect due to the finite spectral width of the source. The coherence is complete for strictly monochromatic radiation, but becomes only partial as other wavelengths are added, giving a finite spectral width to the source. It is never possible to completely separate the two effects, but it is well to name them and point out their significance.

5.3 Propagation of the Mutual Coherence Function

Equations that describe the manner in which the mutual coherence function is propagated can be found by first assuming that the field can be represented by a complex scalar function $u(t)$ that satisfies the scalar wave equation

$$\nabla^2 u(t) = \frac{1}{c^2} \frac{\partial^2 u(t)}{\partial t^2}. \tag{5.29}$$

The mutual coherence function, as we have previously defined it, is

$$\Gamma_{12}(\tau) = \langle u_1(t+\tau) \, u_2^*(t) \rangle. \tag{5.30}$$

The Laplacian, taken at the point Q_1 of eq. (5.30), may be written

$$\nabla_1^2 \Gamma_{12}(\tau) = \langle \nabla_1^2 u_1(t+\tau) \, u_2^*(t) \rangle. \tag{5.31}$$

From eq. (5.29) we can write

$$\nabla_1^2 u_1(t+\tau) = \frac{1}{c^2} \frac{\partial^2 u_1(t+\tau)}{\partial(t+\tau)^2}. \tag{5.32}$$

But

$$\frac{\partial^2 u_1(t+\tau)}{\partial(t+\tau)^2} = \frac{\partial^2 u_1(t+\tau)}{\partial \tau^2},$$

and therefore eq. (5.32) becomes

$$\nabla_1^2 \Gamma_{12}(\tau) = \left\langle \frac{\partial^2 u_1(t+\tau)}{c^2 \partial \tau^2} \, u_2^*(t) \right\rangle. \tag{5.33}$$

The field $u_2(t)$ is independent of τ, and we can therefore make the second partial derivative apply to the whole time-average function,

$$\nabla_1^2 \Gamma_{12}(\tau) = \frac{1}{c^2} \frac{\partial^2}{\partial \tau^2} \langle u_1(t+\tau) \, u_2^*(t) \rangle = \frac{1}{c^2} \frac{\partial^2 \Gamma_{12}(\tau)}{\partial \tau^2}. \tag{5.34}$$

This may be regarded as the fundamental equation describing the propagation of the mutual coherence function.

In the same way, we can take the Laplacian of eq. (5.30) with respect to the coordinates of the point Q_2, and it can be shown that it reduces also to

$$\nabla_2^2 \Gamma_{12}(\tau) = \frac{1}{c^2} \frac{\partial^2 \Gamma_{12}(\tau)}{\partial \tau^2}. \tag{5.35}$$

Each of the eqs. (5.34) and (5.35) contains only four independent variables, while $\Gamma_{12}(\tau)$ contains seven—the three spatial coordinates of each of the two points, plus the delay τ. A complete propagation equation may be obtained by combining (5.34) and (5.35):

$$\nabla_1^2 \nabla_2^2 \Gamma_{12}(\tau) = \frac{1}{c^4} \frac{\partial^4 \Gamma_{12}(\tau)}{\partial \tau^4}. \tag{5.36}$$

Suppose that the source of radiation is a surface of finite but bounded extent. For each pair of points on the surface a mutual coherence function, $\Gamma_{12}(\tau)$, can be specified. We then apply Sommerfeld's radiation condition at infinity, which says that $u_1(t)$ and $u_2^*(t)$ behave asymptotically as point radiators

$$\frac{A_1}{r_1} \exp(ikr_1) \quad \text{and} \quad \frac{A_2}{r_2} \exp(ikr_2) \tag{5.37}$$

as r_1 and r_2 (the distances from a pair of points on the source surface) approach infinity. These conditions may be applied to eqs. (5.34) and (5.35), and these solved as simultaneous wave equations. Since each equation contains four variables, their solutions can be written as four-dimensional Green's functions. The number of variables in each can, however, be reduced to three by first taking the Fourier transform of $\Gamma_{12}(\tau)$, which we will designate by $\hat{\Gamma}_{12}(\omega)$. Now $\Gamma_{12}(\tau)$, as an analytic function, contains only positive frequencies, and we can write

$$\Gamma_{12}(\tau) = \int_0^\infty \hat{\Gamma}_{12}(\omega) \exp(-\omega\tau) \, d\omega \tag{5.38}$$

and

$$\hat{\Gamma}_{12}(\omega) = \begin{cases} \int_{-\infty}^{\infty} \Gamma_{12}(\tau) \exp(\omega\tau) \, d\tau, & \omega > 0 \\ 0, & \omega < 0 \end{cases}. \tag{5.39}$$

By substituting eq. (5.38) into (5.34) and (5.35), and reversing the order of integrations and differentiations, we have the equations

$$\int_0^\infty [\nabla_1^2 + k^2(\omega)]\, \hat{\Gamma}_{12}(\omega)\exp(-\omega\tau)\, d\omega = 0 \tag{5.40}$$

and

$$\int_0^\infty [\nabla_2^2 + k^2(\omega)]\, \hat{\Gamma}_{12}(\omega)\exp(-\omega\tau)\, d\omega = 0. \tag{5.41}$$

These two equations must hold for every value of τ, and therefore

$$[\nabla_1^2 + k^2(\omega)]\, \hat{\Gamma}_{12}(\omega) = 0 \tag{5.42}$$

and

$$[\nabla_2^2 + k^2(\omega)]\, \hat{\Gamma}_{12}(\omega) = 0. \tag{5.43}$$

From these equations, it can be seen that the Fourier transform of the mutual coherence function satisfies the scalar Helmholtz equations.

Now we set up a Green's function $G_1(P_1, P_1', \omega)$ such that

$$[\nabla_1^2 + k^2(\omega)]\, G_1(P_1, P_1', \omega) = -\delta(P_1 - P_1'), \tag{5.44}$$

with the boundary

$$G_1(P_1, P_1', \omega)|_{P'_1 = S_1} = 0, \tag{5.45}$$

where P and S are the position coordinates, and δ is the Dirac delta function.

The solution of eq. (5.42) may be put in terms of this Green's function,

$$\hat{\Gamma}(P_1, S_2, \omega) = -\int_{S_1} \hat{\Gamma}(S_1, S_2, \omega)\left.\frac{\partial G_1(P_1, P_1', \omega)}{\partial n_{S_1}}\right|_{P'_1 = S_1} dS_1. \tag{5.46}$$

With another Green's function, $G_2(P_2, P_2', \omega)$, defined as in eqs. (5.44) and (5.45), and noting the fact that eq. (5.46) provides a boundary condition, we see that the solution of (5.43) becomes

$$\hat{\Gamma}(P_1, P_2, \omega) = -\int_{S_2} \hat{\Gamma}(P_1, S_2, \omega)\left.\frac{\partial G_2(P_2, P_2', \omega)}{\partial n_{S_2}}\right|_{P'_2 = S_2} dS_2. \tag{5.47}$$

By substituting (5.46) into (5.47), we obtain the combined solution

$$\hat{\Gamma}(P_1, P_2, \omega) = \int_{S_2} \int_{S_1} \hat{\Gamma}(S_1, S_2, \omega) \frac{\partial G_1(P_1, P'_1, \omega)}{\partial n_{S_1}}\bigg|_{P'_1 = S_1}$$

$$\times \frac{\partial G_2(P_2, P'_2, \omega)}{\partial n_{S_2}}\bigg|_{P'_2 = S_2} dS_1 \, dS_2. \qquad (5.48)$$

Let us note that $\hat{\Gamma}(P_1, S_2, \omega)$ in eq. (5.46) is the measure of coherence between a point on the source surface and an arbitrary point in space. The phrase *longitudinal direction* is used for space points that lie on perpendiculars to the surface points. The so-called longitudinal coherence is quite different from that between points on the surface. An elaboration of this is beyond the scope of this book, but an interested reader can refer to the excellent texts by Born and Wolf (ref. 5.1, pp. 491–555) and by Beran and Parrent (ref. 5.2, chap. 3).

The Fourier transform $\hat{\Gamma}(P_1, P_2, \omega)$ is a power spectrum, and if P_1 and P_2 are the same, it is the power spectrum of a single point in space. Equation (5.48) then shows that this power spectrum is not determined by the power spectrum of the whole source surface, but rather by $\hat{\Gamma}(S_1, S_2, \omega)$, which is the crosscorrelation between two points of the surface.

According to the Sommerfeld radiation condition, eq. (5.37), the asymptotic behavior of $\hat{u}_1(\omega)$ and $\hat{u}_2^*(\omega)$ as r_1 and r_2 approach infinity should follow

$$\frac{a_1(\theta_1, \phi_1, \omega)}{r_1} \exp(ikr_1) \quad \text{and} \quad \frac{a_2(\theta_2, \phi_2, \omega)}{r_2} \exp(ikr_2). \qquad (5.49)$$

If we assume that the statistics between $u_1(t)$ and $u_2(t)$ are those of a stationary random process, then $\hat{\Gamma}(P_1, P_2, \omega)$ may be expressed as a time average,

$$\hat{\Gamma}(P_1, P_2, \omega) = \lim_{T \to \infty} \frac{1}{2T} \hat{u}_1(T, \omega) \, \hat{u}_2^*(T, \omega), \qquad (5.50)$$

and by (5.49) and (5.50),

$$\hat{\Gamma}(P_1, P_2, \omega) \to a_{12}(\theta_1, \theta_2, \phi_1, \phi_2, \omega) \frac{\exp[ik(r_1 - r_2)]}{r_1 r_2} \qquad (5.51)$$

as $r_1, r_2 \to \infty$.

With $\hat{\Gamma}(P_1, P_2, \omega)$ given by eq. (5.48), the solution for $\Gamma(P_1, P_2, \tau)$ is

$$\Gamma(P_1, P_2, \tau) = \int_0^\infty \hat{\Gamma}(P_1, P_2, \omega) \, e^{-i\omega\tau} \, d\omega. \qquad (5.52)$$

5.4 Some Physical Constraints on Mutual Coherence

The coherence functions have certain limits, which will be discussed in this section. We have already, in discussing eq. (5.23), called attention to a set of limits $[0, 1]$ for the function $|\gamma_{12}(\tau)|$, with 0 representing complete incoherence and 1 representing complete coherence. We have pointed out that the degree of coherence depends upon the value of τ; but it is also true that it depends upon the particular pair of points chosen for comparison. Thus we may expect $|\gamma_{12}(\tau)|$ to be zero for *some* points and *some* delays; but we would not expect it to vanish in general. The question remains, however, whether it is *possible* for an extended field (which might be the source itself) to have the property that $|\gamma_{12}(\tau)| = 0$ or $|\gamma_{12}(\tau)| = 1$ for every pair of points in the field and for any time delay τ. If so, it would seem to be proper to call the entire field incoherent or coherent, respectively.

In this connection we shall quote three well-known theorems, proofs of which the reader can find in the text by Beran and Parrent (ref. 5.2, pp. 47–52).

Theorem 1

An electromagnetic field has a unity degree of coherence (i.e., $|\gamma_{12}(\tau)| = 1$) for every pair of points in the field and every time delay τ, if and only if the field is monochromatic.

Theorem 2

A non-null electromagnetic field, for which $|\gamma_{12}(\tau)| = 0$ for every pair of points in the field and for every time delay τ, cannot exist in free space. Conversely, if $|\gamma_{12}(\tau)| = 0$ for every pair of points on a continuous closed surface, then the surface does not radiate.

Theorem 3

If spectral filtering is used on the radiation from two source points whose degree of coherence is zero, the degree of coherence will remain zero.

With regard to Theorem 1, it must be said that strictly monochromatic fields do not exist in practice, since all fields have some finite spectral bandwidth. However, it is possible for a field to have a spectral bandwidth $\Delta\omega$ that is small compared with the center frequency ω_0 of the radiation. Such a field is called *quasi-monochromatic*. If path differences in the radiation are small, a theory for quasi-monochromatic fields may be developed. Of course the concept of quasi-monochromatic fields is intended to give a practical approach to real monochromaticity, but there are some respects in which they differ widely. For example, although $|\gamma_{12}(\tau)| = 1$ is a requirement for all pairs of points in a strictly monochromatic field, this need not be the case for quasi-monochromaticity. In fact, there may be pairs of points for which $|\gamma_{12}(\tau)|=0$. In order to help the reader obtain a more realistic feeling for the quasi-monochromatic field, I will discuss it a little further.

We take as the condition for quasi-monochromatic radiation

$$\frac{\Delta\omega}{\omega_0}\ll 1. \tag{5.53}$$

It then follows that $\hat{\Gamma}_{12}(\omega)$ is essentially zero for all frequencies outside the band. That is, for an appreciable $\hat{\Gamma}_{12}(\omega)$ we must have

$$|\omega-\omega_0|<\Delta\omega \tag{5.54}$$

From eq. (5.38), the mutual coherence function can then be written

$$\Gamma_{12}(\tau)=\exp(-i\omega_0\tau)\int_0^\infty \hat{\Gamma}_{12}(\omega)\exp[-i(\omega-\omega_0)\tau]\,d\omega. \tag{5.55}$$

By considering only small values of τ, such that $\Delta\omega|\tau|\ll 1$, eq. (5.55) reduces to

$$\Gamma_{12}(\tau)=\Gamma_{12}(0)\exp(-i\omega_0\tau). \tag{5.56}$$

Instead of zero for the value of τ [giving $\Gamma_{12}(0)$], the standard point can be taken as any other value, which we call τ_0, and make $\tau=\tau_0+\tau'$, $\Delta\omega|\tau'|\ll 2\pi$. Then instead of (5.56), we write

$$\Gamma_{12}(\tau_0+\tau')=\Gamma_{12}(\tau_0)\exp(-i\omega_0\tau'). \tag{5.57}$$

The condition $\Delta\omega|\tau'|\ll 2\pi$ is essential in the theory of quasi-monochromatic fields. We can then define coherence in a limited way, by saying that a field is coherent if a τ_0 can be found for every pair of points, such that $|\gamma_{12}(\tau)|=1$ when $\Delta\omega|\tau'|\ll 2\pi$. For very narrow bandwidths ($\Delta\omega/\omega_0$ very small), the field may be considered monochromatic for all values of τ which are practical in a given problem.

In a strictly monochromatic field, the mutual coherence function may be stated as

$$\Gamma_{12}(\tau)=[\Gamma_{11}(0)\,\Gamma_{22}(0)]^{1/2}\exp[i\phi_{12}(\tau)], \tag{5.58}$$

and for $\Delta\omega|\tau'|\ll 1$ this can be reduced to

$$\Gamma_{12}(\tau)=u(P_1)\,u^*(P_2)\exp(-i\omega_0\tau'), \tag{5.59}$$

where u is the field at a particular point (P_1 or P_2).

We do not want to give the impression that all fields with a narrow bandwidth are necessarily coherent, even when τ' is kept small. That is to say, that with $\Delta\omega/\omega_0 \ll 1$ and $\Delta\omega|\tau'| \ll 2\pi$, it still is not necessary that $|\gamma_{12}(\tau)| = 1$. In fact, this quantity can have any value between 0 and 1 under these conditions. However, if the quasi-monochromatic field at every point can be expressed as

$$u(t) = A(t)\{-i[\omega_0 t + \alpha(t)]\}, \tag{5.60}$$

where the functions $A(t)$ and $\alpha(t)$ are slow in their variation with time, compared with $2\pi/\omega_0$, then the field can be said to behave in a coherent manner.

Finally, we note that the treatment of coherent theory in this chapter is by no means complete. For a more intensive treatment of this topic, the reader is referred to the excellent book *Theory of Partial Coherence*, by Beran and Parrent (ref. 5.2).

Problems

5.1 Consider two sources both emitting two discrete frequencies, ω_1 and ω_2. If one source emits radiation of the form

$$u_1 = a_{11} \exp[-i(\omega_1 t + \phi_{11})] + a_{12} \exp[-i(\omega_2 t + \phi_{12})],$$

and the other emits radiation of the form

$$u_2 = a_{21} \exp[-i(\omega_1 t + \phi_{21})] + a_{22} \exp[-i(\omega_2 t + \phi_{22})],$$

(a) determine the degree of coherence $|\gamma_{12}(\tau)|$.

(b) From the result obtained in (a), show that $|\gamma_{12}(\tau)| \leq 1$ for all τ. If $\omega_1 = \omega_2$, so that one may set a_{12} and a_{22} equal to zero, show that the degree of coherence is unity.

5.2 With reference to Young's interference experiment, as shown in fig. 5.3, where $d \ll l_1, l_2$, show that the degree of coherence $|\gamma_{12}| = 1$ when the

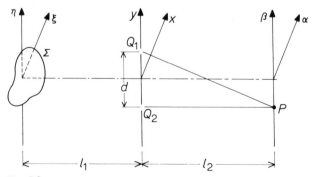

Fig. 5.3

source converges to a point source, and $|\gamma_{12}| = 0$ as the monochromatic surface source Σ becomes infinitely large.

References

5.1 M. Born and E. Wolf, *Principles of Optics*, 2nd rev. ed., Pergamon Press, New York, 1964.

5.2 M. J. Beran and G. B. Parrent, Jr., *Theory of Partial Coherence*, Prentice-Hall, Englewood Cliffs, N. J., 1964.

5.3 M. Françon, *Diffraction Coherence in Optics*, Pergamon Press, New York, 1966.

5.4 E. L. O'Neill, *Introduction to Statistical Optics*, Addison-Wesley, Reading, Mass., 1963.

5.5 J. B. DeVelis, and G. O. Reynolds, *Theory and Applications of Holography*, Addison-Wesley, Reading, Mass., 1967.

5.6 A. M. Yaglom, *An Introduction to the Theory of Stationary Random Functions*, Prentice-Hall, Englewood Cliffs, N.J., 1962.

Part II Information Processing

6 Chapter Fourier Transform Properties of Lenses and Linear Optical Imaging Systems

Prior to discussing linear optical imaging systems, we will treat the important transform properties of lenses. A thorough discussion of the properties of lenses requires a lengthy presentation of the basic theory of geometrical optics, which is beyond the purpose of this text. However, we will use a system theory point of view, which, although not coming directly from the principles of geometrical optics, is quite consistent with these principles in its results. At the beginning of this chapter, we shall discuss in detail the phase transforms of the lenses and their imaging properties. A general analysis of linear optical imaging systems will then be presented.

6.1 Phase Transformation of Thin Lenses

A lens is made of glass or some other transparent material. The index of refraction of a lens is usually greater than that of free space, in which case the velocity of wave propagation inside the lens is lower than the velocity in free space. Prior to going into the general discussion, we should specify what we mean by a *thin lens*. If a light ray entering at a point on one side of a lens emerges at about the same point on the other side of the lens, then the lens may be regarded as thin. That is, the transverse displacement of the ray of light inside a thin lens is negligible. Thus nothing more than a simple phase retardation takes place in a wavefront passing through a thin lens. The amount of retardation is proportional to the thickness of the lens.

Referring to fig. 6.1, we may write the phase retardation across the

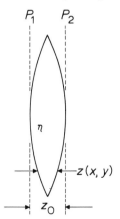

Fig. 6.1 Thickness variation of a convex lens.

wavefront as

$$\phi(x, y) = k[z_0 + (\eta - 1) z(x, y)], \tag{6.1}$$

where $z(x, y)$ is the thickness variation of the lens, z_0 is the maximum thickness of the lens, η is the refractive index of the lens, and k is the wave number. It is clear that $k\eta z(x, y)$ and $k[z_0 - z(x, y)]$ are the phase retardations due to the lens and free space, respectively. Consequently a thin lens can be represented by a general spatial phase transform,

$$T(x, y) = \exp[i\phi(x, y)] = \exp\{ik[z_0 + (\eta - 1) z(x, y)]\}. \tag{6.2}$$

Thus if the thin lens is illuminated by a monochromatic light source in which the complex light field distribution on the plane P_1 of fig. 6.1 is assumed to be $E(x, y)$, then the complex light field at the plane P_2, on on other side of the lens, can be written as

$$E'(x, y) = E(x, y) \, T(x, y). \tag{6.3}$$

In order to better understand the effects of the thickness variation in image formation, we can specify the form of the phase transformation for some practical lenses. The most widely used lenses are those with simple convex and concave surfaces. As an example, for a strictly convex lens (a positive lens, fig. 6.1) the phase transformation may be determined as follows. Let us assume that the radii of curvature of the two spherical surfaces are different and let us divide the lens into left and right halves, as shown in fig. 6.2. The thickness variation of the left half can be written

$$z_1(x, y) = z_{01} - [R_1 - (R_1^2 - \rho^2)^{1/2}] = z_{01} - R_1 \left\{ 1 - \left[1 - \left(\frac{\rho}{R_1} \right)^2 \right]^{1/2} \right\}, \tag{6.4}$$

where z_{01} is the maximum thickness of the left half, R_1 is the radius of the curvature and

$$\rho^2 = x^2 + y^2.$$

Similarly for the right half, the thickness variation can be written

$$\begin{aligned} z_r(x, y) &= z_{0r} - [R_r - (R_r^2 - \rho^2)^{1/2}] \\ &= z_{0r} - R_r \left\{ 1 - \left[1 - \left(\frac{\rho}{R_r} \right)^2 \right]^{1/2} \right\}. \end{aligned} \tag{6.5}$$

The total thickness variation of the lens is the sum of eqs. (6.4) and (6.5):

$$\begin{aligned} z(x, y) &= z_1(x, y) + z_r(x, y) \\ &= z_0 - R_1 \left\{ 1 - \left[1 - \left(\frac{\rho}{R_1} \right)^2 \right]^{1/2} \right\} - R_r \left\{ 1 - \left[1 - \left(\frac{\rho}{R_r} \right)^2 \right]^{1/2} \right\}, \end{aligned} \tag{6.6}$$

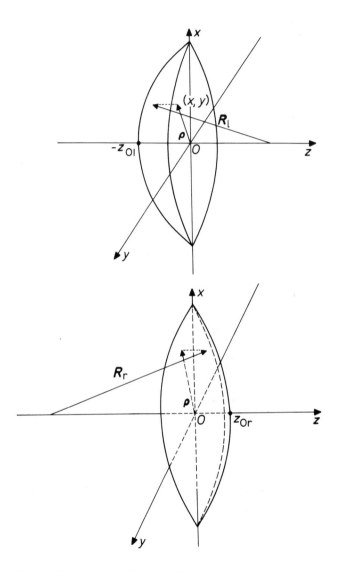

Fig. 6.2 Determination of the lens thickness variation.

where $z_0 = z_{0l} + z_{0r}$.

Equation (6.6) may be simplified if it is confined to a relatively small region of the lens near the optical axis. With this restriction, the following approximations hold:

$$\left[1 - \left(\frac{\rho}{R_1} \right)^2 \right]^{1/2} \simeq 1 - \frac{1}{2} \left(\frac{\rho}{R_1} \right)^2, \tag{6.7}$$

and

$$\left[1 - \left(\frac{\rho}{R_r} \right)^2 \right]^{1/2} \simeq 1 - \frac{1}{2} \left(\frac{\rho}{R_r} \right)^2. \tag{6.8}$$

The thickness variation of the lens can then be approximated by

$$z(x, y) \simeq z_0 - \frac{\rho^2}{2} \left(\frac{1}{R_1} + \frac{1}{R_r} \right). \tag{6.9}$$

It may be noted that the paraxial approximations of eqs. (6.7) and (6.8) give approximately the same result as would be found if the spherical surfaces of the lens were replaced by parabolic surfaces.

From eqs. (6.9) and (6.2) the phase transformation of the convex lens is

$$T(x, y) = \exp(ik\eta z_0) \exp \left[-ik(\eta - 1) \frac{\rho^2}{2} \left(\frac{1}{R_1} + \frac{1}{R_r} \right) \right]. \tag{6.10}$$

Several of the quantities which pertain to the lens itself may be combined as follows:

$$f = \frac{R_1 R_r}{(\eta - 1)(R_1 + R_r)}. \tag{6.11}$$

Although it is not obvious at this point, the quantity f is actually the *focal length* of the lens; i.e., the distance from the lens center, on the axis, at which rays originally parallel to the axis converge to a point. Using (6.11), eq. (6.10) may be written

$$T(x, y) = C_1 \exp \left(-i \frac{k}{2f} \rho^2 \right), \tag{6.12}$$

where $C_1 = \exp(ik\eta z_0)$ is a complex constant.

Similarly, for a concave (i.e., negative) lens (fig. 6.3), the phase transformation can be shown to be

$$T(x, y) = C_2 \exp \left(i \frac{k}{2f} \rho^2 \right), \tag{6.13}$$

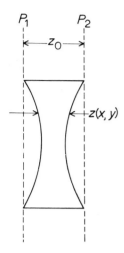

Fig. 6.3 Thickness variation of a concave lens.

where C_2 is a complex constant. Similar derivations may be carried out for the phase transformations for any other type of thin lens. Note that the essential difference in the phase transformations of a convex and concave lens is in the positive and negative quadratic phase delays.

The significance of the plane transformation of lenses can be seen if it is assumed that a monochromatic phase wave field is incident normally on a convex lens. Then the complex light field immediately behind the lens (i.e., on P_2 of fig. 6.1) is

$$E'(x, y) = C \exp\left(-i\frac{k}{2f}\rho^2\right), \tag{6.14}$$

where C is a complex constant. This expression may be interpreted as a quadratic phase transform, which is the approximation to a spherical wavefront. This spherical wavefront converges toward a point of the optical axis at a distance f behind the lens. This confirms the previous statement that f is the focal length. On the other hand, if the lens is concave the wavefront is diverging about a point on the optical axis at the distance f in front of the lens. These two cases of quadratic phase transformation can be illustrated in fig. 6.4. A convex lens is appropriately termed a *converging lens*, while a concave lens is a *diverging* lens.

The quadratic phase transformation has been derived under the paraxial approximation. Without paraxiality, the transformed wavefront will depart from perfect sphericity, and various types of *aberration* will occur.

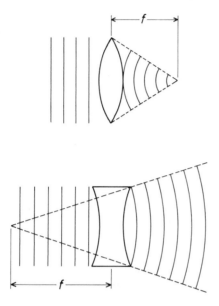

Fig. 6.4 Convergent and divergent effects of positive and negative lenses on an incident plane wave.

Thus practical lenses are often corrected to be free from aberrations, at least to some degree. This correction may be accomplished by grinding the lens surfaces aspherical in order to improve the sphericity of the emerging wavefront.

6.2 Fourier Transform Properties of Lenses

It is very useful that a two-dimensional Fourier transformation may be obtained from a positive lens. Fourier transform operations usually bring to mind complicated electronic spectrum analyzers or digital computers. However, this complicated transform can be performed extremely simply in a coherent optical system; and, because the optical transform is two-dimensional, it has greater information capacity than transforms carried out by means of electronic circuitry.

Prior to the derivation of the Fourier transform properties of a lens, we will use Huygens' principle for the determination of the complex light field on a planar surface P_2 due to a planar light source P_1, as shown in fig. 6.5.

Let us assume that the complex light field at P_1 is $f(x, y)$. Then the complex light distribution at P_2 may be determined by means of Huygens'

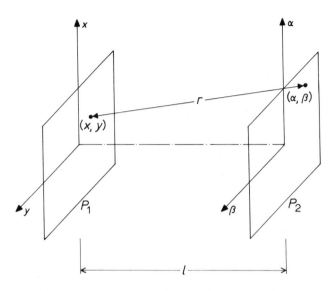

Fig. 6.5 Geometry for the determination of the complex light field.

principle:

$$g(\alpha, \beta) = C \iint_S f(x, y) \exp(ikr) \, dx \, dy, \tag{6.15}$$

where S denotes the surface integral, C is an arbitrary complex constant, $k = 2\pi/\lambda$, and

$$r = [l^2 + (\alpha - x)^2 + (\beta - y)^2]^{1/2}. \tag{6.16}$$

If it is assumed that the separation of the (x, y) and (α, β) coordinate systems is large compared with the spatial dimensions of P_1 and P_2, then r can be paraxially approximated,

$$r \simeq l + \frac{1}{2l} [(\alpha - x)^2 + (\beta - y)^2]. \tag{6.17}$$

The complex light field at P_2 is therefore

$$g(\alpha, \beta) = C \iint_S f(x, y) \exp(ikl) \exp\left\{ i \frac{\pi}{\lambda l} [(\alpha - x)^2 + (\beta - y)^2] \right\} dx \, dy. \tag{6.18}$$

This can be written in the simplified form,

$$g(\alpha, \beta) = C_1 f(x, y) * h_l(x, y), \tag{6.19}$$

where C_1 is an arbitrary complex constant, $*$ denotes the spatial convolution, and

$$h_l(x, y) = \exp\left[i\,\frac{\pi}{\lambda l}(x^2 + y^2) \right]$$

is known as the spatial impulse response. Equations (6.18) or (6.19) define the input-output linear system analog of fig. 6.5, as shown in fig. 6.6.

An additional optical element is required to obtain Fourier transformation by a lens. If the lens is positive, and a point source of monochromatic light is placed at the front focal point, then the light passing through the lens is collimated into a monochromatic plane wave (fig. 6.7). This output wave is the spatial Fourier transform of the point source.

Thus, in fig. 6.5, a point source in the plane P_1 is equivalent to a spatial delta function $\delta(x, y)$. A positive lens is placed in the plane P_2, and the separation of the coordinate systems is made equal to the focal length of the lens. Then the complex light field in front of the lens is

$$g(\alpha, \beta) = C_1 \exp\left[i\,\frac{\pi}{\lambda f}(\alpha^2 + \beta^2) \right]. \tag{6.20}$$

The action of the lens is to make the spherical wave field into a plane wave field. This is merely to say that the lens must introduce a phase transformation

$$T(\alpha, \beta) = \exp\left[-i\,\frac{\pi}{\lambda f}(\alpha^2 + \beta^2) \right], \tag{6.21}$$

so that the complex light field behind the lens is

$$g_1(\alpha, \beta) = C_1, \tag{6.22}$$

a wave field parallel to the (α, β) plane.

Let us consider fig. 6.8, a simple optical system. If the complex light field at P_1 is $f(\xi, \eta)$, then the complex light distribution at P_2 can be written as

$$g(\alpha, \beta) = C\{[f(\xi, \eta) * h_l(\xi, \eta)]\, T(x, y)\} * h_f(x, y), \tag{6.23}$$

where C is again an arbitrary complex constant, $h_l(\xi, \eta)$ and $h_f(x, y)$ are

Fig. 6.6 Input-output linear system analog of the optical setup shown in fig. 6.5.

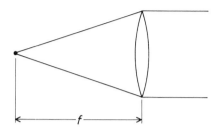

Fig. 6.7 Fourier transform of a monochromatic point source S.

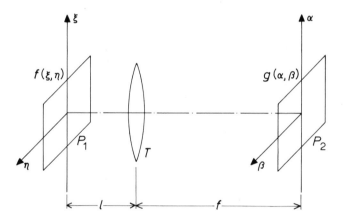

Fig. 6.8 Geometry for the determination of the optical Fourier transformation.

the corresponding spatial impulse responses, and $T(x, y)$ is the phase transformation of the lens.

The linear system analog for eq. (6.23) is shown in fig. 6.9. Equation (6.23) may be written in the integral form

$$g(\alpha, \beta) = C \iint_{S_1} \left[\iint_{S_2} \exp\left(i\frac{k}{2}\,\Delta\right) dx\, dy \right] f(\xi, \eta)\, d\xi\, d\eta, \tag{6.24}$$

where S_1 and S_2 denotes the surface integrals of the light field P_1 and the lens T, respectively, and

$$\Delta \triangleq \left\{ \frac{1}{l}\left[(x-\xi)^2 + (y-\eta)^2\right] + \frac{1}{f}\left[(\alpha-x)^2 + (\beta-y)^2 - (x^2+y^2)\right] \right\}. \tag{6.25}$$

It may be noted that the surface integral of the lens may be assumed to be of infinite extent, since the lens is very large compared to the spatial

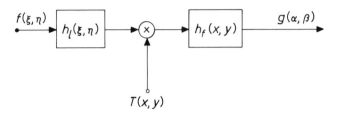

Fig. 6.9 Linear system analog of the optical setup shown in fig. 6.8.

apertures at P_1 and P_2 (paraxiality).

Equation (6.25) may be written as

$$\Delta = \frac{1}{f} [v\xi^2 + vx^2 + \alpha^2 - 2v\xi x - 2x\alpha + v\eta^2 + vy^2 + \beta^2 - 2v\eta y - 2y\beta], \quad (6.26)$$

where $v = f/l$.

By completing the square, eq. (6.26) can be written as

$$\Delta = \frac{1}{f} \left[(v^{1/2}x - v^{1/2}\xi - v^{-1/2}\alpha)^2 - \alpha^2 \left(\frac{1-v}{v} \right) - 2\xi\alpha \right. $$
$$\left. + (v^{1/2}y - v^{1/2}\eta - v^{-1/2}\beta)^2 - \beta^2 \left(\frac{1-v}{v} \right) - 2\eta\beta \right]. \quad (6.27)$$

By using this in eq. (6.24), we have

$$g(\alpha, \beta) = C \exp \left[-i \frac{k}{2f} \left(\frac{1-v}{v} \right) (\alpha^2 + \beta^2) \right]$$
$$\times \int\int_{S_1} f(\xi, \eta) \exp \left[-i \frac{k}{f} (\alpha\xi + \beta\eta) \right] d\xi \, d\eta$$
$$\times \int\int_{S_2} \exp \left\{ i \frac{k}{2f} [(v^{1/2}x - v^{1/2}\xi - v^{-1/2}\alpha)^2 + (v^{1/2}y - v^{1/2}\eta - v^{-1/2}\beta)^2] \right\}$$
$$\times dx \, dy. \quad (6.28)$$

Since the integrations over S_2 are assumed to be taken from $-\infty$ to $+\infty$, we obtain a complex constant which can be incorporated with C. Thus we have

$$g(\alpha, \beta) = C_1 \exp \left[-i \frac{k}{2f} \left(\frac{1-v}{v} \right) (\alpha^2 + \beta^2) \right]$$
$$\times \int\int_{S_1} f(\xi, \eta) \exp \left[-i \frac{k}{f} (\alpha\xi + \beta\eta) \right] d\xi \, d\eta. \quad (6.29)$$

From this it is clear that, except for a spatial quadratic phase variation, $g(\alpha\,\beta)$ is the Fourier transform of $f(\xi, \eta)$. As a matter of fact, the quadratic phase factor vanishes if $l=f$. Evidently if the signal plane P_1 is placed at the front focal plane of the lens, the quadratic phase factor disappears, which leaves an exact Fourier transform relation. Thus eq. (6.29) may be written as

$$G(p, q)=C_1 \iint\limits_{S_1} f(\xi, \eta)\exp[-i(p\xi+q\eta)]\,d\xi\,d\eta \quad \text{for} \quad v=1, \qquad (6.30)$$

where $p=k\alpha/f$ and $q=k\beta/f$ are the spatial frequency coordinates.

It must be emphasized that the exact Fourier transform relation takes place under the condition $l=f$. Under the condition $l\neq f$, a quadratic phase factor will be included. Furthermore, it can easily be shown that a quadratic phase factor also results if the signal plane P_1 is placed behind the lens.

In conventional Fourier transform theory, the transformation from the spatial domain to the spatial frequency domain requires the kernel $\exp[-i(px+qy)]$ and the transformation from the spatial frequency domain to the spatial domain requires the conjugate kernel $\exp[i(px+qy)]$. Obviously a positive lens always introduces the kernel $\exp[-i(px+qy)]$. Therefore, in an optical system one takes only successive transforms, rather than a transform followed by its inverse, as shown in fig. 6.10.

6.3 Optical Image Formation
In this section we will develop some equations for a general optical imaging system. Image formation will be considered for the two extremes; i.e., for the completely incoherent and the completely coherent cases.

In fig. 6.11, a hypothetical optical imaging system is shown. Assume

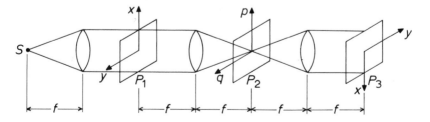

Fig. 6.10 Successive Fourier transformations of lenses. A monochromatic point source is located at S.

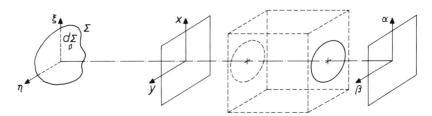

Fig. 6.11 Hypothetical optical imaging system. The black-box optical system is shown lying between the input or signal plane (x, y) and the output or image plane (α, β).

that the light emitted by the source Σ is monochromatic, and suppose that an image of the input signal is formed at the output plane of the optical system. In order to demonstrate the image formation of the optical system, we let $u(x, y)$ be the complex light amplitude distribution on the input signal plane due to an incremental light source $d\Sigma$. If the transmittance of the input plane is $f(x, y)$, then the complex light field immediately behind the signal plane is $u(x, y) f(x, y)$.

If it is assumed that the optical system in the black box is linearly spatially invariant with a spatial impulse response of $h(x, y)$, then the complex light field at the output plane of the system due to $d\Sigma$ can be determined by the convolution equation

$$g(\alpha, \beta) = u(x, y) f(x, y) * h(x, y). \tag{6.31}$$

At this point it may be emphasized that the assumption of linearity of the optical system is generally valid for small amplitude disturbances; however, the spatial-invariance condition may be applicable only over a small region of the signal plane.

From eq. (6.31), the irradiance in the image plane due to $d\Sigma$ is

$$dI(\alpha, \beta) = g(\alpha, \beta) g^*(\alpha, \beta) d\Sigma. \tag{6.32}$$

Therefore the total irradiance of the image due to the whole light source is

$$I(\alpha, \beta) = \int\int_{\Sigma} |g(\alpha, \beta)|^2 \, d\Sigma. \tag{6.33}$$

which can be written out as the convolution integral

$$I(\alpha, \beta) = \int\int\int\int_{-0}^{\infty} \Gamma(x, y; x', y') h(\alpha - x, \beta - y) h^*(\alpha - x', \beta - y')$$

$$\times f(x, y) f^*(x', y') \, dx \, dy \, dx' \, dy', \tag{6.34}$$

where

$$\Gamma(x, y; x', y') = \int\int_\Sigma u(x, y) u^*(x', y') \, d\Sigma. \tag{6.35}$$

Now we choose two points Q_1 and Q_2 on the input signal plane; Q_1 is taken as the origin of the (x, y) plane, and Q_2 is arbitrary. If r_1 and r_2 are the respective distances from Q_1 and Q_2 to $d\Sigma$, then the complex disturbances at Q_1 and Q_2 due to $d\Sigma$ are, respectively,

$$u_1(x, y) = \frac{[I(\xi, \eta)]^{1/2}}{r_1} \exp(ikr_1) \tag{6.36}$$

and

$$u_2(x, y) = \frac{[I(\xi, \eta)]^{1/2}}{r_2} \exp(ikr_2), \tag{6.37}$$

where $I(\xi, \eta)$ is the irradiance across the light source. By substituting eqs. (6.36) and (6.37) in eq. (6.35) we have

$$\Gamma(x, y) = \int\int_\Sigma \frac{I(\xi, \eta)}{r_1 r_2} \exp[ik(r_1 - r_2)] \, d\Sigma. \tag{6.38}$$

In the paraxial case, $r_1 - r_2$ may be approximated by

$$r_1 - r_2 \simeq \frac{2}{r_1 + r_2}(\xi x + \eta y) \simeq \frac{1}{r}(\xi x + \eta y), \tag{6.39}$$

where r is the separation between the light source plane and the signal plane. Then eq. (6.38) can be reduced to

$$\Gamma(x, y) = \frac{1}{r^2} \int\int_\Sigma I(\xi, \eta) \exp\left[i\frac{k}{r}(\xi x + \eta y)\right] d\xi \, d\eta. \tag{6.40}$$

Equation (6.40) is the inverse Fourier transform of the source irradiance.

Now one of the two extreme cases of the hypothetical optical imaging system can be seen by letting the light source become infinitely large. If the irradiance of the source is relatively uniform, that is $I(\xi, \eta) \simeq K$, eq. (6.40) becomes

$$\Gamma(x, y) = K_1 \delta(x, y), \tag{6.41}$$

where K_1 is an appropriate positive constant. This equation describes a completely incoherent optical imaging system.

On the other hand, if the light source is vanishingly small, eq. (6.40) becomes

$$\Gamma(x, y) = K_2,$$ (6.42)

where K_2 is a positive constant. This equation in fact describes a completely coherent optical imaging system.

Referring to eq. (6.34), for the completely incoherent case ($\Gamma(x, y) = K_1\delta(x, y)$), the irradiance at the output is

$$I(\alpha, \beta) = \int \int\limits_{-\infty}^{\infty} \int \int \delta(x' - x, y' - y) h(\alpha - x, \beta - y)$$

$$\times h^*(\alpha - x', \beta - y') f(x, y) f^*(x', y') \, dx \, dy \, dx' \, dy',$$ (6.43)

which can be reduced to

$$I(\alpha, \beta) = \int \int\limits_{-\infty}^{\infty} |h(\alpha - x, \beta - y)|^2 |f(x, y)|^2 \, dx \, dy.$$ (6.44)

It is clear from eq. (6.44) that for the incoherent case the image irradiance is the convolution of the signal irradiance with respect to the impulse response irradiance. In other words, for the completely incoherent case, the optical system is linear in irradiance, i.e.,

$$I(\alpha, \beta) = |h(x, y)|^2 * |f(x, y)|^2.$$ (6.45)

By Fourier transformation, eq. (6.45) can be expressed in the spatial frequency domain:

$$I(p, q) = |H(p, q)|^2 |F(p, q)|^2,$$ (6.46)

where $I(p, q)$, $H(p, q)$, and $F(p, q)$ are the Fourier transforms of $I(\alpha \beta)$, $h(x, y)$, and $f(x, y)$, respectively, and p and q are the spatial frequency coordinates.

In a more convenient from, eq. (6.45) can be written as

$$I(\alpha, \beta) = h_i(x, y) * f_i(x, y),$$ (6.47)

where $h_i(x, y) = |h(x, y)|^2$ and $f_i(x, y) = |f(x, y)|^2$ are the irradiance impulse responses of the system. Then eq. (6.46) may be written as

$$I(p, q) = H_i(p, q) F_i(p, q),$$ (6.48)

where $H_i(p, q)$ and $F_i(p, q)$ are the Fourier transforms of $h_i(x, y)$ and

$f_i(x, y)$ respectively. In fact, $H_i(p, q)$ is given by

$$H_i(p, q) = \int\limits_{-\infty}^{\infty}\int h(x, y) \, h^*(x, y) \, e^{-i(px+qy)} \, dx \, dy. \qquad (6.49)$$

Then by the Fourier multiplication theorem, eq. (6.49) becomes

$$H_i(p, q) = \frac{1}{4\pi^2} \int\limits_{-\infty}^{\infty}\int H(p', q') \, H^*(p'-p, q'-q) \, dp' \, dq', \qquad (6.50)$$

which is the convolution of the complex transfer function with respect to its conjugate. On the other hand, for the completely coherent case $(\Gamma(x, y) = K_2)$, eq. (6.34) becomes

$$I(\alpha, \beta) = g(\alpha, \beta) \, g^*(\alpha, \beta)$$

$$= \int\limits_{-\infty}^{\infty}\int h(\alpha-x, \beta-y) \, f(x, y) \, dx \, dy$$

$$\times \int\limits_{-\infty}^{\infty}\int h^*(\alpha-x', \beta-y') \, f^*(x, y) \, dx' \, dy'. \qquad (6.51)$$

From eq. (6.51) it is obvious that the optical system is linear in complex amplitude, i.e., that

$$g(\alpha, \beta) = \int\limits_{-\infty}^{\infty}\int h(\alpha-x, \beta-y) \, f(x, y) \, dx \, dy. \qquad (6.52)$$

Again, by Fourier transformation eq. (6.52) becomes

$$G(p, q) = H(p, q) \, F(p, q), \qquad (6.53)$$

where $G(p, q)$, $H(p, q)$, and $F(p, q)$ are the corresponding Fourier transforms of $g(\alpha, \beta)$, $h(x, y)$, and $f(x \, y)$, respectively.

A coherence-preserving optical system that makes $\Gamma(x, y) = K$ (a constant) can be achieved by the configuration of fig. 6.12. In this figure, a source of monochromatic light is located at the front focal point of a positive lens. As a result, a collimated plane wave illuminates the signal plane (x, y) of the system.

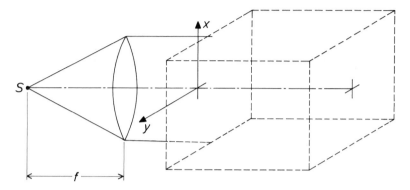

Fig. 6.12 A simple optical configuration to achieve $\Gamma(x, y)=K$. A monochromatic point source is located at S.

Problems

6.1 By means of paraxial approximations, show that the phase transform of the concave lens of fig. 6.3 is indeed given by eq. (6.13).

6.2 Determine the complex light field and the corresponding irradiance at the back focal length of the lens shown in fig. 6.13 where $F(\xi, \eta)$ represents the amplitude transmittance of the transparency, and $l<f$. The incident wave is plane monochromatic.

6.3 Consider a lens cut from a thin section of a cone, as shown in fig. 6.14. By means of the paraxial approximation illustrated in sec. 6.1, determine the corresponding phase transformation. Explain the effect of the lens upon a normally incident monochromatic plane wave.

6.4 The amplitude transparency of a zone-lens grating is given as

$$T(x, y)=\tfrac{1}{2}(1+\cos ax^2) \quad \text{for all} \quad y,$$

where a is an arbitrary constant.
(a) Show that this transparency may be decomposed as the combination of three lenses, i.e., flat, concave, and convex cylindrical lenses.
(b) Determine the corresponding focal lengths of this transparency.
(c) If this transparency is illuminated by a normally incident monochromatic plane wave, determine the effect of the transmitted light field at the back focal plane of the transparency.

6.5 Consider a monochromatic plane wave which is obliquely illumi-

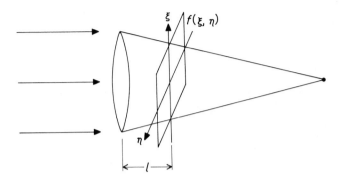

Fig. 6.13

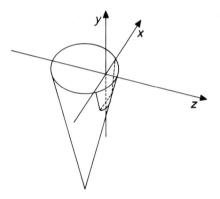

Fig. 6.14

nating a diffraction screen, which consists of an aperture in the shape of a large slit (fig. 6.15). A convex lens is placed behind the aperture as shown. Determine the diffraction pattern at the back focal plane of the lens.

6.6 A monochromatic plane wave is normally incident on an aberration-free objective lens. Due to some alignment error, the image is observed on a plane displaced slightly from the focal plane of the lens. Determine the largest distance error for which the irradiance is still sufficiently accurate to represent the Fraunhofer diffraction pattern.

6.7 The Michelson stellar interferometer is shown in fig. 6.16. A diffraction screen of two parallel slits is placed at the front of an objective lens of a telescope. The light from a distant star enters one of the slits after reflections from mirrors M_{11} and M_{12}, and enters another slit after reflections from mirrors M_{21} and M_{22}. Assume that a filter selecting monochromatic light of wavelength λ is placed at the front of the objective lens. Let a and b be the separations of the slits and mirrors, respectively, and let f be the focal length of the lens. Determine the locations of the maximum and minimum irradiance when
(a) the star lies on the axis of the telescope,
(b) the star lies at a small angle θ to the axis of the telescope, in the direction perpendicular to the slits.

6.8 A spatial filter with time-varying characteristics may be realized by means of the Fabry-Perot interferometric technique. The filter consists of the thin transparent glass plates that are used to support a highly reflective substance coated on the inner surfaces (fig. 6.17). A monochromatic plane wave is normally incident upon the surface of one of the glass plates. If we denote by r and t the respective reflection and transmission coefficients of the reflective substance, and neglect the effect of the glass plates, show that the complex wave field transmitted through the filter is

$$E(\omega) = K \frac{t^2 \exp(-i\omega\, d/c)}{1 - r^2 \exp(-i2\omega\, d/c)},$$

where ω is the circular frequency of the incident plane wave, c is the velocity of the wave propagation, d is the separation of the glass plates, and K is an arbitrary constant.

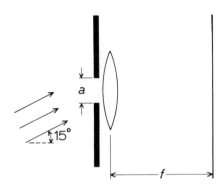

Fig. 6.15

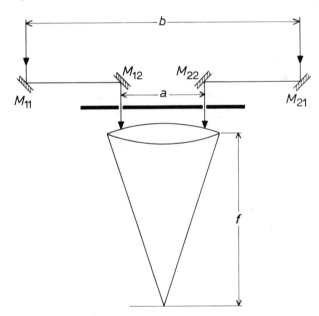

Fig. 6.16

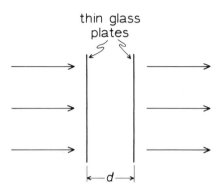

Fig. 6.17

References

6.1 H. H. Hopkins, The Concept of Partial Coherence in Optics, *Proc. Roy. Soc.*, ser. A., **208** (1951), 263.

6.2 H. H. Hopkins, On the Diffraction Theory of Optical Images, *Proc. Roy. Soc.*, ser. A., **217** (1953), 408.

6.3 J. Rhodes, Analysis and Synthesis of Optical Images, *Am. J. Phys.*, **21** (1953), 337.

6.4 L. J. Cutrona et al., Optical Data Processing and Filtering Systems, *IRE Trans. Inform. Theory*, **TT-6** (1960), 386.

6.5 K. Preston, Jr., Use of the Fourier Transformable Properties of Lenses for Signal Spectrum Analysis, in J. T. Tippett et al. (eds.), *Optical and Electro-Optical Information Processing*, MIT Press, Cambridge, Mass., 1965.

6.6 J. W. Goodman, *Introduction to Fourier Optics*, McGraw-Hill, New York, 1968.

Chapter 7   Optical Information Processing and Filtering

We shall not attempt to treat optical information processing in complete detail here. We shall, however, cover a few examples to give a feeling for the principles involved. The reader who wishes to pursue the subject further is referred to the collections edited by Pollack et al. (ref. 7.1), and by Tippett et al. (ref. 7.2). In these references many useful techniques not mentioned here will be found.

Mention must be made of a few important contributors to the field. It was in the early 1950s that the communication theory aspects of optical processing techniques became evident. The most important initial impact was provided by the classic papers "Fourier Treatment of Optical Processes," by Elias, Grey, and Robinson (ref. 7.3), and "Optics and Communication Theory," by Elias (ref. 7.4). However, the very first application of communication theory techniques to modern optics may be O'Neill's paper "Spatial Filtering in Optics" (ref. 7.5). From the broad interest in this new concept, a symposium, "Communication and Information Theory Aspects of Modern Optics," was held in 1960 (ref. 7.6). Since that time the application of communication theory to optics has commanded great interest. The potential applications of coherent spatial filtering were particularly evident in the field of radar signal processing, and it was in this field that Cutrona and his associates at the University of Michigan published "Optical Data Processing and Filtering Systems" in 1960 (ref. 7.7). This article stimulated additional interest in these techniques. The 1964 paper, "Signal Detection by Complex Spatial Filtering," by Vander Lugt (ref. 7.8) introduced the most interesting subject of optical character recognition. Since then, coherent optical information processing has been shown to have a vast variety of applications. It is the combination of communication theory and coherent optics that has stimulated interest in modern optical information processing, and has led to a large amount of interesting research.

7.1 Basic Incoherent Optical Information Processing Systems

In this section we will discuss optical information processing of an incoherent source. Let a signal transparency $f_1(x, y)$ be located at the input signal plane P_1 of an optical system (fig. 7.1). If the transparency is illuminated by an incoherent light source so that the irradiance at P_1 is uniform within the region of interest, then the irradiance immediately behind the transparency is

$$I(x, y) = I_0 f_1(x, y), \tag{7.1}$$

where I_0 is the uniform irradiance which illuminates the signal transparency.

If the input signal transparency is imaged on a one-to-one basis into another signal transparency $f_2(x, y)$ at P_2, then the irradiance immediately behind the second transparency can be written as

$$I(x, y) = I_0 [f_1(x, y) * h_i(x, y)] f_2(x, y). \qquad (7.2)$$

The irradiance impulse responce of the optical system $h_i(x, y)$, is given, as usual, by $h_i(x, y) = |h(x, y)|^2$, where $h(x, y)$ is the complex amplitude spatial impulse response.

If the spatial frequency response of the signal $f_1(x, y)$ is within the limits of the spatial frequency response of the optical system, then the spatial impulse response of the system may be approximated by a Dirac delta function and the irradiance behind P_2 may be approximated by

$$I(x, y) = I_0 f_1(x, y) f_2(x, y), \qquad (7.3)$$

i.e., the output irradiance is, in this approximation, proportional to the product $f_1(x, y) f_2(x, y)$.

The integration of the two processing signals can be accomplished incoherently by focusing the output light on a photodetector, as shown in fig. 7.1. The output voltage from the photodetector is

$$V = K \int\int_S f_1(x, y) f_2(x, y) \, dx \, dy,$$

where K is a proportionality constant, and S denotes surface integration over the output plane P_2.

The integration process of fig. 7.1 can be modified to achieve a multi-

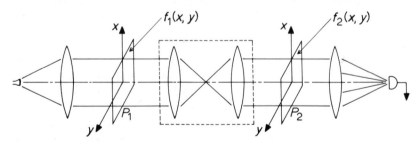

Fig. 7.1 Incoherent processing of signal multiplication and integration. The filament represents an incoherent source, and the imaging system is shown enclosed in dashed lines; a photodiode is shown at the right.

channel multiplication integration, as shown in fig. 7.2. Then the voltages at the outputs of the photodetectors are

$$V(x) = K \int_l f_1(x, y)\, f_2(x, y)\, dy, \tag{7.4}$$

where K is a propositionality constant, and l denotes the line integral in the y direction of the signal for a fixed x.

It may be obvious that if we translate one of the processing signals, for instance, $f_2(x, y)$, along the y axis, then the signal detected by the array of photodetectors may be written,

$$V(x, a) = K \int_l f_1(x, y)\, f_2(x, y-a)\, dy, \tag{7.5}$$

where a is the translation distance.

The processing signals we have considered so far are positive real. This is because a transparency is a passive device whose transmittance T is bounded:

$$0 \le T(x, y) \le 1. \tag{7.6}$$

However, if the processing signal contains positive and negative peaks, its realization can be accomplished by using an appropriate bias level and scaling factor. In this case, the transmittance of the signal is given by

$$T(x, y) = K_0 + K_1 f(x, y), \tag{7.7}$$

where K_0 and K_1 are the bias level and scaling factor, and $T(x, y)$ satisfies the transmittance condition of eq. (7.6).

If the processing signal satisfies eq. (7.7) under the constraint of eq.

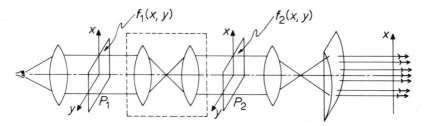

Fig. 7.2 Incoherent processing of multichannel signal multiplication and integration. An array of linear photodetectors is shown at the right.

(7.6), then the output integral can be written as

$$V = K \iint\limits_{S} [K_0 + K_1 f_1(x, y)] [K'_0 + K'_1 f_2(x, y)] \, dx \, dy, \tag{7.8}$$

where K is an appropriate constant. It is apparent that error terms due to the bias levels are introduced into the evaluation of the integration. Even though the values of the bias levels and the scaling factor are known, the removal of these errors may involve variation of the processing signals $f_1(x, y)$ and $f_2(x, y)$. Often, however, a remedy will be obtained by use of the coherent processing to be described in the following section.

7.2 Coherent Optical Information Processing Systems

Let us consider a general coherent optical information processing system (fig. 7.3). If a monochromatic point source is located at the front focal length of a collimating lens L_1, then a monochromatic plane wave will illuminate the input signal plane P_1. A processing signal $f(x, y)$ is inserted at P_1, and the complex light field at the output plane P_3 may be written as

$$g(\alpha, \beta) = K f(x, y) * h(x, y), \tag{7.9}$$

where K is an appropriate constant and $h(x, y)$ is the spatial impulse response of the system.

If we assume that the spatial frequency of the processing signal $f(x, y)$ lies within the spatial frequency limit of the optical system, then the spatial impulse response $h(x, y)$, may be approximated by the Dirac delta function, so that the output light field becomes

$$g(\alpha, \beta) = K f(x, y), \tag{7.10}$$

which is proportional to the input processing signal. Recalling the

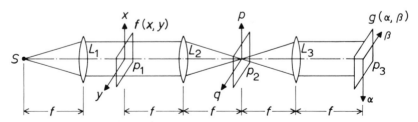

Fig. 7.3 A coherent optical information processing system. Here and in the remaining illustrations of this chapter, S is a monochromatic point source.

Fourier transform properties of lenses (see chap. 6), we see that the complex light amplitude distribution on the plane P_2 is proportional to $F(p, q)$, where $F(p, q)$ is the Fourier transform of the input processing signal $f(x, y)$. The irradiance on P_2 is therefore proportional to $|F(p, q)|^2$.

Several applications of elementary coherent optical systems will be described on the following pages.

COHERENT OPTICAL SPECTRUM ANALYSIS

If the coherent optical system shown in fig. 7.3 is terminated at P_2, as shown in fig. 7.4, then the system is essentially a two-dimensional spectrum analyser. The complex light field at P_2 is

$$E(p, q) = K \int \int_S f(x, y) \, e^{-i(xp+yq)} \, dx \, dy = KF(p, q), \qquad (7.11)$$

where K is a proportionality constant, S denotes surface integration over the input plane P_1, the factor $F(p, q)$ is the two-dimensional Fourier transform of the input processing $f(x, y)$, and p and q are the spatial frequency coordinates. The corresponding irradiance is therefore

$$I(p, q) = E(p, q) \, E^*(p, q) = K^2 \, |F(p, q)|^2, \qquad (7.12)$$

which is proportional to the power spectrum of the processing signal.

The optical spectrum analyser of fig. 7.4 can be modified to yield a multichannel one-dimensional spectrum analyser by the addition of a cylindrical lens, as shown in fig. 7.5. If the input multichannel signal is $f(x_i, y)$, then the complex light field at the output plane of the analyser is

$$E(x_i, q) = K \exp\left[-i \frac{k}{2f} \left(\frac{1-v}{v} \right) q^2 \right] \int f(x_i, y) \, e^{-iqy} \, dy, \quad i = 1, 2, \ldots, n,$$
$$(7.13)$$

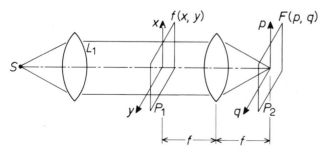

Fig. 7.4 A coherent spectrum analyser.

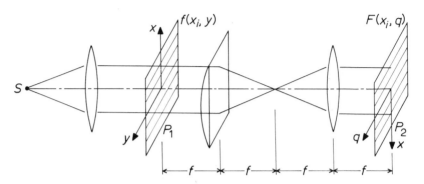

Fig. 7.5 A multichannel coherent spectrum analyser.

where $v = f/l$, and subscript i denotes the corresponding channel in the x axis.

But since $l = 3f$, i.e., $v = 1/3$, eq. (7.13) becomes

$$E(x_i, q) = K \exp\left(-i\frac{k}{f}q^2\right) F(x_i, q), \qquad n = 1, 2, \ldots, n, \tag{7.14}$$

where $F(x_i, q)$ is the one-dimensional Fourier transform of $f(x_i, y)$.

The corresponding irradiance is therefore

$$I(p, q) = E(x_i, q) E^*(x_i, q) = K^2 |F(x_i, q)|^2, \qquad i = 1, 2, \ldots, n. \tag{7.15}$$

SPATIAL FREQUENCY DOMAIN SYNTHESIS

It is possible to synthesize a desired linear-filtering operation in the spatial frequency domain. Such a spatial frequency filter consists of a transparency located at the spatial frequency domain of a coherent optical system, as shown in fig. 7.6. In fact, the desired filter can be synthesized by direct manipulation of the complex amplitude transmittance across the frequency domain. If the processing signal $f(x, y)$ is inserted at the spatial domain P_1, then a Fourier transform of the input signal is distributed at the spatial frequency domain P_2. If a filter of transparency $H(p, q)$ is inserted at P_2, then the complex light field immediately behind P_2 is

$$E(p, q) = KF(p, q) H(p, q), \tag{7.16}$$

where K is a proportionality constant.

The filter has, in general, a complex amplitude transmittance

$$H(p, q) = |H(p, q)|e^{i\phi(p, q)}, \tag{7.17}$$

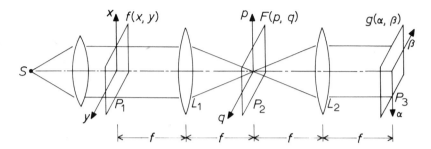

Fig. 7.6 A coherent filtering system.

satisfying the physically realizable conditions,

$$|H(p, q)| \leq 1 \tag{7.18}$$

and

$$0 \leq \phi(p, q) \leq 2\pi. \tag{7.19}$$

Such a transmittance function may be represented by a set of points within or on a unit circle in the complex plane, as shown in fig. 7.7. The amplitude transmission of the filter changes with the optical density, and the phase delay varies with the thickness.

The second lens L_2 of fig. 7.6 takes an inverse Fourier transformation of the complex light field $E(p, q)$ to the output plane P_3, such that the complex-amplitude light distribution across P_3 is

$$g(\alpha, \beta) = K \iint\limits_{S} F(p, q) H(p, q) e^{i(p\alpha + q\beta)} dp \, dq, \tag{7.20}$$

where the surface integration is taken over the spatial frequency domain P_2.

Alternatively by the Fourier multiplication theorem, eq. (7.20) can be written as

$$g(\alpha, \beta) = K \iint\limits_{S} f(x, y) h(\alpha - x, \beta - y) dx \, dy = Kf(x, y) * h(x, y), \tag{7.21}$$

where the integral to taken over the input spatial domain, and $h(x, y)$ is the spatial impulse response of the filter.

It may be pointed out that the frequency domain filter $H(p, q)$ can consist of apertures or slits of any shape. Depending on the arrangement of

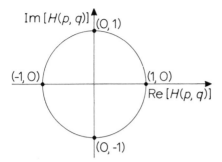

Fig. 7.7 A complex amplitude transmittance function.

apertures, there results a low-pass, high-pass, or band-pass spatial filter. It is clear that any opaque portion in the filter represents a spatial frequency-band rejection. Note also that the inclusion of a phase plate causes a phase delay in the phase-plate portion of the filter. Complex filter synthesis is realizable since one has complete freedom to synthesize the amplitude and the phase responses independently. A technique introduced by Vander Lugt for the synthesis of complex spatial filters will be given in sec. 7.3.

SPATIAL DOMAIN SYNTHESIS
Instead of inserting a complex filter $H(p, q)$, in the spatial frequency plane P_2 (fig. 7.6), put a complex transparency $h(x, y)$ in the input plane P_1. Then the complex light field at P_2 is

$$E(p, q) = K \int\int_S f(x, y) h(x, y) e^{-i(px+qy)} \, dx \, dy, \qquad (7.22)$$

where K is a proportionality constant.

If the processing signal is made to translate in P_1, then eq. (7.22) can be written,

$$E(p, q) = K \int\int_S f(x-\alpha, y-\beta) h(x, y) e^{-i(px+qy)} \, dx \, dy, \qquad (7.23)$$

where α and β are the variables corresponding to the x and y lateral displacements. It is clear now that if the measurement of the light field takes place at $p=q=0$, then the integral of eq. (7.23) can be written

$$g(\alpha, \beta) = K \int\int_S f(x-\alpha, y-\beta) h(x, y) \, dx \, dy, \qquad (7.24)$$

which we recognize as the crosscorrelation function of $f(x, y)$ and $h(x, y)$. Thus,

$$g(\alpha, \beta) = Kf(x, y) \circledast h(x, y). \tag{7.25}$$

On the other hand, if the spatial coordinates of the input processing signal $f(x, y)$ are reversed, and then translated laterally in the input spatial domain, such that the input signal $f(x - \alpha, y - \beta)$ is replaced by $f(\alpha - x, \beta - y)$, then at the origin of P_2 the complex light field is

$$g(\alpha, \beta) = K \int\int_{s} f(\alpha - x, \beta - y) h(x, y) \, dx \, dy. \tag{7.26}$$

This is the convolution integral of $f(x, y)$ and $h(x, y)$. It is identical to eq. (7.21).

In this section, we have discussed the two methods of synthesis that are available for coherent optical information processing, namely,

1. Frequency domain synthesis, in which a complex spatial filter is introduced in the frequency domain, and the filtering operation is performed directly on the complex frequency spectrum.

2. Spatial domain synthesis, in which a complex reference function is introduced in the input spatial domain, and the filtering operation is performed directly on the processing signal.

Theoretically these two techniques of synthesis yield essentially the same result. However, in practice the spatial domain synthesis requires a scanning mechanism, and thus more instrumentation; operations also take longer than in scanner-free frequency domain synthesis. It may be sometimes advantageous to mix these two syntheses together; then the output signal may be written

$$g(\alpha, \beta) = \mathscr{F}^{-1}\{H(p, q)\, \mathscr{F}[f(x, y)\, h(x, y)]\}, \tag{7.27}$$

where $\mathscr{F}$ and $\mathscr{F}^{-1}$ denote the direct and inverse Fourier transforms, $H(p, q)$ is the complex filter function in the frequency domain, and $h(x, y)$ is the complex reference function in the input spatial domain.

MULTICHANNEL FREQUENCY DOMAIN SYNTHESIS

One of the most important features of the coherent information processing system is that of two-dimensionality. This feature is useful when the processing signals are functions of two variables. The signal can be displayed in two spatial dimensions and then both variables can be processed simultaneously. On the other hand, if the same signal were to be processed electronically, some sort of scanning technique would be required.

Occasionally the processing signals are functions of one variable, and the additional capacity of the optical system may not be needed. In this case, the free coordinate of the optical system may be converted into a sequence of one-dimensional channels, such that a finite number of one-variable processing signals may be processed simultaneously. A multi-channel system is shown in fig. 7.8. The input processing signal $f(x, y)$ is composed of a finite number of one-dimensional signals, such that

$$f(x, y) = f(x_i, y), \qquad i = 1, 2, ..., n, \tag{7.28}$$

where x_i represents the ith channel on the x axis, and n is the total number of channels. It may be emphasized that the practical limit of n is the highest number of stripe elements that the optical system can resolve. The complex light field at the front of cylindrical lens L_{c2} is

$$E(x_i, q) = K \exp\left(-i\frac{k}{2f}q^2\right) F(x_i, q), \qquad i = 1, 2, ..., n, \tag{7.29}$$

which is proportional to the Fourier transform $F(x_i, q)$ up to a quadratic phase factor.

If L_{c2} has a phase transform of

$$T(x, q) = \exp\left(i\frac{k}{f}q^2\right), \tag{7.30}$$

and a multichannel filter $H(x_i, q)$ is inserted in the spatial frequency plane P_2, then the complex light field behind P_2 is

$$E'(x_i, q) = K \exp\left(i\frac{k}{f}q^2\right) F(x_i, q) H(x_i, q), \qquad i = 1, 2, ..., n. \tag{7.31}$$

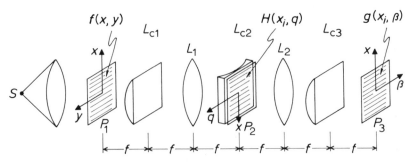

Fig. 7.8 A multichannel coherent filtering system.

Because of the inverse transform operations of lenses L_2 and L_{c3}, the complex light amplitude distribution at the output plane P_3 is

$$g(x_i, \beta) = K \int f(x_i, \beta - y) h(x_i, y) \, dy = K f(x_i, y) * h(x_i, y), \qquad i = 1, 2, \dots, n.$$

$$(7.32)$$

If the input signals are to undergo the same filtering in the spatial frequency domain, then separation into individual channels is not necessary. In this special case, the cylindrical lenses are not needed, and the optical system is the same as that of fig. 7.6. Thus the Fourier transform of the multichannel input signals on P_2 is $F(p, q)$. The filter transparency is $H(p, q)$, which is independent of p. Therefore the complex amplitude distribution across the output plane P_3 is

$$g(x, \beta) = K \int f(x, \beta - y) h(x, y) \, dy.$$

$$(7.33)$$

On the other hand, if the multichannel processing signals are required to operate at the spatial domain, then the input signals $f(x_i, y)$ and the reference function $h(x_i, y)$ can both be inserted in the input plane P_1 of fig. 7.8. The complex light field distribution at the spatial frequency domain is in this case

$$g(x_i, q) = K \exp\left(-i \frac{k}{2f} q^2\right) \int f(x_i, y) h(x_i, y) e^{-iqy} \, dy, \qquad i = 1, 2, \dots, n.$$

$$(7.34)$$

If the input signals $f(x_i, y)$ are translated along the y axis, then the evaluation of eq. (7.34) along the line $q = 0$ is

$$g(x_i, \beta) = K \int f(x_i, \beta + y) h(x_i, y) \, dy, \qquad i = 1, 2, \dots, n.$$

$$(7.35)$$

This can be accomplished by placing a slit along the line $q = 0$. It is clear that eq. (7.35) is the multichannel crosscorrelation, which can be expressed as

$$g(x_i, \beta) = K f(x_i, y) \circledast h(x_i, y), \qquad i = 1, 2, \dots, n, \qquad (7.36)$$

On the other hand, for convolution we could simply reverse the input processing signals such that the output signal is

$$g(x_i, \beta) = K \int f(x_i, \beta - y) h(x_i, y) \, dy = K f(x_i, y) * h(x_i, y), \qquad i = 1, 2, \dots, n,$$

$$(7.37)$$

7.3 Coherent Optical Complex Spatial Filtering

Coherent optical information processing, described in the previous section, permits the synthesis of a vast variety of spatial filters. Such optical filters consist of transparencies which are inserted in the appropriate locations in the optical system. It was in 1963 that Vander Lugt proposed and successfully demonstrated a new method of synthesizing a complex spatial filter (refs. 7.8, 7.9). By means of this technique it is possible to uniquely combine amplitude and phase information in a single filter. Vander Lugt's technique has stimulated great interest in the areas of optical signal detection, and pattern and character recognition. It is because of these interesting applications that we separate this topic in an independent section of this chapter.

The basic concept of matched filtering was introduced in chap. 1. This matched filtering may be easily demonstrated in a simple coherent system diagramed in fig. 7.9. Let us assume that the input processing signal is obscured by some additive Gaussian noise; assume, further, that the matched complex filter $H(p, q)$, has a transmittance which is proportional to the complex conjugate of the Fourier transform of the signal, that is

$$H(p, q) = KF^*(p, q), \tag{7.38}$$

where $F(p, q)$ is the Fourier transform of the input signal $f(x, y)$.

The light amplitude distribution immediately behind the filter is

$$E(p, q) = K_1 [|F(p, q)|^2 + N(p, q) F^*(p, q)], \tag{7.39}$$

where N is the Fourier transform of the random noise.

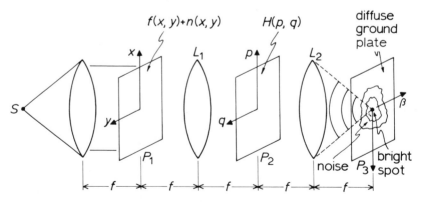

Fig. 7.9 Coherent matched filtering. The signal plus noise is $f(x, y) + n(x, y)$, and the matched filter is $H(p, q)$.

In eq. (7.39), the term $|F(p, q)|^2$ is, of course, real. It represents a plane wave with some amplitude weighting. Thus the complex light field due to the signal term will be imaged into a small region by L_2. It is clear that the effect of the matched filter is to eliminate the phase variation of the Fourier transform signal. It is not intended to restore the signal at the output plane, but rather to compress it into a small bright spot.

On the other hand, the noise term $NF^*(p, q)$ is differently affected. The filter has no compensating effect on the phase spectrum of $N(p, q)$, but it does alternate the amplitude of the noise spectrum in some regions where the signal amplitude spectrum is low. The net effect is that the noise is amplitude-weighted and reproduced at the output plane P_3 essentially unchanged, except that it is attenuated relative to the peak signal.

The complex light field at the output plane P_3 of the matched filter system is therefore

$$g(\alpha, \beta) = K_1 \left[\int \int f(x, y) f(\alpha + x, \beta + y) \, dx \, dy \right.$$
$$\left. + \int \int n(x, y) f(\alpha + x, \beta + y) \, dx \, dy \right]. \qquad (7.40)$$

Alternatively, equation (7.40) can be written

$$g(\alpha, \beta) = K \left[f(x, y) \circledast f(x, y) + n(x, y) * f(-x, -y) \right]. \qquad (7.41)$$

The first term of eq. (7.41) is the autocorrelation of the input signal, and the second term is the convolution of the input noise and the mirror image of the input signal. Nevertheless it is clear that the effect of the matched filter is to compress most of the light diffracted by the signal into a small region; this is the autocorrelation function of the signal. In the case of noise, however, the matched filter does not compress the light into the small region, and no bright spot is observed at the output plane.

So far the basic concept of matched filtering has been discussed; however, to synthesize a two-dimensional matched filter may be difficult in practice. Nevertheless, there exists a stratagem well known to electrical engineers, based on the fact that a band-limited complex function may be represented by a real-valued function of twice the bandwidth of the complex function. In other words, the amplitude and the phase components of the complex function are both modulated by a carrier function whose frequency is greater than the highest frequency component of the complex function. Then by a suitable demodulation technique it is possible to reproduce the complex function from the modulated real function. Using this technique to synthesize a complex spatial filter, we can represent the transparency by means of the real function

$$H(p, q) = K_1 + K_2 |S(p, q)| \cos [\alpha_0 p + \phi(p, q)], \qquad (7.42)$$

where $S(p, q) = |S(p, q)| \exp [-i\phi(p, q)]$ is the complex function, K_1 and K_2 are the appropriate constants such that $0 \leq H(p, q) \leq 1$, p and q are the corresponding spatial frequency coordinates, and α_0 is the carrier frequency of the modulation function. It is clear that the amplitude and phase components of the complex function are properly modulated.

If this complex spatial filter is inserted in the spatial frequency plane of fig. 7.9, then the complex light distribution at the output plane P_3 is

$$g(\alpha, \beta) = K \int \int F(p, q) H(p, q) \exp [-i (\alpha p + \beta q)] \, dp \, dq. \qquad (7.43)$$

This can be written as

$$g(\alpha, \beta) = K K_1 \int \int F(p, q) \exp [-i(\alpha p + \beta q)] \, dp \, dq$$

$$+ \frac{K K_2}{2} \int \int F(p, q) S(p, q) \exp \{-[i(\alpha - \alpha_0) p + \beta q]\} \, dp \, dq$$

$$+ \frac{K K_2}{2} \int \int F(p, q) S^*(p, q) \exp \{-i[(\alpha + \alpha_0) p + \beta q]\} \, dp \, dq. \qquad (7.44)$$

The first term of eq. (7.44) represents a diffraction pattern around the optical axis of the output plane P_3, and the second and third terms represent diffraction around the points at $\alpha = \alpha_0$ and $\alpha = -\alpha_0$ respectively. Therefore the value of the spatial carrier frequency α_0 should be chosen sufficiently large to ensure that the three diffraction patterns will not overlap.

It may be emphasized that the bias level K_1 in eq. (7.42) must be chosen large enough to ensure the physical realization of the filter function. However, when $|S(p, q)|$ has a large dynamic range, the large value of K_1 often leads to poor utilization of the recording medium. This problem can be overcome by using interferometric techniques to synthesize the complex filter.

INTERFEROMETRIC TECHNIQUES IN THE SYNTHESIS OF COMPLEX SPATIAL FILTERS

There are several methods of synthesizing complex spatial filters. In this section we will mention some that make use of interferometric techniques. One interesting method centers around a modification of the Rayleigh interferometer (fig. 7.10; see also ref. 7.2, p. 128). A portion of the collimated monochromatic light illuminates the detecting signal $s(x, y)$, and a portion is converged by lens L_2 to form a reference point source. The complex light field at the back focal plane of L_3 is therefore

$$E(p, q) = K_1 \exp\left(-i\frac{\alpha_0}{2}p\right) + K_2 S(p, q) \exp\left(i\frac{\alpha_0}{2}p\right), \tag{7.45}$$

where $S(p, q)$ is the Fourier transform of the detecting signal.

The corresponding irradiance on the surface of the photographic plate is

$$\begin{aligned}
I(p, q) &= E(p, q)\, E^*(p, q) \\
&= K_1^2 + K_2^2\, |S(p, q)|^2 + K_1 K_2 S(p, q) \exp(ip\alpha_0) \\
&\quad + K_1 K_2 S^*(p, q) \exp(-ip\alpha_0),
\end{aligned} \tag{7.46}$$

where $S(p, q) = |S(p, q)| \exp[i\phi(x, y)]$. If the recording is in the linear region of the photographic emulsion (see chap. 8), then the transparency of the recorded complex spatial filter may be written

$$H(p, q) = K\{K_1^2 + K_2^2\, |S(p, q)|^2 + 2K_1 K_2\, |S(p, q)|\, \cos[\alpha_0 p + \phi(x, y)]\}. \tag{7.47}$$

Equation (7.47) is not only a real-valued function, but is nonnegative as well; i.e., $H(p, q) \geq 0$. Filter synthesis by this interferometric technique

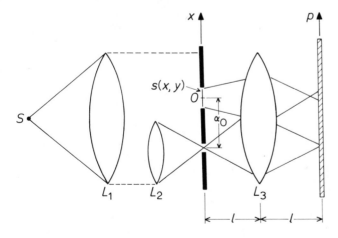

Fig. 7.10 Synthesis of a complex spatial filter; a modified Rayleigh interferometer. The detecting signal is $s(x, y)$; the pinhole behind L_2 provides a point reference source; and a photographic plate P is placed in the (p, q) plane.

offers these advantages at the expense of increased spatial frequency bandwidth required by the amplitude and phase modulation.

If this complex spatial filter is inserted in the spatial frequency plane of
a coherent optical processing system as shown in fig. 7.11, then the field
immediately behind the spatial filter is

$$E(p, q) = F(p, q) H(p, q), \tag{7.48}$$

where $F(p, q)$ is the Fourier transform of the input signal $f(x, y)$. By substitution of eq. (7.47) into eq. (7.48), we have

$$E(p, q) = K[K_1^2 F(p, q) + K_2^2 F(p, q) |S(p, q)|^2 \\ + K_1 K_2 F(p, q) S(p, q) \exp(ip\alpha_0) + K_1 K_2 F(p, q) S^*(p, q) \exp(-ip\alpha_0)]. \tag{7.49}$$

The field at the output plane P_3 is the inverse Fourier transform of eq.
(7.49):

$$g(\alpha, \beta) = \int \int E(p, q) \exp[-i(\alpha p + \beta q)] \, dp \, dq, \tag{7.50}$$

which can be written as the symbolic equation

$$g(\alpha, \beta) = K[K_1^2 f(x, y) + K_2^2 f(x, y) * S(x, y) * S^*(-x, -y) \\ + K_1 K_2 f(x, y) * S(x + \alpha_0, y) + K_1 K_2 f(x, y) * S^*(-x + \alpha_0, -y)]. \tag{7.51}$$

The first and second terms of eq. (7.51) represent the zero order diffraction, which appears at the origin of the output plane; the third and fourth
terms are the convolution and crosscorrelation terms, which are diffracted

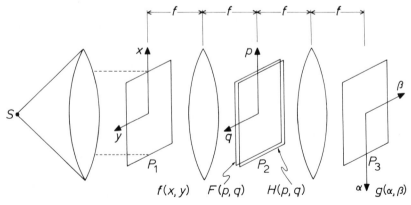

Fig. 7.11 Coherent detection by complex spatial filtering. The input signal is $f(x, y)$; the
filter is $H(p, q)$; and the output signal is $g(\alpha, \beta)$.

in the neighborhood of $\alpha = \alpha_0$ and $\alpha = -\alpha_0$, respectively. The zero order and the convolution terms are of no particular interest here; it is the cross-correlation term that is used in signal detection.

Now if the input signal is assumed to be a detecting signal imbedded in an additive white Gaussian noise n,

$$f(x, y) = S(x, y) + n(x, y), \tag{7.52}$$

then the correlation term of eq. (7.51) can be written

$$R(\alpha, \beta) = K_1 K_2 [S(x, y) + n(x, y)] * S^*(-x + \alpha_0, -y). \tag{7.53}$$

Since the crosscorrelation with respect to $n(x, y)$ and to $S^*(-x + \alpha_0, -y)$ can be shown to be approximately equal to zero, this equation can be written as

$$R(\alpha, \beta) = K_1 K_2 S(x, y) * S^*(-x + \alpha_0, -y), \tag{7.54}$$

which is proportional to the autocorrelation of $S(x, y)$.

In order to ensure that the zero order and the first order diffraction terms of eq. (7.51) will not overlap, the spatial carrier frequency α_0 may be approximated by the inequality

$$\alpha_0 > l_f + \tfrac{3}{2} l_s, \tag{7.55}$$

where l_f and l_s are the spatial lengths in the x direction of the input signal $f(x, y)$ and the detecting signal $S(x, y)$ respectively. To show that this is true, we consider the length of the various output terms of $g(\alpha, \beta)$, as illustrated in fig. 7.12. If l_f and l_s are the respective lengths of input signal $f(x, y)$ and detecting signal $S(x, y)$ in the x direction, then it is clear that lengths of first, second, third, and fourth terms of eq. (7.51) are l_f, $2l_s + l_s$, $l_f + l_s$, and $l_f + l_s$, respectively. From fig. 7.12, we conclude that to achieve complete separation the spatial carrier frequency α_0 must satisfy the inequality (7.55).

Note that the interferometric technique for the complex spatial filter synthesis is similar to the modulation technique in communication theory. The filter synthesis (such as shown in fig. 7.10) is equivalent to single side-band modulation of an electronic communication system. The detecting signal spectrum occupies a bandwidth l_s centered at the spatial carrier frequency α_0.

Following the analogy of spatial filter synthesis and basic communication theory, it is clear that a multiplexed spatial filter may be synthesized. The required complex spatial frequency Fourier spectrum of n detecting signals can be recorded by n carrier frequencies. The minimum value of

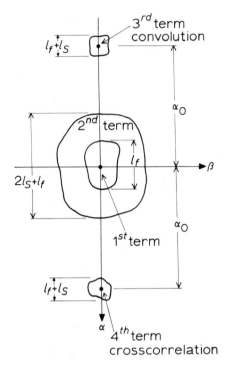

Fig. 7.12 Sketch of the output diffraction of a complex spatial filter.

the carrier frequency for the (i, j)th detecting signal can be approximated by the inequalities:

$$|\alpha_0(i, j+1) - \alpha_0(i, j)| > l_{fx} + \tfrac{1}{2}l_{sx}(i, j+1) + \tfrac{1}{2}l_{sx}(i, j) \tag{7.56}$$

and

$$|\alpha_0(i+1, j) - \alpha_0(i, j)| > l_{fy} + \tfrac{1}{2}l_{sy}(i+1, j) + \tfrac{1}{2}l_{sy}(i, j), \tag{7.57}$$

where l_{fx} and l_{fy} are the spatial lengths of the input signal in the x and y directions, l_{sx} and l_{sy} are the spatial lengths corresponding to the (i, j)th detecting signal, and $\alpha_0(i, j)$ is the vector representation of the spatial carrier frequency with respect to the (i, j)th detecting signal. A sketch of the detecting signal matrix and the reference point source is given in fig. 7.13.

Complex spatial filters are generally sensitive to rotation and to mis-scaling of the input detecting signal. That is, when the input signal is improperly matched to the angular orientation of the detecting signal or is

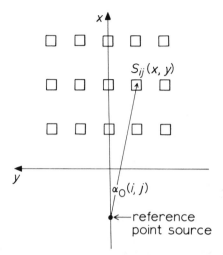

Fig. 7.13 Input plane for multiplex filter synthesis.

given improper magnification, then the response of the filter is greatly decreased from that of a correctly matched one. The degree of sensitivity to the mis-orientation and mis-scaling depends to a large extent on characteristics of the detecting signal. For instance, a detecting signal of other than circular symmetry is more rotationally sensitive than a signal with circular symmetry. Rotational sensitivity may be reduced to some extent by additional filtering. In some cases, the only practical solution to the orientation and scale problems is a manual search operation.

At this point, let us give two alternative coherent optical systems for the synthesis of complex spatial filters (figs. 7.14, 7.15). The optical system shown in fig. 7.14 is a modified Mach-Zehnder interferometer. An oblique plane wave front on the photographic plate is produced by the tilted mirror M_1 and the second beam splitter. The lower part of the interferometer projects the Fourier transform of the detecting signal $S(x, y)$ onto the photographic plate. Hence an interference pattern of the two wavefronts results on the recording medium. The principle of the synthesis shown in fig. 7.15 is essentially the same as the modified Rayleigh interferometric technique, except that a smaller lens L_2 is used to produce the Fourier transformation of the detecting signal, and that a prism produces an appropriate oblique reference wave.

COMPLEX SPATIAL FILTERING IN CHARACTER RECOGNITION

A particularly interesting application of coherent spatial filtering is that

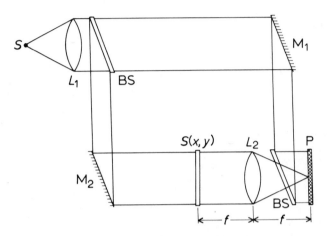

Fig. 7.14 Synthesis of a complex spatial filter; a modified Mach-Zehnder interferometer. BS, beam splitter; $S(x, y)$, detecting signal; P, photographic plate.

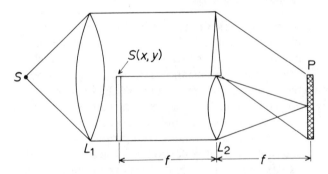

Fig. 7.15 Synthesis of a complex spatial filter. P, photographic plate; $S(x, y)$, detecting signal.

of character recognition. The concept of matched filtering plays an important role in this application. Under the usual assumption of spatial invariance, a complex spatial filter is said to be matched with the detecting signal $S(x, y)$ if and only if the impulse response of the filter is

$$h(x, y) = KS^*(-x, -y). \tag{7.58}$$

The transfer characteristic of the filter is given by

$$H(p, q) = KS^*(p, q), \tag{7.59}$$

where $S^*(p, q)$ is the complex conjugate of the Fourier transform of the detecting signal $S(x, y)$.

To demonstrate that matched filtering provides a means for the recognition of characters, let us assume that the matched filter was designed to detect a particular character $S_i(x, y)$. Now if the input signal $\alpha(x, y)$ consists of a sequence of different nonoverlapping characters,

$$f(x, y) = S_1(x, y) + S_2(x, y) + \cdots + S_n(x, y), \tag{7.60}$$

then it can be shown that the magnitude of the autocorrelation is generally larger than that of the crosscorrelation. That is,

$$|S_i(x, y) \circledast S_i^*(x, y)| \geq |S(x, y) \circledast S_i^*(x, y)|. \tag{7.61}$$

To show that this is true, let us write down the autocorrelation function

$$R_{ii}(\alpha, \beta) = \int \int S_i(x, y) S_i^*(x + \alpha, y + \beta) \, dx \, dy, \tag{7.62}$$

and the crosscorrelation function

$$R_{ji}(\alpha, \beta) = \int \int S_j(x, y) S_i^*(x + \alpha, y + \beta) \, dx \, dy, \qquad i \neq j. \tag{7.63}$$

At the peak $\alpha = \beta = 0$ of the autocorrelation function we can write,

$$|R_{ii}(0, 0)| = \int \int |S_i(x, y)|^2 \, dx \, dy, \tag{7.64}$$

and, similarly,

$$|R_{ji}(0, 0)| = \frac{\left| \int \int S_j(x, y) S_i^*(x, y) \, dx \, dy \right|^2}{\int \int |S_j(x, y)|^2 \, dx \, dy}, \qquad i \neq j. \tag{7.65}$$

But Schwarz's inequality states

$$\left| \int \int S_j(x, y) S_i^*(x, y) \, dx \, dy \right|^2 \leq \int \int |S_i(x, y)|^2 \, dx \, dy \int \int |S_j(x, y)|^2 \, dx \, dy.$$

(7.66)

Therefore it follows that

$$|R_{ji}(0, 0)| \leq |R_{ii}(0, 0)|,$$

(7.67)

the equality holding only when

$$S_j(x, y) = K S_i(x, y).$$

(7.68)

It must be emphasized that character recognition by means of complex spatial filtering is not always a unique process without errors. Nevertheless in most cases it is a workable detection scheme. Examples are shown in figs. 7.16 and 7.17, by Vander Lugt. For more detailed discussion of the complex spatial filtering problem the reader can refer to the excellent article, "Character Reading by Optical Spatial Filtering," by Vander Lugt, Rotz, and Klooster (reprinted as chap. 7 of ref. 7.2).

7.4 Synthetic-Aperture Radar

One of the interesting and unique applications of coherent optical techniques is the processing of synthetic-aperture antenna data. The discussion in this section will follow for the most part that of Cutrona et al. (ref. 7.10).

Let us consider a sidelooking radar system carried by an aircraft in level flight, as shown in fig. 7.18. Suppose that a sequence of pulsed radar signals is directed onto the terrain from the radar system in the plane and that return signals depending on the reflectivity of the terrain across an area adjacent to the flight path are received. Let us define the cross-track coordinate of the radar image as "ground range" coordinate and the along-track coordinate as "azimuth" coordinate. It may be convenient, also, to define the coordinate joining the radar trajectory of the plane and any target under consideration as "slant range" coordinate. If a conventional type of radar system is used, then the azimuth resolution will be (at least in principle) of the order of $\lambda r_1/D$, where λ is the wavelength of the radar signals, r_1 is the slant range, and D is the along-track dimension of the antenna aperture. However, the radar wavelength is several orders of magnitude larger than the optical wavelength, and therefore a very large value of antenna aperture D is required in order to have an angular resolution comparable to that of a photoreconnaissance system. The required

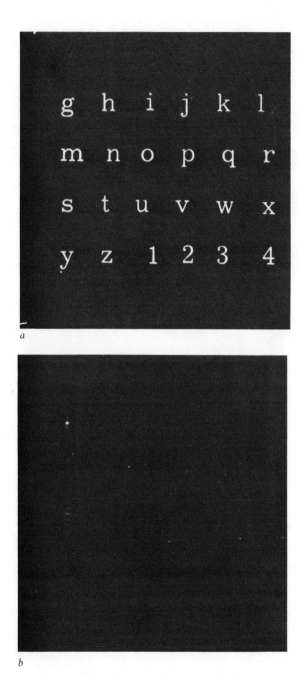

Fig. 7.16 Character recognition: *a*, alphanumerics; *b*, detection of the letter g. (Permission by A. B. Vander Lugt.)

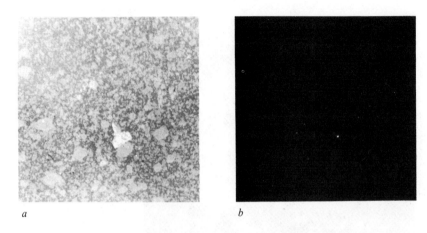

Fig. 7.17 Detection of a signal embedded in random noise: *a*, input scene; *b*, detected signal (Permission by A. B. Vander Lugt.)

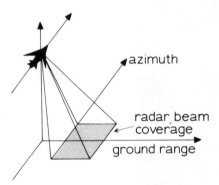

Fig. 7.18 Sidelooking radar.

antenna length may be hundreds, or even thousands, of feet; obviously it is impractical to realize this in an aircraft.

However, this difficulty can be resolved by means of the synthetic-aperture technique. Let us assume that the aircraft carries a small side-looking antenna, and that a relatively broadbeam radar signal scans the terrain by virtue of the aircraft motion. The radar pulses are emitted in a sequence of positions along the flight path, which can be treated as if they were the positions occupied by the elements of a linear antenna array. The return radar signal at each of the respective positions is then recorded "coherently" as a function of time; that is, the radar receiver has available a reference signal such that it is able to record the amplitude and phase information simultaneously. The various recorded complex waveforms are then properly combined to synthesize an effective aperture.

In order to examine in more detail how this synthetic-aperture technique can be accomplished, let us consider first a point-target problem and then extend the results, by superposition, to a more complicated case. We assume a point target located at x_1 in fig. 7.19. The radar pulse is produced by periodic rectangular modulation of a sinusoidal signal of radian frequency ω. This periodic pulsing provides the range information and the fine azimuth resolution, provided that the distance traveled by the aircraft between sample pulses is smaller than $\pi/\Delta p$, where Δp is the spatial bandwidth of the terrain reflections. Now the radar signal returned to the aircraft from a point object may be written as

$$S_1(t) = A_1 \exp\left[i\omega\left(t - \frac{2r}{c} \right) \right], \tag{7.69}$$

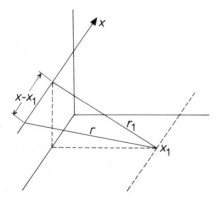

Fig. 7.19 Geometry of sidelooking radar.

where A_1 is an appropriate complex constant. The complex quantity A_1 has contained in it such factors as transmitted power, target reflectivity, phase shift, and the inverse fourth-power propagation law. By the use of the paraxial approximation, the range r may be approximated by

$$r \simeq r_1 + \frac{(x - x_1)^2}{2r_1}. \tag{7.70}$$

Substitution of eq. (7.70) into eq. (7.69) yields

$$S_1(t) = A_1(x_1, r_1) \exp\left\{i\left[\omega t - 2kr_1 - \frac{k}{r_1}(x - x_1)^2\right]\right\}, \tag{7.71}$$

where $k = 2\pi/\lambda$. Equation (7.71) is a function of t and x, in which the aircraft motion links the space variation to the time variable by the relation

$$x = vt, \tag{7.72}$$

where v is the velocity of the aircraft. Now if we assume that the terrain at range r_1 consists of a collection of n points targets, then by superposition the total returned radar signal may be written

$$S(t) = \sum_{n=1}^{N} S_n(t)$$
$$= \sum_{n=1}^{N} A_n(x_n, r_1) \exp\left\{i\left[\omega t - 2kr_1 - \frac{k}{r_1}(vt - x_n)^2\right]\right\}. \tag{7.73}$$

If the returned radar signal of eq. (7.73) is synchronously demodulated, then the demodulation signal may be written,

$$S(t) = \sum_{n=1}^{N} |A_n(x_n, r_1)| \cos\left[\omega_c t - 2kr_1 - \frac{k}{r_1}(vt - x_n)^2 + \phi_n\right], \tag{7.74}$$

where ω_c is the arbitrary carrier frequency and ϕ_n is the arbitrary phase angle.

In order to store the return radar signal of eq. (7.74), a cathode-ray tube is used. The demodulated signal is fed in to modulate the intensity of the electron beam, which is swept vertically across the cathode-ray tube in synchronism with return radar pulses (fig. 7.20). If this modulated cathode-ray display is imaged on a strip of photographic film, which is drawn at a constant horizontal velocity, then the successive range traces will be recorded side by side, producing a two-dimensional format (fig. 7.21). The vertical lines represent the successive range sweeps and the horizontal dimension represents the azimuthal position, which is the sample version

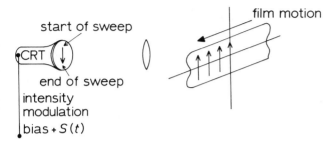

Fig. 7.20 Recording of the return radar signal for subsequent optical processing.

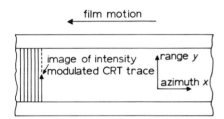

Fig. 7.21 Radar signal recording format.

of $S(t)$. This sampling is carried out in such a way that, by the time the samples have been recorded on the film, the sampled version is essentially indistinguishable from the unsampled version. In this recording, it is clear that the time variable is converted to a space variable defined in terms of the distance along the recorded film. With the proper reading exposure, the transparency of the recorded film represents the azimuthal history of the returned radar signal. Thus, considering only the data recorded along a line $y - y_1$ on the film, the transmittance can be written as

$$T(x, y_1) = K_1 + K_2 \sum_{n=1}^{N} |A_n(x_n, r_1)| \cos \left[\omega_x x - 2kr_1 - \frac{k}{r_1} \left(\frac{v}{v_f} x - x_n \right)^2 + \phi_n \right]$$

$$(7.75)$$

where K_1 and K_2 are bias and proportionality constants, $x = v_f t$ is the film coordinate, v_f is the velocity of the film motion, and $\omega_x = \omega_c / v_f$.

Since the cosine can be written into the sum of two exponential conjugate forms, the summation of eq. (7.75) can be written as the sum of two terms T_1 and T_2, which are given by

$$T_1(x, y_1) = \frac{K_2}{2} \sum_{n=1}^{N} |A_n(x_n, r_1)|$$

$$\times \exp\left\{i\left[\omega_x x - 2kr_1 - \frac{k}{r_1}\left(\frac{v}{v_f}\right)^2\left(x - \frac{v_f}{v} x_n\right)^2 + \phi_n\right]\right\}, \tag{7.76}$$

and

$$T_2(x, y_1) = \frac{K_2}{2} \sum_{n=1}^{N} |A_n(x_n, r_1)|$$

$$\times \exp\left\{-i\left[\omega_x x - 2kr_1 - \frac{k}{r_1}\left(\frac{v}{v_f}\right)^2\left(x - \frac{v_f}{v} x_n\right)^2 + \phi_n\right]\right\}. \tag{7.77}$$

For simplicity in illustration, we restrict ourselves to the one-target problem; thus for $n = j$ eq. (7.76) can be written as

$$T_1(x, y_1) = C \exp(i\omega_x x) \exp\left[-i\frac{k}{r_1}\left(\frac{v}{v_f}\right)^2\left(x - \frac{v_f}{v} x_j\right)^2\right], \tag{7.78}$$

where C is an appropriate complex constant. The first exponent of eq. (7.78) introduces a linear phase function, i.e., a simple tilt of the transmitted wavefront. The oblique angle of this tilt from the plane of the film is given by

$$\sin\theta = \frac{\omega_x}{k_1}, \tag{7.79}$$

where $k_1 = 2\pi/\lambda_1$, with λ_1 the wavelength of the illuminating light source. From the second exponent of eq. (7.78), it can be seen that the transmission function is that of a positive cylindrical lens centered at

$$x = \frac{v_f}{v} x_j, \tag{7.80}$$

with a focal length of

$$f = \frac{1}{2}\frac{\lambda}{\lambda_1}\left(\frac{v_f}{v}\right)^2 r_1, \tag{7.81}$$

where λ_1 is the wavelength of the illuminating light source.

Therefore it can be seen that eq. (7.76), except for a linear phase function, is a superposition of N positive cylindrical lenses, centered at locations given by

$$x = \frac{v_f}{v} x_n, \quad n = 1, 2, \ldots, N. \tag{7.82}$$

Similarly, eq. (7.77) contains a linear phase factor $-\theta$, and represents a superposition of a N negative cylindrical lenses, with centers given by eq. (7.82) and with focal lengths given by eq. (7.81).

In order to reconstruct the image, the transparency yielding eq. (7.75) is illuminated by a monochromatic plane wave, as shown in fig. 7.22. Then by the Fresnel-Kirchhoff theory or Huygens' principle we can show that the real images produced by $T_1(x, y_1)$ and the virtual images produced by $T_2(x, y_1)$ will be reconstructed at the front and back focal planes of the film. The relative positions of the images of the point scatterers are preserved along the line foci because the multiple centers of the lenslike structure of the film are determined by the positions of the point scatterers. However, the reconstructed image will be spread in the y direction; this is because the film, in fact, is a realization of a one-dimensional function along $y = y_1$, and hence exerts no focal power in this direction.

Since it is our aim to reconstruct an image not only in the azimuthal direction but also in the range direction, it is necessary to image the y coordinate directly onto the focal plane of the azimuthal image. In order to accomplish this we recall that from eq. (7.80) the focal length of the azimuthal image distribution is linearly proportional to the range r_1. In turn, the focal distance is linearly related to the y coordinate under consideration. Thus to construct the terrain map, we must image the y coordinate of the transmitted signal onto a tilted plane determined by the focal distances of the azimuthal direction. This imaging procedure may be carried out easily by inserting a positive conical lens immediately behind the recording film, as shown in fig. 7.23. It is clear that if the trans-

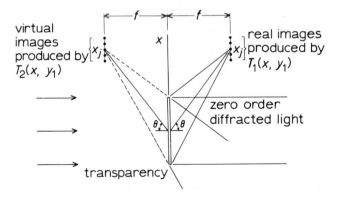

Fig. 7.22 Image reconstruction produced by the film transparency for $y = y_1$.

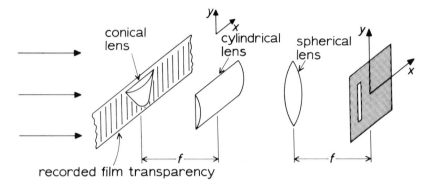

Fig. 7.23 An optical information processing system for imaging synthetic-aperture antenna data.

mittance of the conical lens is

$$T_l(x, y_1) = \exp\left(-i\frac{k_1}{2f}x^2\right),\tag{7.83}$$

where f is a linear function of r_1, as shown in eq. (7.81), then it is possible to remove the entire tilted plane of all the virtual diffraction to the point at infinity, while leaving the transmittance in the y direction unchanged. Thus if the cylindrical lens is placed at the focal distance from the film transparency, it will create a virtual image of the y direction at infinity. Now the azimuthal and the range images (i.e., the x and y directions) coincide, but at the point of infinity. They can be brought back to a finite distance by a spherical lens. In this final operation, a real image of the azimuthal and range coordinates of the terrain will be mapped in focus at the output plane of the system. In practice, however, the desired image may be recorded through a slit in the output plane.

Figure 7.24 is an example of the images obtainable by optical processing of synthetic-aperture radar data. The radar image was obtained in the Monroe area, located south of Detroit, Michigan. This photograph shows a variety of scatterers, including city streets, wooded areas, and farmland. Lake Erie, with some broken ice floes, can be seen on the right.

7.5 Optical Processing of Broadband Signals
An important application of coherent optical systems is in the spectrum analysis of broadband signals. The optical processing technique suggested by Thomas (ref. 7.13) was found to be capable of generating a near real-time spectrum analysis of a large space-bandwidth signal. The applica-

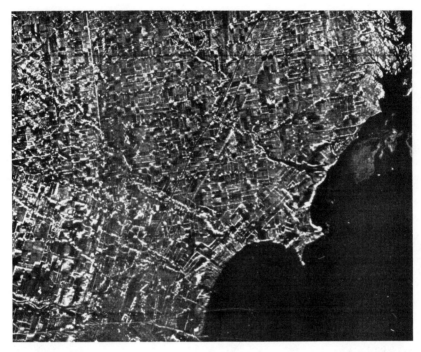

Fig. 7.24 Synthetic-aperture radar image of the Monroe, Michigan, area. (Permission by L. J. Cutrona.)

tion of this technique to the processing of wideband radio signals has been reported by Markevitch (ref. 14).

In this section, we will consider an optical processing technique which is capable of handling broadband signals. Before going into this discussion, we must recognize a basic limitation of the multichannel optical spectrum analyzer (see sect. 7.2): it is that the resolution is limited by the length of the input channel (i.e., the width of the input aperture). However, this basic limitation may be easily overcome by using a two-dimensional optical spectrum analyzer such as sketched in fig. 7.4.

If an input transparency $f(x, y)$ is inserted in the input plane of the analyzer, then the complex light field distributed on the (p, q) plane will be recalled to be

$$F(p, q) = C \int \int f(x, y) \exp[-i(xp + yq)] \, dx \, dy, \qquad (7.84)$$

which is the Fourier transform of the input signal $f(x, y)$, where C is a complex constant.

Now we will see, by means of this basic two-dimensional Fourier transformation, that a large space-bandwidth signal may be processed by means of a slightly different technique. The broadband signal may be recorded, for example, by photographing a cathode ray tube raster pattern without interleaving, as shown in fig. 7.25. The broadband signal is made to modulate the intensity of the CRT beam. The scan rate of the CRT beam is adjusted so that the maximum frequency content of the broadband signal will be properly recorded, without loss of resolution. The required scan velocity of the electron beam is therefore

$$v \geq \frac{f_m}{R}, \tag{7.85}$$

where v is the scan velocity, f_m is the maximum frequency content of the broadband signal, and R is the resolution limit of the optical system. Of course, the return sweep of the CRT beam is assumed to be much faster than $1/f_m$.

The transmittance of the recorded format may be represented by

$$f(x, y) = \sum_{n=1}^{N} f(x) f(y), \tag{7.86}$$

where $N = h/b$. Here

$$f(x) = \begin{cases} f\left[x + (2n-1)\frac{w}{2} \right], & -\frac{w}{2} \leq x \leq \frac{w}{2}, \\ 0, & \text{otherwise} \end{cases} \tag{7.87}$$

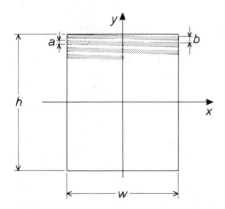

Fig. 7.25 Raster scan input signal transparency.

and

$$f(y) = \begin{cases} 1, & \dfrac{h}{2} - (n-1)\,b - a \le y \le \dfrac{h}{2} - (n-1)\,b \\ 0, & \text{otherwise} \end{cases} \tag{7.88}$$

If this format is inserted in the input plane of the analyzer, then the complex light field distributed on the spatial frequency domain may be written as

$$F(p, q) = C \int \int \sum_{n=1}^{N} f(x)\, f(y) \exp[-i(xp + yq)]\, dx\, dy, \tag{7.89}$$

where C is a complex constant.

Equivalently, eq. (7.89) can be written as

$$F(p, q) = C \sum_{n=1}^{N} \int f(x) \exp(-ixp)\, dx \int f(y) \exp(-iyq)\, dy. \tag{7.90}$$

By substituting eqs. (7.87) and (7.88) in eq. (7.90), we obtain

$$F(p, q) = C_1 \sum_{n=1}^{N} \mathrm{sinc}\left(\frac{qa}{2}\right) \left\{ \mathrm{sinc}\left(\frac{pw}{2}\right) * F(p) \exp\left[i\frac{pw}{2}(2n-1)\right] \right\}$$
$$\times \exp\left\{ -i\frac{q}{2}[h - 2(n-1)\,b - a] \right\}. \tag{7.91}$$

where C_1 is a complex constant, and

$$F(p) = \int f(x') \exp(-ipx')\, dx',$$

$$x' = x + (2n-1)\frac{w}{2}, \quad \text{and} \quad \mathrm{sinc}\, X \triangleq \frac{\sin X}{X}.$$

For simplicity, let us assume that the broadband signal is a simple sinusoid (i.e., $f(x') = \sin p_0 x'$), $w \simeq h$, and $b \simeq a$. Then eq. (7.91) may be written as

$$F(p, q) = C_1 \,\mathrm{sinc}\left(\frac{qa}{2}\right) \sum_{n=1}^{N} \left\{ \mathrm{sinc}\left(\frac{pw}{2}\right) * \frac{1}{2}[\delta(p - p_0) + \delta(p + p_0)] \right.$$
$$\left. \times \exp\left[i\frac{wp_0}{2}(2n-1)\right] \right\} \exp\left\{ -i\frac{q}{2}[w - (2n-1)\,b] \right\}. \tag{7.92}$$

To further simplify the analysis, let us consider only one of the components, say $\delta(p-p_0)$, thus

$$
F_1(p, q) = C_1 \operatorname{sinc}\left[\frac{w}{2}(p-p_0)\right] \operatorname{sinc}\left(\frac{qa}{2}\right) \exp\left(-i\frac{qw}{2}\right)
$$
$$
\times \sum_{n=1}^{N} \exp\left[i\frac{1}{2}(2n-1)(wp_0+bq)\right]. \tag{7.93}
$$

The corresponding irradiance may be written as

$$
I_1(p, q) = |F_1(p, q)|^2 = C_1^2 \operatorname{sinc}^2\left[\frac{w}{2}(p-p_0)\right] \cdot \operatorname{sinc}^2\left(\frac{qa}{2}\right)
$$
$$
\times \sum_{n=1}^{N} \exp[i2n\theta] \sum_{n=1}^{N} \exp[-i2n\theta], \tag{7.94}
$$

where

$$
\theta = \frac{1}{2}(wp_0+bq). \tag{7.95}
$$

But (ref. 7.15)

$$
\sum_{n=1}^{N} e^{i2n\theta} \sum_{n=1}^{N} e^{-2n\theta} = \left(\frac{\sin N\theta}{\sin \theta}\right)^2. \tag{7.96}
$$

Therefore, $I_1(p, q)$ may be written as

$$
I_1(p, q) = C_1^2 \operatorname{sinc}^2\left[\frac{w}{2}(p-p_0)\right] \cdot \operatorname{sinc}^2\left(\frac{qa}{2}\right) \cdot \left(\frac{\sin N\theta}{\sin \theta}\right)^2. \tag{7.97}
$$

It may be seen from eq. (7.97) that the first sinc factor represents a relatively narrow spectral line in the p direction, located at $p=p_0$, which is derived from the Fourier transform of a pure sinusoid truncated within the width w of the input transparency. The second sinc factor represents a relatively broad spectral band in the q direction, which is derived from the Fourier transform of a rectangular pulse of width a (i.e., the channel width). The last factor deserves special mention, in that for large values of N it approaches a sequence of narrow pulses. The locations of the pulses, (i.e., the peaks), may be obtained by letting $\theta=n\pi$, which gives

$$
q = \frac{1}{b}(2\pi n - wp_0), \quad n=1, 2, \dots. \tag{7.98}
$$

Thus this factor yields the fine spectral resolution in the q direction.

To continue the interpretation of eq. (7.97), we note that irradiance in the p direction is confined within a relatively narrow region, which essentially depends on w. The half-width of the spectral spread (i.e., from the center to the first zero), is seen to be

$$\Delta p = \frac{2\pi}{w},\tag{7.99}$$

which is the resolution limit of an ideal transform lens. In the q direction, the irradiance is first confined within a relatively broad spectral band, (which primarily depends on the channel width a) centered at $q=0$, and then modulated by a sequence of narrow periodic pulses. The half-width of the broad spectral band is

$$\Delta q = \frac{2\pi}{a}.\tag{7.100}$$

The separation of the narrow pulses is obtained from a similar equation, i.e.,

$$\Delta q_1 = \frac{2\pi}{b}.\tag{7.101}$$

It may be seen from eqs. (7.100) and (7.101) that there will be only a few pulses located within the spread of the broad spectral band for each p_0, as shown in fig. 7.26.

The actual location of any of the pulses is determined by the signal frequency. Thus, if the signal frequency changes, the position of the pulses also changes, in accordance with

$$dq = \frac{w}{b}\,dp_0.\tag{7.102}$$

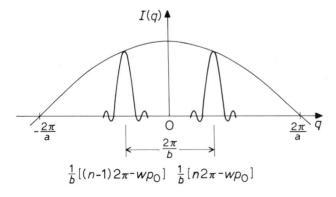

Fig. 7.26 A broad spectral band modulated by a sequence of narrow pulses.

The displacement in the q direction is proportional to the displacement in the p direction. But $N=h/b \simeq w/b$ is the number of scan lines. Therefore the output spectrum includes the required frequency discrimination, which is equivalent to analysis of the entire signal in one continuous channel.

In order to avoid the ambiguity in reading the output plane, all but one of the periodic pulses should be ignored. This may be accomplished by masking all of the output plane except the region

$$\frac{\pi}{b} \leq q \leq -\frac{\pi}{b}, \quad 0 \leq p < \infty, \tag{7.103}$$

as shown in fig. 7.27. The periodic pulses are $2\pi/b$ apart; therefore, as the input signal frequency advances, one pulse leaves the open region defined by eq. (7.103) at $q = -\pi/b$ while another pulse enters the region at $q = \pi/b$. As a result of eq. (7.102), a single bright spot would be scanned out diagonally on the output plane. The frequency locus of the input signal is also determined in cycles per second. Finally, to remove the nonuniform characteristic of the second sinc factor of eq. (7.97), a weighting transparency may be placed at the output plane of the processor.

So far we have assumed the input signal has been recorded on a linear photographic film (a subject upon which we will elaborate in the next

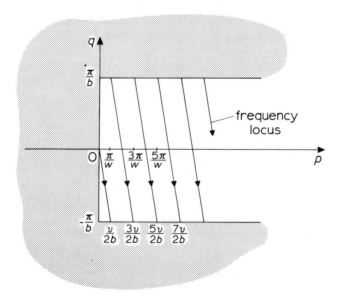

Fig. 7.27 Frequency locus at the output spectral plane.

chapter). The analysis has been carried out on a one-frame basis. However, the operation could also be performed on a continuously running tape basis. Thus it is possible to synthesize a near real-time optical spectrum analyzer.

A convenient mode of near real-time operation would be to use a continuous strip of photographic film and a rapid film developer, as diagramed in fig. 7.28. A modulating signal is fed into the z-axis of a cathode ray tube. The modulated electron beam is then swept across the CRT by means of an internal sawtooth generator. The output of the scanned CRT is then imaged on a strip of slow-moving photographic film. The exposed film then proceeds to a rapid film developer. This rapid film developer is commercially available (refs. 7.13, 7.14); developing and fixing take a few seconds to a minute. After developing, the film is transported to the input plane spectrum analyzer. The output spectrum may be displayed on a television screen as shown in fig. 7.28. The continuous time varying spectrum of the signal will be observed. This output spectrum can also be photographed on a continuous strip of film (by means of a movie camera) for later use.

The access time from the input processing signal to the output spectrum display is generally limited by the transit time of the film traveling from the CRT to the optical processor. This access time depends somewhat on

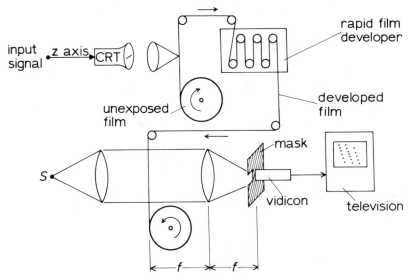

Fig. 7.28 Synthesis of a near real-time spectrum analyser.

the physical configuration of the system; however, the most important limiting factor is the film development time. It may be feasible to reduce the access time to something between a few seconds and a minute.

The wideband optical spectrum analyzer offers considerable flexibility in data handling. To name just one application, the output time-varying spectrum may be directly connected to an appropriate analog-digital converter and then to a digital computer for signal processing. This application may be important to the future of automatic signal detection, synthesis, and recognition. A near real-time continuous optical spectrum analyzer with a space-bandwidth product capability greater than 10^6 is within the current state of the art (refs. 7.13 and 7.14).

Problems

7.1 Given the coherent optical processor shown in fig. 7.29. Denote the transmittance of the input transparency by $f(x, y)$ and the output complex light field by $F(p, q)$ [the Fourier transform of $f(x, y)$]. The output irradiance is recorded on a linear photographic plate. If this recorded transparency is inserted in the input plane, in place of the original transparency, show that the output light distribution is proportional to the autocorrelation function of $f(x, y)$. S is a monochromatic point source.

7.2 Suppose the transmittance $T(x, y)$ of a recorded photographic plate is given as

$$T(x, y) = \tfrac{1}{2} + \tfrac{1}{4} \sin\sqrt{x^2 + y^2}.$$

Design a coherent optical processor and an appropriate spatial filter so that the output irradiance will be of the highest contrast.

7.3 A transparency consists of a signal imbedded in an additive white

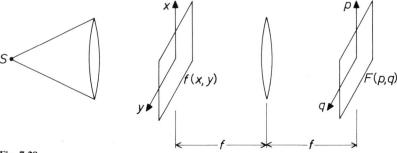

Fig. 7.29

Gaussian noise. Synthesize an appropriate coherent optical information processor for the signal filtering. If the signal is assumed spatially limited, i.e., if

$$f(x, y) = \begin{cases} A(x, y), & 0 \le x^2 + y^2 \le a^2 \\ 0, & \text{otherwise} \end{cases},$$

then construct a simple amplitude filter such that the output signal-to-noise ratio will be appreciably increased.

7.4 For the system of fig. 7.3, design an appropriate spatial filter such that the output signal is the spatial derivative of the input transparency.

7.5 Let the input transparency of the detector sketched in fig. 7.11 be given by

$$f(x, y) = \exp[-(x^2 + y^2)] \exp[i\beta(x^2 + y^2)^{1/2}],$$

where β is a positive real constant, and the additive spatial noise is assumed to be white Gaussian.
(a) Synthesize a matched filter such that the signal-to-noise ratio at the output plane will be optimum.
(b) Determine the corresponding light amplitude distribution on the output plane of the processor.

7.6 Certain relationships between the real and imaginary parts of a linear system function are necessary. These relationships may be expressed by means of Hilbert transform pairs,

$$f_r(x) = -\frac{1}{\pi} \int_{-\infty}^{\infty} \frac{f_i(\alpha)}{x - \alpha} d\alpha,$$

and

$$f_i(x) = \frac{1}{\pi} \int_{-\infty}^{\infty} \frac{f_r(\alpha)}{x - \alpha} d\alpha,$$

where $f(x) = f_r(x) + if_i(x)$, a complex-valued function, with $f_r(x)$ and $f_i(x)$ the respective real and imaginary parts. Design a coherent optical system which is able to perform the respective Hilbert transformations.

7.7 The optical processing technique of prob. 7.6 may be extended to a two-dimensional Hilbert transform pair,

$$f_r(x, y) = \frac{1}{\pi^2} \int\limits_{-\infty}^{\infty} \int\limits_{-\infty}^{\infty} \frac{f_i(\alpha, \beta)}{(x - \alpha)(y - \beta)} \, d\alpha \, d\beta,$$

and

$$f_i(x, y) = \frac{1}{\pi^2} \int\limits_{-\infty}^{\infty} \int\limits_{-\infty}^{\infty} \frac{f_r(\alpha, \beta)}{(x - \alpha)(y - \beta)} \, d\alpha \, d\beta,$$

where $f(x, y) = f_r(x, y) + i f_i(x, y)$. Design an optical system which can perform the two-dimensional transformations.

7.8 Given a method of synthesizing a complex spatial filter as shown in fig. 7.30a, a signal transparency $s(x, y)$ is inserted at a distance d behind the transform lens. The amplitude transmittance of the recorded spatial filter is linearly proportional to the signal transparency. If this resulting spatial filter is used for signal detection, determine the appropriate location of the input plane of the system, shown in fig. 7.30b.

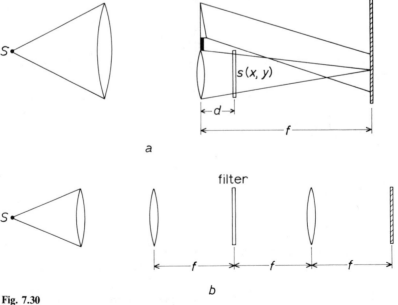

Fig. 7.30

7.9 The complex spatial filter of the previous problem is inserted in the input plane of the optical processor shown in fig. 7.31. What is the complex light distribution at the output plane?

7.10 Two input transparencies of amplitude transmittances $f_1(x, y)$ and $f_2(x, y)$ are placed in the front focal plane of the transform lens of an optical processor, as shown in fig. 7.32. The transparencies are centered in the (x, y) coordinate plane at $(a, 0)$ and $(-a, 0)$ respectively. Compute the irradiance at the back focal plane of the lens.

7.11 The output irradiance of the system of fig. 7.32 is recorded on a linear photographic plate. If this transparency is placed at the input plane of the optical system, as shown in fig. 7.33,
(a) Determine the complex light distribution at the back focal plane of the transform lens.

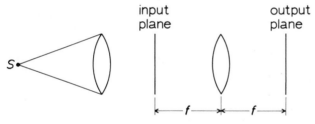

Fig. 7.31

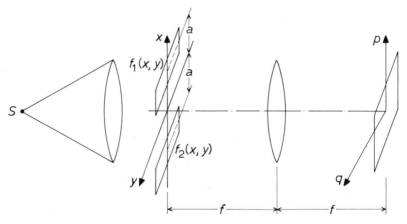

Fig. 7.32

(b) Discuss briefly the important functional operations, with respect to $f_1(x, y)$ and $f_2(x, y)$, at the output plane (α, β).

7.12 Consider the coherent optical processing system with an incident monochromatic plane wave (fig. 7.34). Two signal transparencies of transmittance $f_1(x, y)$ and $f_2(x, y)$ are inserted in the front focal plane of the first transform lens, with centers at $(a, 0)$ and $(-a, 0)$. If there is a sinusoidal grating with amplitude transmittance

$$T(p, q) = \tfrac{1}{2} + \tfrac{1}{2}\cos ap \quad \text{for every } q,$$

determine the light amplitude distribution at the output plane of the optical system.

7.13 Repeat prob. 7.12, if a sinusoidal grating of transmittance

$$T(p, q) = \tfrac{1}{2} + \tfrac{1}{2}\sin ap \quad \text{for every } q,$$

replaces the grating of the previous problem. Show the effect on output functional operations.

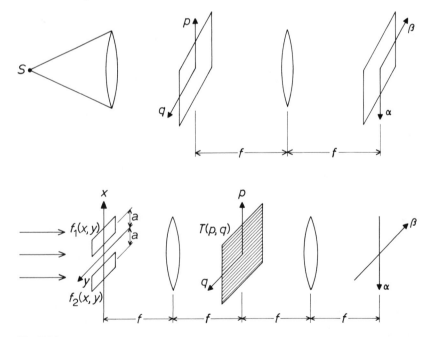

Fig. 7.34

7.14 Consider a spatial-multiplex transparency, which is the sequential record of two images $A_1(x, y)$ and $A_2(x, y)$, made through a sinusoidal grating in contact with the film. The corresponding transmittance may be written,

$$T(x, y) = K + A_1(x, y) \cos p_0 x + A_2(x, y) \cos q_0 y,$$

where K is an arbitrary constant, and p_0 and q_0 are the respective carrier frequencies in the (p, q) spatial frequency coordinates. The images $A_1(x, y)$ and $A_2(x, y)$ are assumed to be spatially bandlimited within a circular region of radius $r = r_0$, where $r_0 < \frac{1}{2} p_0$, and $r_0 < \frac{1}{2} q_0$. This transparency is placed at the input plane P_1 of the optical processor in fig. 7.11.

(a) Determine the complex light distribution on the spatial frequency plane P_2.

(b) Design a spatial filter so that the output light field will reimage $A_1(x, y)$ and $A_2(x, y)$, one at a time.

7.15 With reference to eq. (7.96) and the broadband optical spectrum analyzer described in sec. 7.5:

(a) Determine the corresponding frequency resolution at the output spectral plane. (Hint: The frequency resolution, that is the 3db point, in the q direction may be determined by setting eq. (7.96) equal to half the comb function peak amplitude, $N^2/2$).

(b) If the number of scan lines N becomes very large, show that the number of resolution elements at the output plane is linearly proportional to the number of scan lines.

References

7.1 D. K. Pollack, C. J. Koester, and J. T. Tippett (eds.), *Optical Processing of Information*, Spartan Books, Baltimore, Md., 1963.

7.2 J. T. Tippett et al. (eds.), *Optical and Electro-Optical Information Processing*, The MIT Press, Cambridge, Mass., 1965.

7.3 P. Elias, D. S. Grey, and D. Z. Robinson, Fourier Treatment of Optical Processes, *J. Opt. Soc. Am.*, **42** (1952), 127.

7.4 P. Elias, Optics and Communication Theory, *J. Opt. Soc. Am.*, **43** (1953), 229.

7.5 E. L. O'Neill, Spatial Filtering in Optics, *IRE Trans. Inform. Theory*, **IT-2** (1956), 56.

7.6 E. L. O'Neill (ed.), *Communication and Information Theory Aspects of Modern Optics*, General Electric Co., Electronics Laboratory, Syracuse, N.Y., 1962.

7.7 L. J. Cutrona et al., Optical Data Processing and Filtering Systems, *IRE Trans. Inform. Theory*, **IT-6** (1960), 386.

7.8 A. Vander Lugt, Signal Detection by Complex Spatial Filtering, *IEEE Trans. Inform. Theory*, **IT-10** (1964), 139.

7.9 A. Vander Lugt, *Signal Detection by Complex Spatial Filtering* (Radar Laboratory Report No. 4594-22-T), Institute of Science and Technology, University of Michigan, Ann Arbor, 1963.

7.10 L. J. Cutrona et al., On the Application of Coherent Optical Processing Techniques to Synthetic-Aperture Radar, *Proc. IEEE*, **54** (1966), 1026.

7.11 J. W. Goodman, *Introduction to Fourier Optics*, McGraw-Hill, 1968.

7.12 E. N. Leith et al., Introduction to Optical Data Processing, course notes, University of Michigan Engineering Summer Conference, Ann Arbor, 1967.

7.13 C. E. Thomas, Optical Spectrum Analysis of Large Space-Bandwidth Signals, *Applied Optics*, **5** (1966), 1782.

7.14 B. V. Markevitch, Optical Processing of Wideband Signals, 3rd Annual Wideband Recording Symposium, Rome Air Development Center, April, 1969.

7.15 R. V. Churchill, *Fourier Series and Boundary Value Problems*, McGraw-Hill, New York, 1941, p. 32.

Chapter 8 Basic Properties of Photographic Film

Photographic film (or plate) occupies an important place as an optical element in modern coherent processing systems. Although it serves primarily as a recording medium, it can also be used in the synthesis of complex spatial filters and signal transparencies. There are several other optical elements (photochromic film, thermoplastic tape, etc.), some of whose optical properties are similar to those of photographic film. However, these materials at the present time are still in the research and development stage. It is doubtful whether these new materials can ever totally replace photographic film and plate. For this reason we will devote this chapter to a discussion of the basic properties of photographic film. However, the presentation will by no means be complete. More thorough treatments of the photographic process will be found in the books by Mees and James (refs. 8.1, 8.2).

8.1 Photographic Film as a Recording Medium

We discuss first photographic film considered as a recording medium. A photographic detector is generally composed of a base, which may be made of a transparent glass plate or acetate film, and a layer of photographic emulsion (fig. 8.1). The photographic emulsion consists of a large number of tiny photosensitive silver halide particles, which are suspended more or less uniformly in a gelatin for support. When the photographic emulsion is exposed to light, some of the silver halide grains absorb optical energy and undergo a complex physical change. Some of the grains that absorb sufficient light energy are immediately reduced, forming tiny metallic silver particles. (These are the so-called development centers.) The reduction to silver is completed by the chemical process of *development*. Those grains which were not exposed or which have not absorbed sufficient optical energy will remain unchanged. If this developed film is then subjected to some chemical *fixing* process, then the unexposed silver halide grains will be removed, leaving only the metallic silver particles in the gelatin. These metallic silver grains remaining in the emulsion are largely opaque at optical frequencies. The transmittance of developed film therefore depends on the density of the metallic silver grains (i.e., the exposed and developed grains) in the gelatin. The relation of the intensity transmittance and the density of the developed grains was first demonstrated in a classic article written in 1890 by F. Hurter and V. C. Driffield (ref. 8.3). They showed that the area density of the metallic silver particles of a developed film should be proportioned to $-\log T_i$; thus they defined the

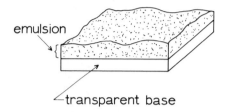

Fig. 8.1 Section of a photographic film. The emulsion is composed of silver halide particles embedded in gelatin.

photographic density by

$$D = -\log T_i, \tag{8.1}$$

where T_i is the *intensity transmittance*. The intensity transmittance is in turn defined as

$$T_i(x, y) = \left\langle \frac{I_o(x, y)}{I_i(x, y)} \right\rangle, \tag{8.2}$$

where $\langle \ \rangle$ denotes the localized ensemble average and $I_i(x, y)$ and $I_o(x, y)$ denote the input and output irradiance, respectively, at the point (x, y).

One of the most commonly used descriptions of the photosensitivity of a given photographic film is that given by the Hurter-Driffield curve (or for short called "H and D" curve), as shown in fig. 8.2. It is the plot of the density of the developed grains versus the logarithm of the exposure E. The plot shows that if the exposure is below a certain level, then the photographic density is quite independent of the exposure; this minimum density is usually referred as *gross fog*. As the exposure increases beyond the *toe* of the curve, the density begins to increase in direct proportion to $\log E$. The slope of the straight-line portion of the H and D curve is usually referred to as the *film gamma* γ. If the exposure is further increased beyond the straight line portion of the H and D curve, then the density saturates, after an intermediate region called the *shoulder*. In the saturated region, there is no further increase of developed-grain density with the increase of exposure.

It is the straight-line portion of the H and D curve which is used in conventional photography. However, we will show in chap. 12 that, for the linear optimization in wavefront recording and in complex spatial filter synthesis, the optimum recording is not restricted to the linear region of the H and D curve.

A photographic film with a large value of γ is generally referred to as high-contrast film. Conversely, a film which has a low value of γ is referred

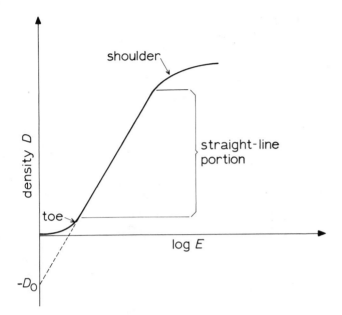

Fig. 8.2 The Hurter-Driffield ("H and D") curve.

to as low-contrast film. The value of γ is generally affected not only by the type of photographic emulsion used, but also by the chemical of the developer and the time of the developing process. Therefore, in practice, it is possible to achieve a prescribed value of γ, with a fair degree of accuracy, if suitable type of film, developer, and developing time are used.

In some cases, particularly in incoherent processing, the intensity transmittance of a developed film is a more appropriate parameter than the film density. We will first consider the properties of the photographic film under incoherent illumination, and then move toward the coherent case.

If a given film is recorded in the straight-line region of the H and D curve, then the photographic density may be written as

$$D = \gamma_n \log E - D_0, \tag{8.3}$$

where the subscript n means that a negative film is being used, and $-D_0$ is the intercept of the projection of the straight-line portion of the H and D curve with the density ordinate. By substitution of eq. (8.1) in eq. (8.3) we have

$$\log T_{in} = -\gamma_n \log(It) + D_0, \tag{8.4}$$

where I is the incident irradiance, and t is the exposure time. (Note that the energy delivered to the film is given by $E = It$.) Equivalently, eq. (8.4) can be written,

$$T_{in} = K_n I^{-\gamma_n}, \tag{8.5}$$

where $K_n = 10^{D_0} t^{-\gamma_n}$, a positive constant.

From eq. (8.5), it is clear that the intensity transmittance is nonlinear with respect to the incident irradiance. However, it may be possible to achieve a positive power-law relation between the intensity transmittance and the incident irradiance. To do so, it may require the two-step process called *contact printing*.

In the first step, a negative film is recorded. Then in the second step a negative film is laid under the developed film and an incoherent light is transmitted through the first film to expose the second. By developing the second film so as to obtain the prescribed value of γ, a positive transparency with a linear relationship between the intensity transmittance and the incident irradiance may be obtained. In order to illustrate this two-step process, let the resultant intensity transmittance of the second developed film (i.e., the positive transparency) be

$$T_{ip} = K_{n2} I_2^{-\gamma_{n2}}, \tag{8.6}$$

where K_{n2} is a positive constant. The subscript n2 denotes the second-step negative. The irradiance I_2 is that incident on the second film; this may be written as

$$I_2 = I_1 T_{in}, \tag{8.7}$$

where

$$T_{in} = K_{n1} I^{-\gamma_{n1}},$$

with the subscript n1 meaning the first-step negative. The illuminating irradiance of the first film during contact printing is I_1, and I is the irradiance originally incident on the first film. By substitution of (8.7) into (8.6) we have

$$T_{ip} = K I^{\gamma_{n1}\gamma_{n2}}, \tag{8.8}$$

where

$$K = K_{n2} K_{n1}^{-\gamma_{n2}} I_1^{-\gamma_{n2}}$$

is a positive constant. A linear relationship between the intensity transmittance of the positive transparency and the incident recording irradiance may be achieved if the overall gamma is made to be unity, i.e., if we make $\gamma_{n1}\gamma_{n2} = 1$.

If the film is used as an optical element in a coherent system, then it is more appropriate to use the complex amplitude transmittance rather than the intensity transmittance. The complex amplitude transmittance is defined by

$$T(x, y) = [T_i(x, y)]^{1/2} \exp[i\phi(x, y)], \tag{8.9}$$

where T_i is the intensity transmittance, and $\phi(x, y)$ represents random phase retardations. Such phase retardations are primarily due to emulsion thickness variations. These thickness variations are of two sorts. One is the coarse "outer scale" variation, which is a departure from optical flatness of the emulsion and base. The fine "inner scale" variation is a result of random fluctuations in the density of developed silver grains, which has been found to produce nonuniform swelling of the surrounding gelatin. This fine scale variation of emulsion thickness obviously depends upon the exposure of the film.

In most practical applications, phase retardations due to emulsion thickness variations can be removed by means of an *index-matching liquid gate* (fig. 8.3). Such a gate consists of two parallel optically flat glass plates with the space between them filled with refractive index–matching liquid, i.e., a liquid whose refractive index is very close to that of the film emulsion. If a developed film is submerged in the liquid gate, then the overall complex amplitude transmittance may be written as

$$T(x, y) = [T_i(x, y)]^{1/2}, \tag{8.10}$$

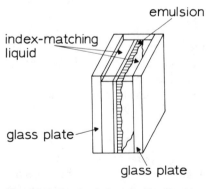

emulsion

index-matching liquid

glass plate

glass plate

Fig. 8.3 Refractive index-matching liquid gate. The outward-facing surfaces of the glass are ground optically flat.

which is a real function. The random phase retardation is therefore removed.

Recall the negative and positive transparencies of the two-step process, i.e., eqs. (8.5) and (8.8). The amplitude transmittances of these two transparencies can be written as

$$T_n = [T_{in}]^{1/2} = K_n^{1/2} I^{-\gamma_n/2} = K_n^{1/2} (uu^*)^{-\gamma_n/2}, \qquad (8.11)$$

and

$$T_p = [T_{ip}]^{1/2} = K^{1/2} I^{(\gamma_{n1}\gamma_{n2})/2} = K^{1/2} (uu^*)^{(\gamma_{n1}\gamma_{n2})/2}, \qquad (8.12)$$

where u is the complex amplitude of the incident recording field. Thus the amplitude transmittance of the two-step contact process can be written as

$$T = K_1 |u|^\gamma, \qquad (8.13)$$

where K_1 is a positive constant and $\gamma = \gamma_{n1}\gamma_{n2}$. Therefore, a linear relationship between the amplitude transmittance and the amplitude of the recording light field may be achieved by making the overall gamma equal to unity. This is identical to the case for incoherent illumination. It may be pointed out that, for convenience, the first gamma is frequently chosen to be less than unity (e.g., $\gamma_{n1} = 1/2$), and the second gamma is then chosen to be larger than *two* (e.g., $\gamma_{n2} = 4$). The overall gamma is therefore equal to two. This provides a square-law relation rather than linearity for the intensity transmittance versus incident irradiance. By (8.10), however, it becomes a linear relation for amplitude transmittance.

In most coherent information processing it is more convenient to use the transfer characteristic directly, rather than the H and D curve. This direct transfer characteristic is frequently referred to as the T-E curve (fig. 8.4). It can be seen from fig. 8.4 that if the film is properly biased at an operating point which lies well within the linear region of the transfer characteristic of the T-E curve, then within a certain dynamic range the film will offer the best linear transfer amplitude transmittance. If E_Q and T_Q ("quiescent") denote the corresponding bias exposure and amplitude transmittance, then within the linear region of the transfer characteristic the amplitude transmittance may be written as

$$T \simeq T_Q + \alpha(E - E_Q) = T_Q + \alpha' |\Delta u|^2, \qquad (8.14)$$

where α is the slope measured at the quiescent point of the T-E curve, $|\Delta u|$ is the incremental amplitude variation, and $\alpha' = \alpha t$, where t is the exposure time. For further details of the use of the T-E curve and the direct transfer description, see ref. 8.4, p. 48, and ref. 8.5.

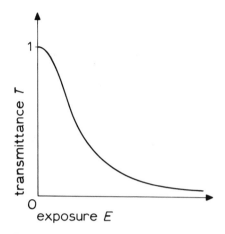

Fig. 8.4 Amplitude transmittance versus exposure of a photographic negative (T-E curve).

To conclude this section, it may be emphasized that the constraint of the signal recorded within the linear region of the T-E curve may not be necessary in some coherent optical systems, particularly in wavefront reconstructions. The removal of this linear constraint in wavefront reconstruction will be treated in chap. 12. It will be shown that for optimal linear image reconstruction, the wavefront recording is actually required to exceed the linear region of the T-E curve!

8.2 A Stochastic Approach to the Problem of Noise in Photographic Film

Studies of the granularity of photographic emulsions have appeared in a number of papers. In particular, we refer to those by Jones (ref. 8.6), Zweig (refs. 8.7–8.9), Picinbono (ref. 8.10), and Savelli (ref. 8.11). The work of Jones and Zweig was based on empirical data relating rms density to average density of a photographic film which had been uniformly exposed and developed. Picinbono and Savelli assumed that the probability of reduction of photographic grains follows a stationary Poisson distribution. In this section we will study photographic noise as a very special type of stochastic process. We will show that the noise behavior of photographic emulsions can be interpreted as a continuous-parameter Markov chain (ref. 8.12; ref. 8.13, p. 288). This model predicts a new relationship between rms density and average density. Furthermore, the distribution of the number of developed grains is shown to be not a Poisson distribution, but one which corresponds to a spatial nonstationary probability.

In order to determine the probability distribution of the undeveloped

grains (as well as developed grains), the following assumptions are made.
1. Grain size is uniform and the grains are uniformly distributed throughout the film.
2. A given particle is either perfectly opaque, if the grain has been developed, or perfectly transparent, if the grain has remained undeveloped.
3. The minimum resolvable image (Altman and Zweig, ref. 8.14, call this
   a cell) contains a large number of grains.

In the following, a conditional-probability distribution of the undeveloped grains will be calculated. The corresponding mean and variance will
be determined, and the application to a realistic photographic film will be
illustrated.

From the assumptions stated, it is possible to derive a conditional-probability distribution of the undeveloped grains defined in the volume
covered by the minimum resolvable point image. Let us define the conditional probability as

$$P_m(t) = P[x(t) = m \mid x(0) = M], \quad x(t) = 0, 1, 2, ..., M, \tag{8.15}$$

where t is the exposure time, $x(t)$ is the number of undeveloped grains, and
$P_m(t)$ is the conditional probability that m undeveloped grains exist at
time t. The quantity M is the total number of grains enclosed in the
minimum resolvable point image. If the probability of development of
more than one grain in an interval Δt is assumed to be the order of $(\Delta t)^2$,
then a set of nonhomogeneous Kolmogorov differential equations can be
obtained (ref. 8.13):

$$\frac{dP_M(t)}{dt} = -Mq(t) P_M(t) \quad \text{for} \quad m = M, \tag{8.16}$$

$$\frac{dP_m(t)}{dt} = -mq(t) P_m(t) + (m+1) q(t) P_{m+1}(t) \quad \text{for} \quad m < M, \tag{8.17}$$

with the initial conditions

$$P_M(0) = 1, \tag{8.18}$$

$$P_m(0) = 0 \quad \text{for} \quad m < M, \tag{8.19}$$

where $mq(t) \, dt$ is the probability that one of the m grains will be developed
at time t, and $mq(t)$ is the rate of development, which is also called the
transition intensity. The solution of eqs. (8.16) and (8.17) can be shown to
be (see app. A)

$$P_m(t) = \frac{M!}{(M-m)! \, m!} [1 - e^{-\rho(t)}]^{M-m} e^{-m\rho(t)} \quad \text{for} \quad m \leq M, \tag{8.20}$$

where $\rho(t) = \int_0^t q(t')\, dt'$. $\qquad\qquad\qquad\qquad\qquad\qquad\qquad\qquad$ (8.21)

The corresponding mean and variance of $P_m(t)$ are

$$\bar{m} = Me^{-\rho(t)}, \qquad\qquad\qquad\qquad\qquad\qquad\qquad\qquad (8.22)$$

$$\sigma^2 = Me^{-\rho(t)}\left[1 - e^{-\rho(t)}\right]. \qquad\qquad\qquad\qquad\qquad (8.23)$$

The expression relating rms density fluctuation to the average density is therefore

$$\sigma = \left[\bar{m}\left(1 - \frac{\bar{m}}{M}\right)\right]^{1/2} = \left[\bar{n}\left(1 - \frac{\bar{n}}{M}\right)\right]^{1/2}, \qquad\qquad (8.24)$$

where $\bar{n} = M - \bar{m}$ is the average density of the developed grains.

The conditional-probability distribution of the developed grains can be obtained by simply substituting the random variable $m = M - n$ in eq. (8.20), where n is the number of developed grains included in the minimum resolvable point image.

In order to determine the conditional probability $P_m(t)$ of a typical photographic film, we must know $q(t)$. To determine $q(t)$, we can start from the H and D curve of a given photographic film. The density of developed grains increases with exposure (that is, the irradiance times exposure time). For a given irradiance, we can plot the density of the developed grains against the exposure time t. In fig. 8.5, a set of D versus t curves for different values of irradiance I are plotted. The corresponding slopes of the D versus t curves are also plotted in dashed lines in the same figure. Assuming that the dashed lines in this figure follow the Rayleigh probability distribution law, then the corresponding slopes can be written as

$$\frac{dD}{dt} = \begin{cases} M \dfrac{t}{\alpha^2} \exp\left[-\dfrac{1}{2}\left(\dfrac{t}{\alpha}\right)^2\right], & t \geq 0 \\[2mm] 0, & \text{otherwise} \end{cases}, \qquad (8.25)$$

where the parameter α corresponds to the exposure time when dD/dt is maximum.

We may consider that the photographic density is equal to the number of developed grains,

$$D = M - \bar{m}. \qquad\qquad\qquad\qquad\qquad\qquad\qquad\qquad (8.26)$$

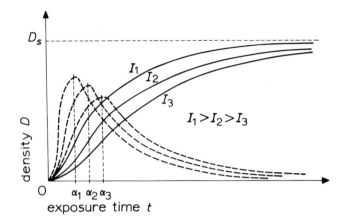

Fig. 8.5 Solid curves: density of developed grains as a function of exposure time for various values of irradiance I. Dashed curves: slope of the solid curves.

By substituting eq. (8.22) in the above equation, and then differentiating with respect to t, we have

$$\frac{dD}{dt} = Mq(t)\, e^{-\rho(t)}.$$ (8.27)

By comparing eqs. (8.27) and (8.25), we conclude that

$$q(t) = \frac{t}{\alpha^2}$$ (8.28)

and

$$\rho(t) = \frac{1}{2}\left(\frac{t}{\alpha}\right)^2.$$ (8.29)

From eq. (8.24), we can see that the random behavior of the photographic noise is dependent on the average density while the probability of grain development is nonstationary (see ref. 8.15, p. 300).

Furthermore, from eq. (8.20), the conditional-probability distribution of the undeveloped (as well as the developed) grains is more similar to a continuous-parameter binomial distribution than to a Poisson distribution.

Accordingly, from eqs. (8.22), (8.23), and (8.29), we can conclude that the mean of undeveloped (developed) grains is a monotonically decreasing (increasing) function of t, as shown in fig. 8.6. However the variance is not a monotonic function of t (fig. 8.7). The maximum value of noise, $\sigma^2 = M/4$,

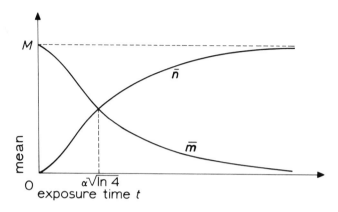

Fig. 8.6 Mean of developed and undeveloped grains as a function of exposure time.

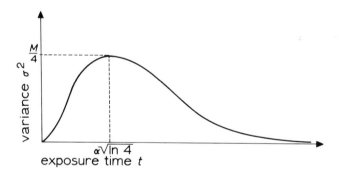

Fig. 8.7 Mean-square fluctuation of developed grains as a function of exposure time.

occurs at $t = \alpha(\ln 4)^{1/2}$ and the minimum values, $\sigma^2 = 0$, occur at $t = 0$ and $t = \infty$. This figure shows that a variation of the standard deviation is quite compatible with the experimental results in the range obtained in fig. 6 of ref. 8.16. Equation (8.24) also shows that the standard deviation of density increases with respect to the value of M. That is, for an emulsion of smaller grain size, the rms fluctuation in density is higher. Moreover, it will be shown in the following that the informational sensitivity is also an increasing function of M.

INFORMATIONAL SENSITIVITY

The signal of interest in conventional photography may be specified by the linear gradient (ref. 8.2, p. 540)

$$g = \frac{dD}{dE}, \tag{8.30}$$

where E is the exposure.

Then from eqs. (8.22) and (8.26), the signal is

$$g = \frac{M}{I} \frac{t}{\alpha^2} \exp\left[-\frac{1}{2}\left(\frac{t}{\alpha}\right)^2 \right], \tag{8.31}$$

where I is the irradiance of the minimum resolvable image. Therefore, the informational sensitivity for the effectiveness of information transmission is defined (see ref. 8.2, p. 540) by

$$\text{Informational sensitivity} \triangleq \frac{g}{\sigma} = \frac{\sqrt{M}}{I} \frac{t}{\alpha^2} \left[\frac{1}{e^{\rho(t)} - 1} \right]^{1/2}. \tag{8.32}$$

The informational sensitivity as a function of exposure time given by eq. (8.32) is sketched in fig. 8.8. The informational sensitivity is not a monotonic function of t, but has a maximum at some finite exposure time. The variation of the informational sensitivity is compatible with the shape of the graphs in fig. 8 of ref. 8.16. Equation (8.32) also shows that the informational sensitivity increases as the square root of M. That is, for any particular resolvable cell, a smaller-grain emulsion would be expected to have a higher informational sensitivity than one of larger grain size.

In addition, from eq. (8.31) and the corresponding dashed curves of fig. 8.5, we can readily see that the variation of signal as a function of average

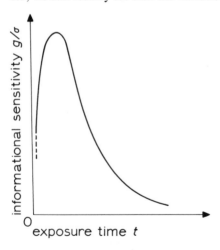

Fig. 8.8 Informational sensitivity as a function of exposure time.

density is quite similar to the experimental results obtained in fig. 7 of ref. 8.16.

This stochastic model predicts a new relation between rms noise and average developed-grain density. It shows the probability of grain development to be nonstationary and the distribution of developed grains to be non-Poisson. If the effects due to random variations in grain size and the stochastic behavior of its sensitivity are considered, then the rms noise will be higher than we have obtained.

8.3 Film-Grain Noise and Signal-to-Noise Ratio

In the last section, the noise behavior of photographic emulsions was studied as a continuous-parameter Markov process. This stochastic model predicts a new relation between rms fluctuation and average density of the developed photographic grains. In this section, the rms fluctuation and average transmittance will be derived from a well-known approximation in the theory of functions of a random variable (see ref. 8.17). From these new results, the transmittance signal-to-noise ratio will be determined, and the dynamic range of a photographic film will also be defined.

FILM-GRAIN NOISE

The formulas for the mean and variance of the developed photographic grains have been derived in the previous section. They are repeated here:

$$\bar{n} = M\{1 - \exp[-\rho(E)]\} \tag{8.33}$$

and

$$\sigma^2 = M \exp[-\rho(E)]\{1 - \exp[-\rho(E)]\}, \tag{8.34}$$

where $\bar{n}$ and σ^2 are the mean and variance, respectively, of the developed photographic grains, M is the total number of grains in the minimum resolvable image area, E is the exposure, $\rho(E) = \frac{1}{2}(E/\alpha)^2$, and α is an appropriate constant corresponding to the maximum derivative of the H and D curve.

By using the well-known formula for the relation between the transmittance and the density of the developed photographic grains (ref. 8.18, p. 117),

$$T = 10^{-D/2}, \tag{8.35}$$

where T is the amplitude transmittance and D is the density of the developed photographic grains in terms of a defined unit volume, we find that the corresponding average and mean square fluctuation of the transmittance can be approximated by the following equations (ref. 8.15, p. 115):

$$T \simeq 10^{-\bar{n}/2} \tag{8.36}$$

and

$$\sigma_T^2 \simeq \left(\frac{\partial T}{\partial D}\right)^2 \sigma^2. \tag{8.37}$$

By substituting eq. (8.33) into (8.36), we have

$$T \simeq \exp\{-\beta M[1 - \exp(-\rho(E)]\}. \tag{8.38}$$

By using eqs. (8.34), (8.35), and (8.38) in eq. (8.37), we obtain

$$\sigma_T^2 \simeq M\beta^2 \exp[-\rho(E)] \exp\{-2\beta M[1 - \exp(-\rho(E))]\} \times \{1 - \exp[-\rho(E)]\}, \tag{8.39}$$

where $\beta = \frac{1}{2}\ln 10$, and $\rho(E) = \frac{1}{2}(E/\alpha)^2$.

The average and mean-square fluctuation of the transmittance as functions of E/α are plotted in figs. 8.9 and 8.10, respectively, for different values of M. It can be seen from fig. 8.9 that the average transmittances are monotonic functions which decay rapidly with increasing values of M. It also can be seen that the curves in fig. 8.9 are compatible with eq. (8.35). From fig. 8.10, it is quite clear that the noise transmittance (i.e., the variance of the transmittance) is a spatially nonstationary stochastic process which is inherently attached to the signal recorded.

SIGNAL-TO-NOISE RATIO

The signal-to-noise ratio for a photographic emulsion may be defined either with respect to the developed grains in the emulsion, or with respect to the corresponding transmittance. However, the former definition is not of particular interest in coherent optical information processing and holography. Therefore, we will restrict ourselves to the definition of the signal-to-noise ratio with respect to the transmittance.

The definition of signal and noise adopted in this book are essentially those of the input-output communication system point of view. In contrast, the conventional definition of the signal is based on the differences in the localized photographic density (see, e.g., ref. 8.2). The distinction between these two definitions is essential; the former is primarily based on how much of the input signal (i.e., the information input) a photographic film can convey, while the latter is based on the constraint of a physical detector (e.g., the eye). For instance, if a film has been ex-

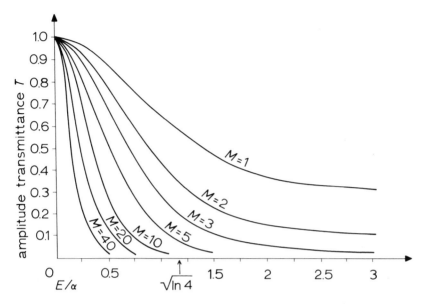

Fig. 8.9 Amplitude transmittance as a function of E/α, for different values of M.

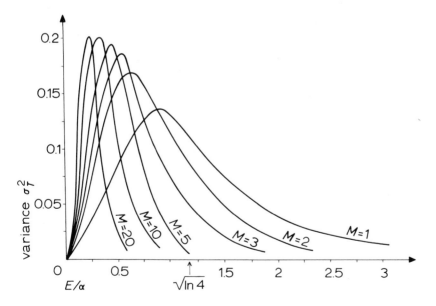

Fig. 8.10 Mean-square fluctuation in transmittance as a function of E/α, for different values of M.

posed to a uniform photographic density, then by conventional definition it contains no informaion (i.e., no signal), since the detector cannot distinguish one region from another. However, if the spatial location on the film has been predetermined, then a uniform film could be set to represent a code word (i.e., a signal exists). Thus the information is nonzero. This may be more clearly seen in the following. With regard to fig. 8.11a, the information is apparently zero, since we cannot distinguish the resolvable regions. However, if the same figure is predivided (for example, by means of a spatial coordinate system) into $m \times n$ regions, as shown in fig. 8.11b, then the possible information content of this figure certainly cannot be regarded as zero, although the density is uniform. Indeed, fig. 8.11b may be represented by a coded signal in that the transmittance of each resolvable cell is $T(x_i, y_j) = k$, $i = 1, 2, ..., n$, $j = 1, 2, ..., m$, where k is a constant smaller than unity. Therefore, at least from this simple coded photographic point of view, it is clear that a uniform film transparency should not be regarded as carrying zero information.

Furthermore, in holography (chap. 10), the transmitted light fields are responsible for the filtering operation and image reconstruction. It is very important that the film densities at the definite coordinate points be determined by the recorded signal $f(x, y)$. That is, each of the resolvable cells (x, y) represents a specific transmitted light wave (i.e., signal), which will ultimately contribute to the filtering operation and in image formation. Therefore, it is more convenient to define the signal in terms of the transmittance, rather than by the difference in localized density, as in the conventional definition. In other words, each resolvable point (x, y) can be represented as an input-output subcommunication channel as

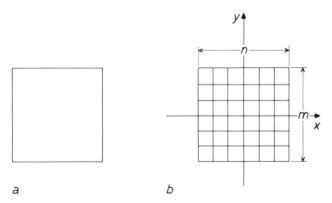

a   b

Fig. 8.11 A uniformly exposed photographic film.

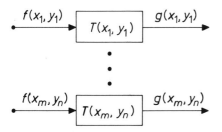

Fig. 8.12 Photographic film as an array of $m \times n$ input-output subcommunication channels. The drawing represents the operations $g(x_i, y_j) = f(x_i, y_j) \, T(x_i, y_j)$.

shown in fig. 8.12. It is logical that, in coherent processing and wave-front reconstruction, the signal defined for the filtering operation and image reconstruction should not be measured in terms of the localized density difference (in which the spatial coordinate is frequently disregarded), but that it should be defined by means of the film transparency at each of the spatial coordinate points (x, y). To emphasize, it is the transmittance $T(x, y)$, i.e., the transmitted light field, which is responsible for the information processing operation and for the image formation, not the transmittance spatial derivatives, $\partial^2 T(x, y)/\partial x \, \partial y$.

Therefore, from eqs. (8.38) and (8.39), the signal-to-noise ratio may be written

$$\frac{S}{N} \triangleq \frac{T^2}{\sigma_T^2} = \frac{1}{M\beta^2 \exp[-\rho(E)] \{1 - \exp[-\rho(E)]\}} . \tag{8.40}$$

From eq. (8.40), the transmittance signal-to-noise ratio is inversely proportional to M. Plots of the signal-to-noise ratio as a function of E/α for different values of M are given in fig. 8.13. In this figure the curves of the transmittance signal-to-noise ratio are clearly not monotonic functions. The optimum values of signal-to-noise ratio, (i.e., $S/N = \infty$) occur at $E/\alpha = 0$ and $E/\alpha = \infty$ and the minimum value of signal-to-noise occurs at $E/\alpha = (\ln 4)^{1/2}$. It may be noted that high signal-to-noise ratios occur on the upper scale of the plots (i.e., $E/\alpha > (\ln 4)^{1/2}$), but this is not of practical interest, because we see from fig. 8.9 that the usefulness of the transmittances lies in the lower scale of the E/α axis (i.e., $0 \le E/\alpha < (\ln 4)^{1/2}$). It can also be seen from fig. 8.9 that the transmittance decreases rapidly as the value of M increases. Therefore, it may be concluded that for better signal-to-noise ratio, the signal should be recorded in the lower region of the E/α axis.

For example, if a specified lower bound of signal-to-noise ratio is

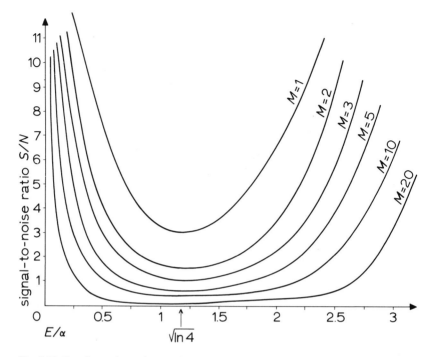

Fig. 8.13 Signal-to-noise ratio as a function of E/α, for different values of M.

required, then it can be shown from eq. (8.40) that

$$\frac{E}{\alpha} \le \ln \left\{ \frac{2}{1 + \left[1 - \dfrac{4}{M\beta^2 (S/N)_0} \right]^{1/2}} \right\}^{1/2}, \qquad (8.41)$$

where $(S/N)_0$ is the specified lower-bound signal-to-noise ratio. For instance, if the lowest required signal-to-noise is assumed to be unity, then we have

$$\frac{E}{\alpha} \le \ln \left\{ \frac{2}{1 + \left(1 - \dfrac{4}{M\beta^2} \right)^{1/2}} \right\}^{1/2}, \quad \text{for} \quad \left(\frac{S}{N} \right)_0 = 1, \quad M \ge 3. \qquad (8.42)$$

It may be emphasized however, that, if the photographic film is recorded at the on-off saturated extremes (i.e., at the two limits $E = 0$ and $E = \infty$), then a high signal-to-noise ratio is to be expected. This maximum signal-to-noise ratio in on-off binary recording has been observed by many investigators, particularly in the application of computer generated holograms. Such observations support the theory that we have developed.

It may be of interest to show plots of noise-to-signal ratio as functions of E/α, as given in fig. 8.14. From this figure, the noise-to-signal ratio is linearly proportional to M, the maximum value of N/S occurs at $E/\alpha = (\ln 4)^{1/2}$, and the minimum values of N/S occur at $E/\alpha = 0$ and $E/\alpha = \infty$.

Although the signal-to-noise ratio is inversely proportional to M, the dynamic range, which will next be discussed, is directly proportional to M.

DYNAMIC RANGE

In conventional photography, the dynamic range of a photographic film is defined in the linear region of the H and D curve. However, in linear optimization for wavefront recording, which we will consider in Chapter 12, optimum recording does not imply restriction to the linear region of the H and D curve, but instead requires recording across a greater range of the T-E transfer characteristic. Therefore, we will define the *dynamic range* as the ratio of the maximum transmitted irradiance to the minimum

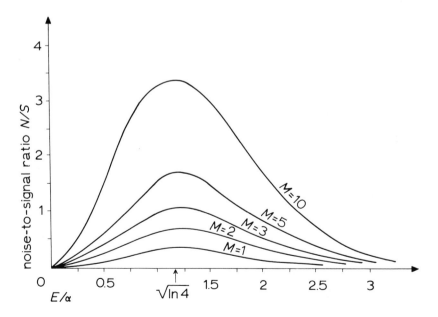

Fig. 8.14 Noise-to-signal ratio as a function of E/α, for different values of M.

transmitted irradiance of a given photographic plate:

$$\frac{T^2(0)}{T^2(\infty)} = \frac{1}{\exp(-2\beta M)}.$$ (8.43)

In terms of decibels, the dynamic range may be written

$$DR \triangleq \left.\frac{T^2(0)}{T^2(\infty)}\right|_{db} = 10M.$$ (8.44)

From eq. (8.44) it is clear that the dynamic range of a photographic film is directly proportional to M. Theoretically, the dynamic range is not bounded from above. However, in practice, the upper bound is limited by the finite grain size in the minimum resolvable image area, that is

$$DR \leq 10M_{max},$$ (8.45)

where M_{max} is the maximum number of the grains which can be accumulated in the defined unit volume. Thus the minimum number of grains needed to have a specified dynamic range for a practical film is

$$M_{min} = DR/10.$$ (8.46)

As an example, if it is required to have a 50 db dynamic range of a film, then the minimum number of grains is 5 grains per minimum resolvable image area. Therefore to have a dynamic range of 50 db for a practical film, there must be a least 5 grains per unit volume, that is $M \geq 5$. It may be noted that the dynamic range is not a continuous function, but discrete, since M is discrete in nature.

8.4 Information Channel Capacity of Photographic Film

The formula for the information capacity of photographic film was first derived by Fellgett and Linfoot (ref. 8.19), and was later applied by Jones (ref. 8.20). However, the derivation of this formula was based on the assumption of an additive stationary Gaussian photographic channel. In this section (based on ref. 8.21), the approach toward the information channel capacity of photographic film or plate will instead follow from the multiplicative nonstationary discrete stochastic model which was introduced in sec. 8.2.

Since the noise in the film is inherently associated with the recorded signal, the distinguishable levels of amplitude transmittance can be quantized with respect to standard variations of noise. A photographic film may be considered to be a four-dimensional orthogonal Euclidean space (two spatial coordinates, plus amplitude and phase coordinates). However, it will be shown later that the information channel capacity can be fully expressed in a three-dimensional orthogonal coordinate system (two spatial coordinates, plus the amplitude coordinate).

INFORMATION CHANNEL CAPACITY OF A CARTESIAN DIAGRAM

Before going into the determination of the information channel capacity of a photographic film, the information theory view of a Cartesian diagram (ref. 8.22) will be discussed. On the diagram of fig. 8.15a, the information content (i.e., the entropy), is obviously zero, since it is impossible for us to distinguish the different regions in the diagram. However, if the diagram is divided into two distinct but otherwise arbitrary regions, as is shown in the examples of fig. 8.15b, then the information content I of each of the diagrams can be determined from the probabilistic random selection of the regions:

$$I = - \sum_{n=1}^{2} P_n \log_2 P_n \quad \text{bits}, \tag{8.47}$$

where P_n is the probability of selecting the nth region. If the random-selection probabilities are equal, then the information content of the diagrams of fig. 8.15b is

$$I = \log_2 2 = 1 \quad \text{bit}.$$

That is, the entropy is maximum when the probabilities are equal.

The information content for a many-celled diagram follows immediately. If a two-dimensional orthogonal diagram contains $m \times n$ distinguishable regions (fig. 8.16), then the information content is

$$I = - \sum_{i=1}^{m} \sum_{j=1}^{n} P_{ij} \log_2 P_{ij} \quad \text{bits}, \tag{8.48}$$

where P_{ij} is the random-selection probability of the ijth region, and

$$\sum_{i=1}^{m} \sum_{j=1}^{n} P_{ij} = 1.$$

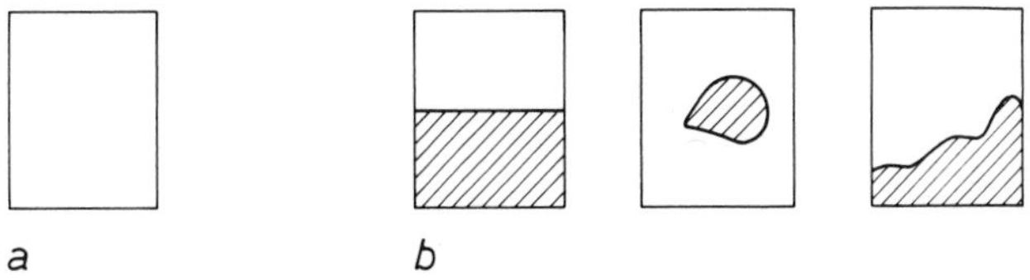

a   b

Fig. 8.15 One-dimensional Cartesian diagram: a, with a single nondistinguishable region; b, with two distinguishable regions.

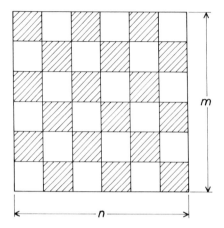

Fig. 8.16 Two-dimensional Cartesian diagram with $m \times n$ distinguishable regions.

If the random selections of the distinguishable cells are equiprobable, then the information content of this diagram becomes

$$I = \log_2 mn \quad \text{bits}. \tag{8.49}$$

We can generalize the foregoing. For example, if each of the distinguishable regions can assume q distinguishable color levels, the resulting system is describable in a three-dimensional space, and the information content is given by

$$I = -\sum_{i=1}^{m} \sum_{j=1}^{n} \sum_{k=1}^{q} P_{ijk} \log_2 P_{ijk} \quad \text{bits}, \tag{8.50}$$

where P_{ijk} is the probability of selecting the ijkth cell, and

$$\sum_{i=1}^{m} \sum_{j=1}^{n} \sum_{k=1}^{q} P_{ijk} = 1.$$

Again, for equiprobable random selection the information content is

$$I = \log_2 mnq \quad \text{bits}. \tag{8.51}$$

Based on the total number of possible combinations of the distinguishable regions, the *information channel capacity*, i.e., the maximum amount of information of the diagram that can be transmitted by means of a code word (ref. 8.23), of the diagram is

$$C = mn \log_2 q \quad \text{bits per diagram}. \tag{8.52}$$

Without loss of generality, the foregoing concepts can be extended to Cartesian diagrams in spaces of more than three dimensions.

SPATIAL INFORMATION CONTENT OF PHOTOGRAPHIC FILM

We can determine the spatial information content of a photographic film by considering it to be a Cartesian diagram, i.e., by counting the number of distinguishable or detectable regions in the film. Let twice the Rayleigh distance (see sec. 4.6) be used for the minimum resolvable distance d on the photographic film, i.e., let

$$d = \frac{1.22 \lambda L}{R}, \tag{8.53}$$

where λ is the wavelength of the source, L is the distance between the lens and the film, and R is the radius of the lens. It may be noted that this is an approximation inferior to that used by Fellgett and Linfoot (ref. 8.19). The practical lower limit of d is a few orders of λ. Then by means of a pattern of squares (fig. 8.17) the upper limit of spatial information content I_S of a rectangular film is seen to be

$$I_S = \log_2 \frac{a}{d} + \log_2 \frac{b}{d} \quad \text{bits}, \tag{8.54}$$

where a and b are the dimensions of the film. The fractions a/d and b/d will not be integral in general; in such cases the nearest integer may be used without noticeable error.

For simplicity of calculation, the irradiance of the minimum resolvable point object will be assumed to be uniformly projected onto a circular region of diameter d (i.e., the minimum resolvable distance) in the film. The equations of mean and variance of the amplitude transmittance (i.e., the signal) in the minimum resolvable cell were derived in the previous section [eqs. (8.38) and (8.39)]. It has been shown that noise is inherently attached to the recorded signal. Therefore, it is theoretically possible

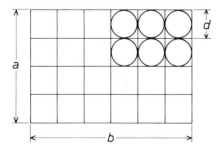

Fig. 8.17 A pattern of squares for the determination of the spatial information content of a rectangular photographic film.

(assuming an ideal detector) to quantize the amplitude transmittance into $k+2$ distinguishable levels such as

$$T(E_1) - \sigma_T(E_1) \simeq 1,$$
$$T(E_2) - \sigma_T(E_2) = T(E_1) + \sigma_T(E_1),$$
$$T(E_3) - \sigma_T(E_3) = T(E_2) + \sigma_T(E_2),$$
$$\vdots$$
$$0 \simeq T(E_k) + \sigma_T(E_k), \tag{8.55}$$

where $E_1 < E_2 < \cdots < E_k$.

We assume that these discrete levels are completely independent of each other, and that an arbitrary succession of levels is capable of representing a certain message. Therefore, if the distinct messages (i.e., $k+2$ discrete levels) of the amplitude transmittance are considered to be equally probable a priori, then the information content of the amplitude transmission is

$$I_T = \log_2(k+2) \quad \text{bits.} \tag{8.56}$$

For phase delay information, we could start from the phase delay equation (ref. 8.24),

$$\phi = KD, \quad 0 \leq \phi \leq 2\pi, \tag{8.57}$$

where K is an appropriate constant and D is the density of photographic grains. The corresponding mean and variance of the phase delay in the minimum resolvable cell can be approximated by

$$\phi \simeq KM\{1 - \exp[-\rho(E)]\}, \tag{8.58}$$

and

$$\sigma_\phi^2 \simeq K^2 M \exp[-\rho(E)] \{1 - \exp[-\rho(E)]\}. \tag{8.59}$$

It is therefore theoretically possible that signals defined with respect to the phase delay could be quantized into $r+1$ levels,

$$\phi(E_1) - \sigma_\phi(E_1) \simeq 0,$$
$$\phi(E_2) - \sigma_\phi(E_2) = \phi(E_1) + \sigma_\phi(E_1),$$
$$\phi(E_3) - \sigma_\phi(E_3) = \phi(E_2) + \sigma_\phi(E_2), \tag{8.60}$$
$$\vdots$$
$$2\pi \simeq \phi(E_r) + \sigma_\phi(E_r),$$

where $E_1 < E_2 < \cdots < E_r$.

Again, if the quantized phase delays are assumed to be equiprobable, then the information contained in the phase delay is

$$I_\phi = \log_2(r+1) \quad \text{bits.} \tag{8.61}$$

The overall information content of a photographic film may now be written as

$$I = I_S + I_T + I_\phi, \tag{8.62}$$

where I_S, I_T, and I_ϕ give the spatial, amplitude transmittance, and phase delay information, respectively. However, from eq. (8.57) the phase delay is uniquely related to the density of the developed grains, which in turn is related to the amplitude transmittance. Thus the information carried by the phase delay could be regarded as redundant to the information carried by the amplitude transmittance, or vice versa. Moreover, from the constraint $0 \le \phi \le 2\pi$ on the phase delay, it may be shown that

$$I_T \ge I_\phi. \tag{8.63}$$

The information content of the film is then

$$I = I_S + I_T. \tag{8.64}$$

The corresponding information channel capacity is therefore

$$C = \frac{ab}{d^2} \log_2(k+2) \quad \text{bits per given film.} \tag{8.65}$$

INFORMATION CAPACITY OF A HOLOGRAM*

A rigorous determination of the information capacity of a hologram is difficult and remains to be accomplished. However, a naïve, qualitative determination will be presented in this section.

In holography a resolvable point object will be recorded in the form of Fresnel zones [eq. (10.9)], and the upper limit of spatial information in a hologram will be taken as the maximum number of minimum resolvable Fresnel zone arrays that it can contain. Again we use twice the Rayleigh distance for the theoretical minimum resolvable distances [eq. (10.9)], i.e.,

$$l_x = \frac{2\lambda L_1}{H_x}, \tag{8.66}$$

and

---

* The subject of holography or wavefront reconstruction will be treated in chaps. 10–13. The reader who is not familiar with the subject of holography can refer to chap. 10 before reading this section.

$$l_y = \frac{2\lambda L_1}{H_y}, \tag{8.67}$$

where l_x and l_y are the minimum resolvable distances in the xy coordinate system, H_x and H_y are the dimensions of the (rectangular) hologram, L_1 is the separation between the point object and the hologram, and λ is the recording wavelength. If the hologram is circular in shape, eqs. (8.66) and (8.67) become

$$l = 2.44 \frac{\lambda L_1}{H}, \tag{8.68}$$

where l is the minimum resolvable distance and H is the diameter of the hologram. However, it can be easily demonstrated that the spread of a reconstructed point object becomes broader if the object is recorded off from the optical axis of the hologram. The minimum resolvable distance increases rapidly as the point objects are located further away from the optical axis of the recording medium. Therefore, the upper limit of spatial information content of a rectangular hologram may be approximated by

$$I_S = \log_2 \frac{H_x}{l_x} + \log_2 \frac{H_y}{l_y} \quad \text{bits}. \tag{8.69}$$

One might question why *longitudinal* information does not contribute to the spatial information content of a hologram, since a holographic image is three-dimensional in nature. The answer to this question is that the film is for all practical purposes a two-dimensional recorder, the interposition of which suppresses the longitudinal information.

The amplitude information in holography is found to be more difficult to obtain. However, a naive approach proceeds as follows. If we disregard the speckle noise* and the dc build-up** we may quantize the amplitude transmittance of the hologram into $S+2$ levels:

$$T(E_0) + \sigma_T(E_0) \simeq T(E_0 - E_1) - \sigma_T(E_0 - E_1),$$
$$T(E_0 - E_1) + \sigma_T(E_0 - E_1) \simeq T(E_0 - E_2) - \sigma_T(E_0 - E_2),$$
$$\vdots$$
$$T(E_0 - E_S) + \sigma_T(E_0 - E_S) \simeq 1, \tag{8.70}$$

where E_0 is the dc bias, and the E_S's are the signal amplitudes.

---

* A spatial noise, which is mainly due to the optical roughness of the surface of the object and the high coherent illumination.
** The quiescent bias exposure of the holographic recording.

Again, if these quantized amplitude transmittances of the hologram are assumed to be equiprobable, then the amplitude transmittance information content may be expressed as

$$I_{Th} = \log_2 (S+2) \quad \text{bits.} \tag{8.71}$$

The corresponding channel capacity is therefore

$$C_h = \frac{H_x H_y}{l_x l_y} \log_2 (S+2) \quad \text{bits per given hologram.} \tag{8.72}$$

If we compare the quantization in eq. (8.55) to that in eq. (8.70) it is clear that the total number of the quantized levels in the holographic process is at the most equal to one-half of the number for the direct imaging photographic process, i.e.,

$$S \leq \frac{k}{2}. \tag{8.73}$$

Thus, for the same recording medium, the upper limit of amplitude information content in a direct imaging photographic process is higher than that in a holographic process, or

$$I_T > I_{Th}. \tag{8.74}$$

Furthermore, if the recording films for the photographic and holographic processes are equal in size, and the minimum resolvable distances are the same then, from eqs. (8.65) and (8.72), it is clear that the information channel capacity of a photographic film is greater than that of a hologram,

$$C > C_h. \tag{8.75}$$

The reader may be surprised at this result, since it has been stated by many writers that a hologram can contain more information than a photograph. However, it is the encoding (i.e., recording) process that makes the holographic process superior to ordinary photography. The distinction between the information channel capacity of a direct imaging photographic process and of a holographic process must be clearly perceived. The information-channel capacity in photography is defined as the upper limit of the amount of information which can be *photographically* recorded. In other words, there must exist an encoding process such that the amount of information recorded can approach the information channel capacity of the photographic film. The information channel capacity of a hologram is defined as the upper limit of the amount of information which can be holographically recorded. That is, there must exist a coding process such that the amount of information recorded can approach the information channel capacity of the hologram. The light amplitude distribution from the objects are, in general, complex. Present-day direct imaging photographic processes are not able to record the phase information. However, the holographic process is able to record the phase information as well as the amplitude. Thus, in practice, a hologram contains more information than a direct image photograph. Nevertheless, the information content of a hologram may never reach that of a maximally encoded photograph. In fact, a hologram *is* an encoded photographic plate; therefore it is impossible for the information channel capacity of a hologram to be greater than that of a photograph.

Problems

8.1. Convert the H and D curve of fig. 8.18 to one giving amplitude transmittance as a function of exposure (i.e., construct a T-E curve).

8.2 If the *contrast* of the amplitude transmittance is defined as

$$\frac{T_{max} - T_{min}}{T_{max} + T_{min}},$$

then from the T-E transfer characteristic obtained in prob. 8.1, determine an approximate linear exposure in which the contrast is optimized.

8.3 Consider a certain photographic plate which is recorded by a low

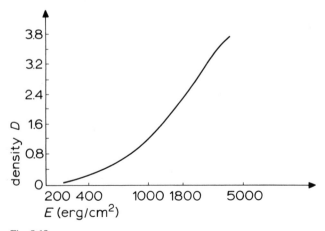

Fig. 8.18

contrast exposure over the plate. The exposure is written as

$$E(x, y) = E_Q + E_1(x, y), \quad \text{with } |E_1(x, y)| \ll E_Q,$$

where E_Q is the bias exposure and $E_1(x, y)$ is the variable incremental exposure. If the recorded photographic plate is biased in the linear region of the H and D curve, show that the contrast of the amplitude transmittance is linearly proportional to $E_1(x, y)$, provided $|E_1(x, y)| \ll E_Q$.

8.4 Suppose that the T-E transfer characteristic of a certain photographic emulsion is as shown in fig. 8.19, and the input exposure that given by a sinusoidal grating,

$$E(x, y) = 1000 + 800 \cos(100\pi x) \quad \text{erg/cm}^2, \quad \text{for every } y.$$

Use a graphical method to plot the corresponding output transmittance $T(x, y)$.

8.5 By using fig. 8.13, explain why a hard-clipped grating gives the optimum transmission irradiance-to-noise ratio.

8.6 Referring to sec. 8.4, evaluate the information content and informa-

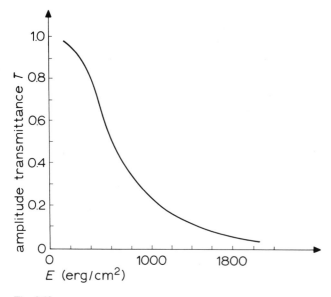

Fig. 8.19

tion channel capacity of a binary (i.e., hard-clipped) photographic plate.

References

8.1 C. E. K. Mees, *The Theory of the Photographic Process*, rev. ed., Macmillan, New York, 1954.

8.2 C. E. K. Mees and T. H. James, *The Theory of the Photographic Process*, 3rd ed., Macmillan, New York, 1966.

8.3 F. Hurter and V. C. Driffield, Photochemical Investigations and a New Method of Determination of the Sensitiveness of Photographic Plates, *J. Soc. Chem. Ind.*, **9** (1890), 455.

8.4 E. L. O'Neill (ed.), *Communication and Information Theory Aspects of Modern Optics*, General Electric Company, Electronics Laboratory, Syracuse, N.Y., 1962.

8.5 A. Kozma, Photographic Recording of Spatially Modulated Coherent Light, *J. Opt. Soc. Am.*, **56** (1966), 428.

8.6 R. C. Jones, New Method of Describing and Measuring the Granularity of Photographic Materials, *J. Opt. Soc. Am.*, **45** (1955), 799.

8.7 H. J. Zweig, Autocorrelation and Granularity. Part I. Theory, *J. Opt. Soc. Am.*, **46** (1956), 805.

8.8 H. J. Zweig, Autocorrelation and Granularity. Part II. Results on Flashed Black-and-White Emulsions, *J. Opt. Soc. Am.*, **46** (1956), 812.

8.9 H. J. Zweig, Autocorrelation and Granularity. Part III. Spatial Frequency Response of the Scanning System and Granularity Correlation Effect Beyond the Aperture, *J. Opt. Soc. Am.*, **49** (1959), 238.

8.10 M. B. Picinbono, Modèle Statistique Suggéré par la Distribution de Grains D'argent Dans les Films Photographiques, *Compt. Rend.*, **240** (1955), 2206

8.11 M. M. Savelli, Résultats Pratiques de l'étude d'un Modèle à Trois Paramètres pour la Représentation des Propriétés Statistiques de la Granularité des Films Photographiques Notamment des Propriétés Spectrales, *Compt. Rend.*, **246** (1958), 3605.

8.12 F. T. S. Yu, Markov Photographic Noise, *J. Opt. Soc. Am.*, **59** (1969), 342.

8.13 E. Parzen, *Stochastic Processes*, Holden-Day Publishing Company, San Francisco, 1962.

8.14 J. H. Altman and H. J. Zweig, Effect of Spread Function on the Storage of Information on Photographic Emulsions, *Phot. Sci. Eng.*, **7** (1963), 173.

8.15 A. Papoulis, *Probability, Random Variables, and Stochastic Processes*, McGraw-Hill, New York, 1965.

8.16 H. J. Zweig, G. C. Higgins, and D. L. MacAdam, On the Information-Detecting Capacity of Photographic Emulsions, *J. Opt. Soc. Am.*, **48** (1958), 926.

8.17 F. T. S. Yu, Film-Grain Noise and Signal-to-Noise Ratio, *J. Opt. Soc. Am.*, **60** (1970), 1547.

8.18 E. L. O'Neill, *Introduction to Statistical Optics*, Addison-Wesley, Reading, Mass., 1963.

8.19 P. B. Fellgett and E. H. Linfoot, On the Assessment of Optical Images, *Trans. Roy. Soc.*, ser. A, 247 (1955), 369.

8.20 R. C. Jones, Information Capacity of Photographic Films, *J. Opt. Soc. Am.*, **51** (1961), 1159.

8.21 F. T. S. Yu, Information Channel Capacity of a Photographic Film, *IEEE Trans. Inform. Theory*, **IT-16** (1970), 477.

8.22 F. T. S. Yu, Information Content of a Sound Spectrogram, *J. Au. Eng. Soc.* **15** (1967), 407.

8.23 C. E. Shannon and W. Weaver, *The Mathematical Theory of Communication*, University of Illinois Press, Urbana, Ill., 1949.

8.24 D. G. Falconer, Role of the Photographic Process in Holography, *Phot. Sci. Eng.*, **10** (1966), 133.

Chapter 9   Image Restoration and Information

The purpose of this chapter is to present some basic techniques and physical constraints of image restoration from the information theory point of view. *Information* may be defined as a measure of the degree of uncertainty of some message or process; i.e., the more uncertain we are about a message we have received, the more information it contains. From the information theory viewpoint, a perfectly dishonest person is as good an informant as a perfectly honest person, provided we have the prior knowledge of which person is perfectly dishonest. (This is by no means to say that if one cannot be a perfect gentleman, then he should be a perfect rogue!) It may be clear from this simple example that information theory, is in fact, a probability theory.

A communication system may be represented by the block diagram shown in fig. 9.1. For example, in speech communication, I have a message (i.e., the source) to transfer to you by means of a spoken language, such as English, French, German, Chinese, etc. Then I have to select a suitable language (i.e., a source encoder) which will be appropriate to our conversation. After selection of the language, the message still cannot be conveyed unless it is converted into an appropriate acoustical wave (at the transmitter). This acoustical wave is the information (message) carrier. When this acoustical signal reaches the receiver (your ears), then a proper decoding (translating) process takes place to make the message recognizable by the user (i.e., your mind). From this simple daily communication process, it is easily seen that a suitable encoding operation is not adequate unless an

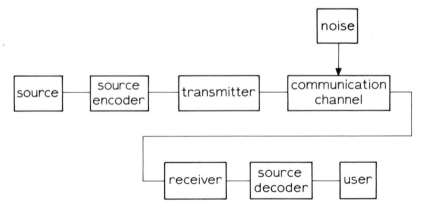

Fig. 9.1 Block diagram of a communication system.

appropriate decoding procedure takes place. For instance, if I used a wrong encoding (e.g., the Chinese language instead of English) you might not be able to decode my message, even if the speech communication channel is perfect (i.e., contains no noise). This is because proper decoding requires some previous knowledge of the encoding scheme (i.e., an appropriate information storage), such as prior knowledge of the Chinese language. Therefore, the decoding process can also be regarded as a recognition process.

Communication or information theory is a large subject, and we can investigate it only in an informal way here. However, the basic viewpoint to be developed will serve our own purposes. The reader who is interested in a rigorous development of the subject can refer to the excellent books by Fano (ref. 9.1) and by Brillouin (ref. 9.2).

Communication theory may be categorized into two general disciplines. One is Wiener's communication theory (ref. 9.3, 9.4), and the other is Shannon's information theory (ref. 9.5). Although both of these share a probabilistic basis, there is a basic distinction between them.

The significance of Wiener's communication theory is that if a signal (i.e., the information), is corrupted by some physical means (e.g., noise, system distortion, etc.), it may be possible to recover the signal partially or completely from the corrupted signal. It is for this reason that he advocates correlation detection, prediction, optimum filtering, etc.

However, in Shannon's information theory, this is carried a step further, for Shannon shows that the signal can be processed both before and after transmission. That is, before the message is transmitted, it is possible to encode it in a form appropriate to combat the disturbances in the communication channel. Then by using an appropriate decoding process, it is possible to retrieve the message optimally in the receiving end. This is the primary reason Shannon advocates information measure, channel capacity, coding processes, etc. The major interest in Shannon's theory is the efficient utilization of the channel capacity.

A fundamental theorem proposed by Shannon is the most important and most surprising result of information theory. It may be stated approximately thus:

Given that a stationary finite-memory communication channel has a channel capacity C. If the binary transmission rate R of the message is smaller than C, then there exist channel encoding and decoding processes in which the probability of error in transmission per digit can be made arbitrarily small. Conversely, if the transmission rate R is larger than C, then there do not exist encoding and decoding processes with this property, and the probability of error in transmission cannot be made arbitrarily small.

In other words, the presence of random disturbances in a communication channel does not, by itself, set any limit to the transmission accuracy. Rather, it sets a limit to the transmission rate for which arbitrarily high transmission accuracy may be accomplished.

In summary of this brief introduction to information theory, I would like to point out again the distinction between the Wiener and Shannon disciplines. Wiener assumes, in effect, that the signal in question can be processed only after it has been corrupted by noise. Shannon, on the other hand, suggests that the signal can be processed both before and after transmission through the communication channel. Nevertheless, the main objective of these two branches of communication theory are basically the same, namely, the faithful reproduction of the original signal.

9.1 Image Restoration

The restoration, by means of coherent techniques, of the image from a smeared or out-of-focus photographic film has been demonstrated. (refs. 9.6, 9.8). However, as will be shown in the next section, the restored image will not be better than the photograph taken under the maximum allowable exposure time.* In other words, the restored image theoretically can only approach the case when the picture is taken without letting it smear or get out of focus.

In general the problem of image restoration can be divided into two categories:
1. Restoration of the image from the distorted signal.
2. Restoration of images by superimposing the signals distorted by smearing or being out of focus.

In the first category, it is possible for the image to be reconstructed; however, one does not use the extra energy (e.g., due to smearing) recorded on the film. In the second category, it is implied that it is not only possible to extract the signal, but also to use the extra energy recorded on the film.

If one disregards the existence of the random noise in the film, then

---

* The *maximum allowable exposure time* is defined as the longest possible exposure time in which the recorded image will not significantly smear. It depends upon the size of the object and the details of the scene. The smaller the size or the finer the detail, the shorter the exposure time demanded.

the above two categories are seen to be essentially the same, since the same amount of information will be extracted in each. On the other hand, if the inherent random noise of the photographic film is considered, then the second category will have a higher information content. However, it can be shown that the second category is a physically unrealizable procedure (ref. 9.9).

In the following, a few techniques of restoration because of linear image motion will be illustrated. Some of the physically realizable factors will be given in the next section.

If an image is distorted by some physical means, then the distorted image, in general, can be expressed in its Fourier transform form,

$$G(p) = S(p)\,D(p), \tag{9.1}$$

where $G(p)$ is the distorted image function, $D(p)$ is the distorting function, $S(p)$ is the image function, and p is the spatial frequency.

It is well known that the image function $S(p)$ can be recovered if we know the distorting $D(p)$, that is,

$$S(p) = G(p)\,\frac{1}{D(p)}. \tag{9.2}$$

Disregarding the realizability of the inverse filter $1/D(p)$, we may ask: Does this inverse filtering process increase the information content of the distorted signal $G(p)$? The answer to this question is no, because we know exactly how it is distorted. In other words, the additional bits of information we gain from the inverse filtering are diminished by the bits of information from the knowledge of $D(p)$. One might argue: If this method of recovering the signal does not increase the information content, why do we do it? The answer is that it is a recognition problem. Since we recognize $G(p)$ as well as $S(p)$, for us they are essentially no different. But, for those who do not know how the signal was distorted, certainly a realistic signal $S(p)$ contains more information than a distorted one. Moreover, the information gain for one who does not recognize $G(p)$ comes from the expenditure of energy in converting $G(p)$ to $S(p)$.

Suppose that the image of a resolvable point object of constant irradiance I is projected onto the film of a camera. If the point object is moving at a constant velocity v, and the time of the exposure of the film is t, then the moving point image recorded on the photographic film will be a straight line of length $\Delta x = mvt$, with m a proportionality constant. The corresponding transmittance $f(x)$ may be expressed by

$$f(x) = \begin{cases} A, & -\frac{1}{2}\Delta x \le x \le \frac{1}{2}\Delta x \\ 0, & \text{otherwise} \end{cases}, \tag{9.3}$$

where A is a positive constant proportional to I. The corresponding Fourier transform is

$$F(p) = A\Delta x \frac{\sin(p\Delta x/2)}{p\Delta x/2}. \tag{9.4}$$

To restore this point image, we apply the inverse filtering process, giving

$$F(p)\frac{A}{F(p)} = A\mathscr{F}[\delta(x)], \tag{9.5}$$

where $\mathscr{F}$ denotes the Fourier transform, and $\delta(x)$ is the Dirac delta function.

An inverse filter $A/F(p)$ may not be realizable in practice. Nonetheless, if one assumes that the inverse filter can be completely or partially realized, then the restored image is at best equal to $A\delta(x)$, the minimum resolvable point image. In other words, this inverse filtering process might be able to restore the image from the distorted image, but it will not increase the overall signal spectrum over that obtained from a maximum allowable exposure. However, one might ask: Can we have a physical filter $H(p)$ such that

$$F(p)\,H(p) = B\mathscr{F}[\delta(x)], \tag{9.6}$$

where B is a positive real constant greater than A? The answer to this question is, of course, no, since the coherent optical information processor we are dealing with is a passive system. Furthermore, because of the practical complications involved in filter syntheses, we are led to another approach.

Again, suppose a moving image $f(x)$ is recorded on the film, such that the length is Δx. If the smearing is controlled by some exposure modulation, such as that composed of a finite sequence of identical functions, then the transmittance of the recorded function may be expressed as

$$\begin{aligned} g(x) &= f(x) + f(x - \Delta l) + f(x - 2\Delta l) + \cdots + f(x - N\Delta l) \\ &= \sum_{n=0}^{N} f(x - n\Delta l), \end{aligned} \tag{9.7}$$

where $N = \Delta x/\Delta l$, with Δl the incremental translation of $f(x)$.

The corresponding Fourier transform is

$$G(p)= \sum_{n=0}^{N} F(p)\, e^{-ipn\Delta l}, \tag{9.8}$$

where $G(p)$ and $F(p)$ are the corresponding Fourier transforms of $g(x)$ and $f(x)$ respectively. From eq. (9.8), the image restoration may be attempted by

$$G(p)\, H(p)=F(p), \tag{9.9}$$

where the filter is

$$H(p)=\frac{1}{\sum_{n=0}^{N} e^{-ipn\Delta l}}. \tag{9.10}$$

It may be emphasized that if we properly translate the x axis of eq. (9.7), then, from Lagrange's identity (ref. 9.10), the denominator of eq. (9.10) may be written as

$$\sum_{n=-N/2}^{N/2} e^{-inp\Delta l}=\frac{\sin\left[\left(\dfrac{N}{2}+\dfrac{1}{2}\right)p\Delta l\right]}{\sin\left(\dfrac{1}{2}p\Delta l\right)}. \tag{9.11}$$

For large N and small Δl, the above equation can be approximated by

$$\sum_{n=-N/2}^{N/2} e^{-inp\Delta l}\simeq\frac{\sin\dfrac{p\Delta x}{2}}{\dfrac{1}{2}\dfrac{p\Delta x}{N}}. \tag{9.12}$$

Obviously, $H(p)$ is not a physically realizable function. It may be emphasized again that if we assume $H(p)$ is physically realizable, by the same interpretation, the restored image is at best equal to that in the case where the image is recorded over the maximum allowable exposure time.

The significance of this process is that the filter is not an inverse function of $F(p)$. In some cases, it might be more convenient to approximate a filter of $H(p)$, rather than its inverse.

Because of the physical unrealizability of the filter, we must modify the exposure modulating process. We note from eqs. (9.9) and (9.10) that the filter $H(p)$ may be made physically realizable if unity is added to the denominator of eq. (9.10),

$$H(p)=\frac{1}{1+\sum_{n=0}^{N} e^{-ipn\Delta l}}. \tag{9.13}$$

It is clear that if the filter of eq. (9.13) is used, the physical realizability condition is

$$|H(p)| \leq 1, \text{ for all } p.$$ (9.14)

A method of making $H(p)$ physically realizable is to control the film exposure properly, so that the resulting transmittance is

$$g(x) = 2f(x) + f(x - \Delta l) + f(x - 2\Delta l) + \cdots + f(x - N\Delta l)$$
$$= f(x) + \sum_{n=0}^{N} f(x - n\Delta l),$$ (9.15)

where $N = \Delta x / \Delta l$ and $\Delta x = mvt$.

The corresponding Fourier transform then has the form

$$G(p) = F(p) \left[1 + \sum_{n=0}^{N} e^{-ipn\Delta l} \right].$$ (9.16)

Obviously, a filter such as

$$H(p) = \frac{1}{1 + \sum_{n=0}^{N} e^{-inp\Delta l}} \simeq \frac{1}{1 + N \frac{\sin(p\Delta x/2)}{p\Delta x/2}},$$ (9.17)

is physically realizable. However, it can be shown that the restored image spectrum is equal to $F(p)$, which is only one-half of the first signal recorded in eq. (9.15).

MODULATING CAMERA

If the recorded image can be modulated such that each of the resolvable objects can be imaged by itself,* then it may be possible to reconstruct the signal by means of a coherent optical system. The smearing of the image recorded will not only be corrected; in addition, the overall signal-to-noise ratio may be increased.

Suppose a resolvable point image is recorded on the film by a modulating process so that the amplitude transmittance is

$$T(x) = \frac{1}{2} - \frac{1}{2} \cos \frac{\pi}{\lambda f} x^2 = \frac{1}{2} - \frac{1}{4} \exp\left(i \frac{\pi}{f\lambda} x^2\right) - \frac{1}{4} \exp\left(-i \frac{\pi}{f\lambda} x^2\right),$$ (9.18)

where λ is the wavelength of the coherent source and f is the focal length

---

* If each of the resolvable point images is recorded by means of an appropriate shutter-modulating function, then these modulated point images may be reconstructed (i.e., re-imaged) by means of wavefront reconstruction.

of the one-dimensional Fresnel zone lens (refs. 9.9, 9.11, 9.12). It may be noted that a zone lens is similar to a zone plate, as described in sec. 4.5, except that the transmittance is a sinusoidal function of x^2.

If this modulated film of eq. (9.18) is illuminated by a monochromatic plane wave as shown in fig. 9.2, then the complex light field at the focal plane of $T(x)$ may be determined by the Fresnel-Kirchhoff theory or Huygens' principle (ref. 9.13):

$$E(\alpha; k) = \int_{-\infty}^{\infty} T(x)\, E_l^+(\alpha - x; k)\, dx, \tag{9.19}$$

where

$$E_l^+(x; k) = B\left(-\frac{i}{\lambda f}\right) \exp\left[i\frac{2\pi}{\lambda}\left(f + \frac{x^2}{2f}\right)\right]$$

is the spatial impulse response, and B is a complex constant.

By substituting eq. (9.18) in eq. (9.19), the solution is

$$E(\alpha; k) = -iB\left[\frac{1}{2}\left(\frac{1}{f\lambda}\right)^{1/2} \exp\left(i\frac{\pi}{4}\right) - \frac{1}{4}\left(\frac{1}{2f\lambda}\right)^{1/2} \exp\left(i\frac{\pi}{2f\lambda}\alpha^2 + i\frac{\pi}{4}\right)\right.$$
$$\left. - \frac{1}{4}\exp\left(i\frac{\pi}{f\lambda}\alpha^2\right)\delta(\alpha)\right]. \tag{9.20}$$

It is clear that the first term of eq. (9.20) is the zero order (i.e., dc) term; the second term is the divergent diffraction term; the last term represents the convergent diffraction term. In other words, a zone lens such as the one of eq. (9.18), is equivalent to three lenses (dc, divergent and convergent). However, if we examine eq. (9.20) a bit more carefully, we see that the convergent image term is reduced by a factor of four (factor of sixteen with respect to irradiance) compared with the one used in the maximum

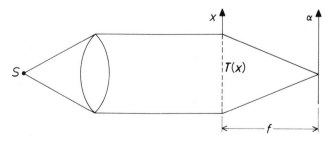

Fig. 9.2 Coherent illumination. A monochromatic point source is at S.

allowable exposure time. Obviously, the additional energy recorded on the film due to the modulation was not effectively used. On the contrary, this extra modulation energy, i.e., bits of information, has been converted into useless information in the dc and divergent terms. From this simple analysis we are led to doubt that there exists a realizable shutter modulation process in which the extra recording can be fully or partially converted into useful information. The complete restoration of smeared images may be regarded as an energy and time problem for which a physically realizable procedure involving the expenditure of a finite amount of energy does not exist.

In general, a smeared image can be represented in a three-dimensional orthogonal space with a closed time interval $(t_1 \le t \le t_2)$. The smeared image can therefore be regarded as a time and space problem. Nevertheless, if we claim to be able to restore the image due to smearing by piling up the extra energy recorded on the film, it is equivalent to saying it is possible to separate the times and the spaces from the smeared image. Of course, from the basic theory of relativity (ref. 9.4), it is impossible for us to do so without expending an infinite amount of energy.

9.2 Uncertainty and Information

In order to demonstrate that the restoration of a smeared image is a problem related to Heisenberg's uncertainty principle (ref. 9.15), we may start from the basic inequality of a photographic film,

$$It \ge E_0, \tag{9.21}$$

where E_0 is the minimum energy per unit area of the film required for the resolvable image, and t is the exposure time. In other words, if the recorded image is below the minimum required energy E_0, it is impossible for the image to be resolved. This minimum energy E_0 can be called the threshold level of the photographic film.

To show that the foregoing inequality is equivalent to Heisenberg's uncertainty relation, one can simply substitute $E_0 = h\nu$; thus

$$Et \ge h, \tag{9.22}$$

where $E = I/\nu$, h is Planck's constant, and ν is the frequency.

These inequalities give the theoretical limit of the film. The inequalities (9.22) or (9.21) give the *detectability* (or *recordability*) of the film. If the situation does not satisfy the detectability, it is impossible for the recorded images to be resolved. In the following, we will use this inequality to relate to the theory of information.

For stationary objects and film, if the signal-to-noise ratio of the image is high, there is no essential difficulty in satisfying the inequalities (9.21) or (9.22). However, for moving objects, film or both, there exists a maximum allowable exposure time so that the recorded image will not be distorted due to smearing:

$$t \leq t_{max}, \tag{9.23}$$

where t_{max} is the maximum allowable exposure time. In general, t_{max} is linearly related to the size of the recorded image.

From the inequality (9.22), we note that a trade of energy and exposure time is possible; for high irradiance, the film can record smaller objects. However, if the irradiance from the object is small, it may be impossible to reduce the maximum allowable exposure time without violating the detectability condition.

If we are tempted to satisfy the detectability condition by letting the exposure time become greater than the allowable one, then the recorded image will be further degraded (due to smear). Some of the information recorded in the allowable time interval $0 \leq t \leq t_{max}$ is partially or completely destroyed by this excessive recording. This information loss can also be interpreted as an increase of the entropy of the film (ref. 9.16). Hence, a smeared image is expected to contain less information than that of the unsmeared one under the maximum allowable time condition. This loss of information due to additional recording is also the physical basis of the unrealizability of the inverse filter discussed in the previous section. In order to bring back the information loss due to smearing, the expenditure of an infinite amount of energy is required.

In coherent optical processing we have a lot of energy available from the source. Can we use it? The answer to this question appears to be no, because the energy in coherent systems is used to convey the information; it is not convertible into information. For example, suppose there is a printed page which contains a certain amount of information. If it is located in a dark room, the information has no way of reaching us; we need a certain quantity of light in order to make the information observable. This quantity of light is the energy used for the information transmission. Moreover, if this page contains a number of misprinted words, and these misprinted words are independent of the correct ones, then it will require an enormous number of bits of information (or energy) to restore the intended words.

We conclude that the extra energy (or bits of information) which we intended the film to absorb to correct the smear actually further degrades the information content of the image. On the other hand, to say that we may reconstruct the original image completely from film degraded by smear or modulation is equivalent to saying that it is possible to make the recording instrument record the unrecordable object after the recording. The contradiction is apparent.

But, one can still argue, if an object is embedded in noise (such as turbulence), why is it possible to restore the image by using complex spatial filtering? Before we attempt to answer this question, we might ask if the source irradiance and the exposure time do not satisfy the detectability condition. If they do not, can we restore the imaging object? Certainly we cannot. Furthermore, if they do satisfy the detectability condition, can we restore the imaging object to better than in the case without noise? Again, we cannot. At best, we can only approach the case without noise as a limit. Therefore, it can be concluded that the imaging object can be restored (with a probability of error) from the degraded random noise if, and only if, the intensity from the object and the exposure time of the film have fulfilled the basic detectability condition. Furthermore, the restored image, at best, can only be equal to the image taken under the condition without random noise (such as turbulence, etc.).

A few questions can still be raised; is there a coding process (such as a modulating camera) that can improve the smearing correction? The answer to this question is no. To improve the information transmission of a communication channel, the coding process should be done at the transmitting end, but not at the receiving end.

How about coherent detection methods, such as correlation detection, sampling, etc? Can it improve the image restoration? The answer is again no, because we have no prior knowledge about the object (i.e., the recorded image). Therefore, no way exists to correlate the recorded smear image. We may summarize these considerations as follows:

1. A smeared image is basically correctable. However, the corrected results can only approach the case in which the image is taken under the maximum allowable exposure time criterion $t = t_{max}$ of the film. In practice, the corrected result is far below the t_{max} criterion, which is due to the unrealizability of the filter.

2. The modulating camera method or any other modulating process is unable to improve the image resolution over that of the maximum allowable time criterion, because the information content of the image is further degraded by this modulating process.

3. The smear correction problem is basically a time and space problem. It is physically impossible to restore the image by piling up partially or

completely the extra energy (i.e., bits of information), recorded due to smear.

4. If the irradiance from the object and the exposure time of the film do not satisfy the detectability condition of the instrument (i.e., the film), it is physically impossible to restore the recorded image. To do so would violate the uncertainty principle. It is possible to restore the recorded image to a point where it equals the image under the maximum allowable exposure time criterion, if and only if the detectability condition is satisfied.

5. The unrealizability of an inverse filter can be interpreted from the information theory point of view: The information lost due to the degradation of the recorded image (such as smearing) can only be recovered with the expenditure of an infinite amount of energy.

6. Any existing coding processes and coherent detection techniques cannot improve image restoration; because the efficient coding should be processed at the transmitting end, but not at the receiving end. In coherent detection, there is a lack of prior knowledge of the signals (objects) recorded.

However, *some* image restoration after smearing is of course possible. The optimum image restoration that may be obtained can be found by means of a mean-square error approximation such as

$$\overline{\varepsilon^2(x, y)} = \lim_{\substack{x_1 \to \infty \\ y_1 \to \infty}} \frac{1}{x_1 y_1} \int_{-x_1/2}^{x_1/2} \int_{-y_1/2}^{y_1/2} [f_0(x, y) - f_d(x, y)]^2 \, dx \, dy, \qquad (9.24)$$

where $f_d(x, y)$ is the desired image, $f_0(x, y)$ is the restored image, and the evaluation of the spatial filter is subject to the physical constraint

$$|H(p, q)| \leq 1, \qquad (9.25)$$

where $H(p, q)$ is the complex transmittance of the correcting filter.

It may be emphasized that the restoration of a smeared image has application where the image was recorded without intention of image motion, i.e., where minimum exposure choice could not be made at the time the picture was recorded.

9.3 Optical Resolving Power and Physical Realizability

The possibility of resolving power beyond the classical limit of an idealized optical system has been recognized by Coleman and Coleman (ref. 9.17), Toraldo (ref. 9.18), Ronchi (refs. 9.19, 9.20), and Harris (ref. 9.21). In this section (based upon ref. 9.22), however, some of the physical limitations, besides the inevitable noise of optical systems, will be discussed.

We will state two long-established theorems from analytic function theory, which have been found to be useful in studies of resolution beyond the classical limit. For proof of these theorems refer to refs. 9.23 and 9.24.

Theorem 1
The Fourier transform of a spatially bounded function is analytic throughout the entire domain of the spatial frequency plane.

Theorem 2
If a function of a complex variable is analytic in a region R, it is possible, from the knowledge of the function in an arbitrary small region within R, to determine the whole function within R by means of analytic continuation.

Corollary (Identity or Uniqueness Theorem)
If any two analytic functions have values which coincide over arbitrary small region in the region of analyticity, then the values of these two functions must be equal everywhere throughout their common region of analyticity.

From these two theorems, it is clear that if the spatial transfer characteristic (i.e., the spatial frequency and phase response) of an optical system is known, then for a bounded object it is possible to resolve the object with an infinite precision by means of analytic continuation. Furthermore, from the corollary of the second theorem, the ambiguity of two close objects does not exist, and the resolved object is therefore unique.

The infinite precision of object resolution is, however, a mathematical ideal. In the following we will see that infinitely precise resolution is physically unrealizable. In fact, the amount of information gained from the extension of the complex spectral density of the object, which is limited by the spatial cutoff frequency of the optical system, must come from the effort of proceeding with the analytic continuation. We will also show that the degree of precision depends on the accuracy of the functional spectrum expansion. The smaller the regions of the spectral density known, the larger is the effort required for better precision of the object resolution.

Let the function $f(z)$ represent the complex spatial frequency spectrum of a bounded object (where $z = x + iy$, with x and y the spatial frequency coordinates). If $f(z)$ is analytic throughout a given region R, and the functional value is assumed given over an arbitrary small region (or an arc) within R, then, by analytic continuation, $f(z)$ can be uniquely determined throughout the region R.

Let A denote an arc within R on which the function $f(z)$ is given (fig. 9.3). Since $f(z)$ is analytic in R, the value of the derivative $f'(z)$ is independent of the manner in which Δz tends to zero. Suppose z_0 denotes a point on A. Then because of the analyticity of $f(z)$, this function can be expanded by Taylor series about z_0, at all points interior to a circle having its center at z_0 and lying within R. Thus,

$$f(z) = f(z_0) + \sum_{n=1}^{\infty} \frac{f^n(z_0)}{n!}(z - z_0)^n, \quad \text{for} \quad |z - z_0| \leq r_0, \tag{9.26}$$

where $f^n(z_0)$ denotes the nth order derivative of $f(z)$ at z_0, and r_0 is the corresponding radius of the circle of convergence.

Thus, in order to determine the functional value of $f(z)$ within $|z - z_0| \leq r_0$, we must know all the functional derivatives $f^n(z_0)$. For the most general f, of course, it is not possible to write down the infinite number of derivatives required. If we assume that eq. (9.26) converges to $f(z)$ with little error when the series is truncated at some finite $n = N$ and also that we can find or approximate the associated derivatives $f^n(z_0)$ (which is difficult to achieve in practice; see ref. 9.25), then the functional value of $f(z)$ within the convergent circle C_0 can be approximated by

$$f(z) \simeq f(z_0) + \sum_{n=1}^{N} \frac{f^n(z_0)}{n!}(z - z_0)^n. \tag{9.27}$$

Now consider any other point P in R. Let A_1 denote a curve connecting

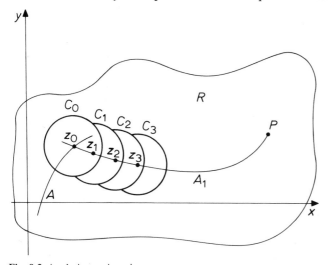

Fig. 9.3 Analytic continuation.

z_0 and P and lying in the interior of R (fig. 9.3), and let the radius r_0 be equal to the shortest distance between the curve A_1 and the boundary R. Let z_1 be a point on curve A_1 within the circle C_0, and let C_1 be the circle with the center at z_1 and radius r_0. Then the function $f(z)$ within C_1 is approximated by

$$f(z) \simeq f(z_1) + \sum_{n=1}^{N} \frac{f^n(z_1)}{n!} (z - z_1)^n. \tag{9.28}$$

Continuing this process, the curve A_1 can be covered by a finite sequence of circles C_0, C_1, C_2, ..., C_m, each of radius r_0; consequently the functional value of $f(z)$ over the whole of R can be determined. However, from the well-known theorem that a spatially bounded object cannot be spatially frequency bounded (ref. 9.26), the complete extension of $f(z)$ requires an infinite sequence of circles to cover the entire spatial frequency plane. Therefore, this analytic extension of $f(z)$ will also require an infinite amount of effort, i.e., it is not physically realizable.

Eq. (9.28) is the analytic continuation of eq. (9.27). The error of this approximate extension of $f(z)$ increases as the steps of the analytic continuation proceed.

If we allow the complex spatial frequency spectrum to be for all practical purposes bounded,* then by an analytic continuation similar to the foregoing we can see that the degree of precision of object resolution increases with the increase of the spatial frequency bandwidth of the optical system.

Furthermore, the amount of information gained by the analytic continuation of the spatial frequency spectrum clearly comes from the effort (i.e., energy) of obtaining the derivatives $f^n(z_0)$ (ref. 9.16).

We have shown in this section that the infinite precision of object resolution is not possible to achieve in practice. However, considering the uniqueness theorem cited earlier, one might wonder why we need to expand the spatial frequency spectrum, since an arbitrary small region of the spectrum is sufficient to represent the object. Unfortunately, to utilize this fact we need to have a priori knowledge of the objects and their corresponding spatial frequency spectra. If the objects to be detected are finite in number, this may be possible. However, in practice the objects are generally not finite in number; in fact, there may be uncountably many objects. In this case, an infinitely large volume of information storage (i.e., the dictionary of their corresponding spatial frequency

---

* Although mathematically the spatially bounded object cannot be spatially frequency bounded, this assumption of boundedness can often be made in practice.

spectra) must be available. Such information capacity is, of course, physically unrealizable.

9.4 Restoration of Blurred Photographic Images

The restoration of an image from a smeared or defocused photographic image by means of coherent optical spatial filtering has been briefly mentioned. Some of the physical constraints discoverable from the standpoint of information theory have also been discussed in the preceding sections. However, the synthesis of a complex spatial filter to reduce the effect of blurring has not been given. As we have pointed out in chap. 7, a complex spatial filter may be realized by the combination of an amplitude filter and a thin-film phase filter. Such a synthesis, however, is generally difficult to achieve in practice (ref. 9.27). By means of the holographic techniques, which will be considered in the following chapter, the desired phase filter may be easy to realize. The preparation of such a phase filter has been studied by Stroke and Zech (ref. 9.28) for the restoration of blurred images, and by Lohmann and Paris (ref. 9.29) for optical data processing.

In this section we will consider the synthesis of a phase filter that, when combined with an amplitude filter, can be used for the restoration of an image which has been blurred. The complex filtering process that we will discuss may be able to correct some of the blurred images, but it is by no means optimum; the constraints of sec. 9.2 still apply.

We recall the Fourier-transform expression of a linearly distorted (blurred) image, eq. (9.1),

$$G(p) = S(p) D(p), \tag{9.29}$$

where $G(p)$ is the distorted-image function, $S(p)$ is the non-distorted-image function, $D(p)$ is the distorting function of the imaging system, and p is the spatial frequency. Then the corresponding inverse filter transfer function for the restoration is

$$H(p) = \frac{1}{D(p)}. \tag{9.30}$$

As we have noted in sec. 9.1, the inverse filter function is generally not physically realizable, particularly for blurred images due to linear motion or defocusing. If we are willing, however, to accept a certain degree of error, then an approximate inverse filter might be possible to realize. For example, let the transmission function of a linear smeared point image [i.e., eq. (9.3)] be

$$f(\xi) = \begin{cases} 1, & -\tfrac{1}{2}\varDelta\xi \leq \xi \leq \tfrac{1}{2}\varDelta\xi, \\ 0, & \text{otherwise} \end{cases}, \tag{9.31}$$

where $\Delta\xi$ is the smear length. If a transparency satisfying eq. (9.31) is inserted in the input plane P_1 of a coherent optical data processor, as shown in fig. 9.4, the resultant complex light field on the spatial frequency plane is

$$F(p) = \Delta\xi \, \frac{\sin(p\Delta\xi/2)}{p\Delta\xi/2}, \tag{9.32}$$

which is essentially the Fourier transform of the smeared point image. A plot of the Fourier spectrum given by eq. (9.32) is shown in fig. 9.5. It can be seen that the Fourier spectrum is bipolar.

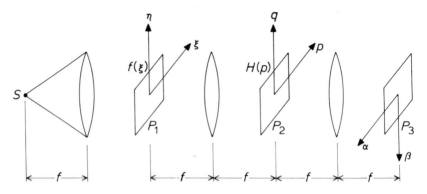

Fig. 9.4 Coherent data processing system.

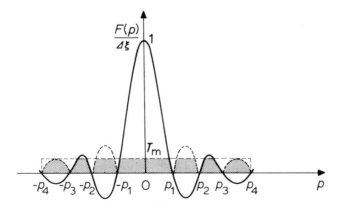

Fig. 9.5 The solid curve represents the Fourier spectrum of a linear smeared point image. The shaded area represents the corresponding restored Fourier spectrum of a point image. The zeroes are given by $p_n = 2n\pi/\Delta\xi$, $n = 1, 2, 3, \dots$.

In principle, the smeared image may be corrected by means of inverse filtering (sec. 9.1). A suitable inverse filter function is

$$H(p) = \frac{p\Delta\xi/2}{\sin(p\Delta\xi/2)}.$$ (9.33)

It is noted that the inverse filter function itself is not only a bipolar but also an infinite poles function. Thus the filter is not physically realizable. However, if we are willing to sacrifice some of the resolution, then an approximate filter function may be realized. In order to do so, we will combine the amplitude filter of fig. 9.6 with the independent phase filter of fig. 9.7. The transfer function of this combination is

$$H(p) = A(p)\, e^{i\phi(p)}.$$ (9.34)

If this approximated inverse filter is inserted in the spatial frequency plane of the data processor of fig. 9.4, the restored Fourier transfer function will be

$$F_1(p) = F(p)\, H(p).$$ (9.35)

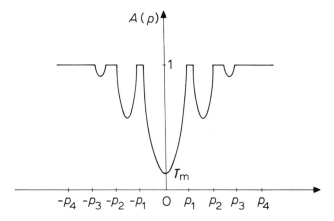

Fig. 9.6 Amplitude filter function.

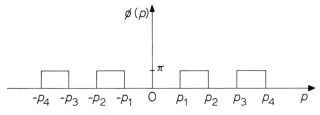

Fig. 9.7 Phase filter function.

If we let T_m be the minimum transmittance of the amplitude filter, then the restored Fourier spectrum of the point image is the shaded spectrum shown in fig. 9.5. We can now define the relative degree of image restoration (ref. 9.30):

$$\mathscr{D}(T_m) \text{ (percent)} = \frac{1}{T_m \Delta p} \int_{\Delta p} \frac{F(p) H(p)}{\Delta \xi} \, dp \times 100, \tag{9.36}$$

where Δp is the spatial bandwidth of interest. In fig. 9.5, for example, $\Delta p = 2p_4$. From eq. (9.36) we can plot degree of image restoration as a function of T_m (fig. 9.8). We see that perfect restoration is approached as T_m approaches zero. However, at the same time the restored Fourier spectrum is also vanishing, and so no image will be reconstructed. Thus perfect restoration cannot be achieved in practice. These considerations aside, it seems that noise (film granularity and speckling) is the major limiting factor in the image restoration. To achieve a high degree of restoration, a lower transmittance T_m is required, and the restored Fourier spectrum is therefore weaker. In turn, a lower signal-to-noise ratio of the restored image will result. Therefore, considering the noise problem, it

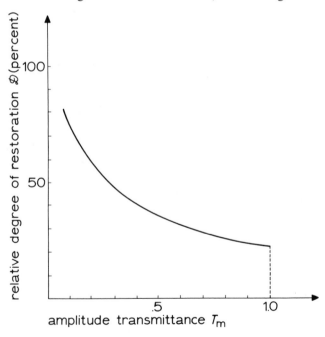

Fig. 9.8 Relative degree of image restoration as a function of T_m, for linear image motion.

is clear that our optimum value T_m must be obtained, at least in practice, for optimum image restoration.

As was mentioned earlier in this section, a phase filter may be synthesized by means of a holographic technique (perhaps by a computer-generated hologram). We will now see how such a phase filter works in image restoration. Let us assume that the transmittance of a holographic phase filter is

$$T(p) = \tfrac{1}{2}\{1 + \cos[\phi(p) + \alpha_0 p]\}, \tag{9.37}$$

where α_0 is an arbitrarily chosen constant, and

$$\phi(p) = \begin{cases} \pi, & p_n \leq p \leq p_{n+1}, \ n = \pm 1, \ \pm 3, \ \pm 5, \dots \\ 0, & \text{otherwise} \end{cases}. \tag{9.38}$$

With the amplitude filter added, the complex filter function can be written as

$$H_1(p) = A(p) \, T(p) = \tfrac{1}{2} A(p) + \tfrac{1}{4}[H(p) \exp(i\alpha_0 p) + H^*(p) \exp(-i\alpha_0 p)], \tag{9.39}$$

where $H(p)$ is the approximate inverse filter function of eq. (9.34). Note also that $H(p) = H^*(p)$, because of eq. (9.38).

Now, if this complex filter $H_1(p)$ is inserted in the spatial frequency plane P_2 of fig. 9.4, then the complex light field behind P_2 is

$$F_2(p) = \tfrac{1}{2} F(p) \, A(p) + \tfrac{1}{4}[F(p) \, H(p) \exp(i\alpha_0 p) + F(p) \, H^*(p) \exp(-i\alpha_0 p)]. \tag{9.40}$$

It is clear that the first term of eq. (9.40) is the restored Fourier spectrum due to the amplitude filter alone, which will be diffracted on the optical axis at the output plane P_3 of fig. 9.4. The second and third terms are the restored Fourier spectra of the smeared image, in which the restored images due to these terms will be diffracted away from the optical axis at the output plane, and centered at $\alpha = \alpha_0$ and $\alpha = -\alpha_0$ respectively. As an illustration, fig. 9.9 shows the calculated irradiance of a restored point image blurred by linear motion, after transmission through an amplitude filter ($T_m = 0.1$), a phase filter ($p \leq p_4$), and the combination of the amplitude and phase filters. The results of an experiment using such filters are shown in fig. 9.10. It may be worth pointing out that the restoration of defocused images can be accomplished by a similar procedure, except that the complex spatial filter in this latter case must have rotational symmetry.

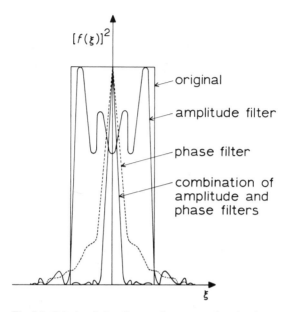

Fig. 9.9 Calculated irradiance of a restored point image blurred by linear motion. (Permission by J. Tsujiuchi.)

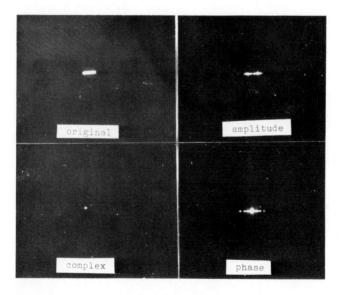

Fig. 9.10 Restored images. (Permission by J. Tsujiuchi.)

It should be emphasized that the relative degree of restoration is with respect to the spatial bandwidth of interest. It is clear that the ultimate limit of Δp is restricted by the diffraction limits of the optical imaging and processing systems, whichever comes first (ref. 9.31). Therefore, it does not follow that a high degree of restoration implies restoration beyond the diffraction limit.

Problems

9.1 Equations (9.18) and (9.20), when applied to the modulating camera described in sec. 9.1, show that the real restored image is interfered with by the zero-order and the divergent diffractions. Design a truncated modulated function based on eq. (9.18) such that the restored point image does not suffer from this unwanted interference.

9.2 Repeating prob. 9.1, and designing for an image of diameter d, determine the truncation of eq. (9.18) required to completely separate the restored from the dc and divergent diffractions.

9.3 An image of stationary irradiance from a distant object is projected on the recording medium of a camera. The object moves at a variable speed during the time the shutter is open, so that the image is severely smeared on the recording medium. The transmission function of the recorded film is given by

$$T(x, y) = A(x - x', y) \exp(-x^2/2), \quad -2 \le x' \le 2,$$

where $A(x, y)$ is the unsmeared image of the object. Design a coherent optical processor and an appropriate inverse spatial filter to enhance the image.

9.4 If the object of prob. 9.3 is moving at a constant acceleration, the transmission function of the recorded film may be written,

$$T(x, y) = A(x - x', y)\left(1 - \frac{x}{2a}\right), \quad 0 \le x' \le a,$$

where a is an arbitrary constant. Synthesize a complex inverse filter to enhance this image.

9.5 Consider the transmission function of a linear smeared image to be

$$T(x, y) = \begin{cases} A(x - x', y), & -\dfrac{\Delta x}{2} \le x' \le \dfrac{\Delta x}{2} \\ 0, & \text{otherwise} \end{cases},$$

where $A(x, y)$ is the unsmeared image, and Δx is the corresponding smear distance. For some reason, the minimum transmittance T_m of the filter is not to go below 25%. Using the techniques presented in sec. 9.4, (a) design an optimum complex spatial filter, and (b) compute the relative degree of image restoration.

References

9.1 R. M. Fano, *Transmission of Information*, MIT Press, Cambridge, Mass., 1961.

9.2 L. Brillouin, *Science and Information Theory*, Academic Press, New York, 1956.

9.3 N. Wiener, *Cybernetics*, Technology Press and Wiley, New York, 1948; 2nd MIT Press edition, MIT Press, Cambridge, Mass., 1965.

9.4 N. Wiener, *Extrapolation, Interpolation, and Smoothing of Stationary Time Series*, Technology Press and Wiley, New York, 1949; MIT Press paperback edition, MIT Press, Cambridge, Mass., 1964.

9.5 C. E. Shannon and W. Weaver, *The Mathematical Theory of Communication*, University of Illinois Press, Urbana, Ill., 1962.

9.6 J. Tsujiuchi, Correction of Optical Images by Compensation of Aberrations and Spatial Frequency Filtering, in E. Wolf (ed.), *Progress in Optics*, vol. II, North-Holland Publishing Company, Amsterdam, 1963.

9.7 J. L. Horner, Optical Spatial Filtering with the Least Mean-Square-Error Filter, *J. Opt. Soc. Am.*, **59** (1969), 553.

9.8 G. W. Stroke, F. Furrer, and D. R. Lamberty, Deblurring of Motion-Blurred Photographs Using Extended-Range Holographic Fourier-Transform Division, *Opt. Commun.*, **1** (1969), 141.

9.9 F. T. S. Yu, Image Restoration, Uncertainty, and Information, *J. Opt. Soc. Am.*, **58** (1968), 742; *Appl. Opt.* **8** (1969), 53.

9.10 R. V. Churchill, *Fourier Series and Boundary Value Problems*, McGraw-Hill, New York, 1941.

9.11 O. Bryngdahl and A. Lohman, Holographic Compensation of Motion Blur by Shutter Modulation, *J. Opt. Soc. Am.*, **59** (1969), 1175.

9.12 O. Bryngdahl, Holographic Encoding with Completely Incoherent Light, *J. Opt. Soc. Am.*, **60** (1970), 510.

9.13 M. Born and E. Wolf, *Principles of Optics*, 2nd ed., Pergamon Press, New York, 1964.

9.14 J. L. Synge, *Relativity*, North-Holland Publishing Company, Amsterdam, 1955.

9.15 J. L. Powell and B. Crasemann, *Quantum Mechanics*, Addison-Wesley, Reading, Mass., 1961.

9.16 L. Brillouin, *Science and Information Theory*, 2nd ed., Academic Press, New York, 1962.

9.17 H. S. Coleman and M. F. Coleman, Theoretical Resolution Angles for Point and Line Test Objects in the Presence of a Luminous Background, *J. Opt. Soc. Am.*, **37** (1947), 572.

9.18 G. Toraldo di Francia, Resolving Power and Information, *J. Opt. Soc. Am.*, **45** (1955), 497.

9.19 V. Ronchi, *Optics, The Science of Vision*, New York University Press, New York, 1957.

9.20 V. Ronchi, Resolving Power of Calculated and Detected Images, *J. Opt. Soc. Am.*, **51** (1961), 458.

9.21 J. L. Harris, Diffraction and Resolving Power, *J. Opt. Soc. Am.*, **54** (1964), 931.

9.22 F. T. S. Yu, Optical Resolving Power and Physical Realizability, *J. Opt. Soc. Am.*, **59** (1969), 497, and *Opt. Commun.*, **1** (1970), 319.

9.23 E. T. Whittaker, and G. N. Watson, *A Course of Modern Analysis*, 4th ed., Cambridge University Press, New York, 1940.

9.24 E. A. Guilleman, *The Mathematics of Circuit Analysis*, Wiley, New York, 1951.

9.25 J. F. Steffesen, *Interpolation*, Chelsea Publishing Company, New York, 1950.

9.26 R. E. A. Paley and N. Wiener, Fourier Transform in the Complex Domain, *Am. Math. Soc. Colloq.*, **19** (1934), 16.

9.27 T. M. Halladay and J. D. Gallatin, Phase Control by Polarization in Coherent Spatial Filtering, *J. Opt. Soc. Am.*, **56** (1966), 869.

9.28 G. W. Stroke and R. G. Zech, A Posteriori Image-Correcting Deconvolution by Holographic Fourier-Transform Division, *Phys. Letters*, ser. A, **25** (1967), 89.

9.29 A. W. Lohmann and D. P. Paris, Computer Generated Spatial Filters for Coherent Optical Data Processing, *Appl. Opt.*, **7** (1968), 651.

9.30 J. Tsujiuchi, T. Honda, and T. Fukaya, Restoration of Blurred Photographic Images by Holography, *Opt. Commun.* **1** (1970), 379.

9.31 F. T. S. Yu, Coherent and Digital Image Enhancement, Their Basic Differences and Constraints, *Opt. Commun.* **3** (1971), 440.

Part  **Holography**

Chapter 10 Introduction to Linear Holography

The theory of wavefront reconstruction was introduced by D. Gabor in 1948 (ref. 10.1), who further developed it in a series of classic articles (refs. 10.2, 10.3). At the time he encountered two difficulties. The main difficulty was that a high-intensity coherent source suitable for wavefront recording was not available; the second difficulty was that the virtual and real images could not be separated. Nevertheless, he set down the basic foundation for modern three-dimensional photography or holography. As a matter of fact, the word "holography" was first applied to this process by Gabor. The word is derived from the combination of two Greek words: "holos," meaning "whole," and "graphein," meaning "to write." * Thus, holography means a complete writing (i.e., recording).

This new imaging technique was at first received with only mild interest. In the 1950s, a number of investigators, including G. L. Rogers (ref. 10.4), H. M. A. El-Sum (ref. 10.5), and A. Lohmann (ref. 10.6), significantly extended the theory and understanding of this new imaging technique. With the invention of the laser, coherent light sources of adequate intensity were at last available. The difficulty of separating the overlapping real and virtual images was overcome by Leith and Upatnieks, who added a high-spatial-frequency carrier to the recording wavefront (refs. 10.7–10.9). Since then numerous articles on holography and its engineering applications have been published. The purpose of this chapter is to approach holography from an elementary engineering point of view. That is to say, we will employ impulse-exitation and other concepts from system theory. It is well known that optical instruments exhibit similarities with some electrical systems. It follows that holography may be approached from an input-output system analog.

In the following sections, wavefront construction and reconstruction will be demonstrated, and holographic magnifications, resolution limits, and bandwidth requirements will be calculated. Third order holographic aberrations, spatially incoherent holography, and color holography will also be discussed.

10.1 Wavefront Construction and Reconstruction

In this section, on- and off-axis wavefront constructions and reconstructions of a simple point object will be demonstrated, extension toward a more complicated object will be given, and a system analog of holographic recording and reconstruction will be illustrated.

---

* The same term had long been applied to documents written wholly by hand.

To show the wavefront construction of a point object, let a monochromatic point radiator be located at a distance R away from a photographic plate, as shown in fig. 10.1. The complex light amplitude at a distance r from the radiator is

$$u = \frac{A}{r} \exp[i(kr - \omega t)], \tag{10.1}$$

where A is a complex constant, k is the wave number, and ω is the radian frequency of the light source. Let the *reference wave*, a monochromatic plane wavefront* of the same frequency, travel perpendicular to the recording medium. The complex light amplitude of the reference wave is given by

$$v = B \exp\{i[k(R + z) - \omega t]\}, \tag{10.2}$$

where B is again a complex constant. At this point, and without loss of generality, the time-varying factor $\exp(i\omega t)$ will be dropped from the calculation.

At the surface of the recording medium ($z = 0$), eqs. (10.1) and (10.2) become

$$u(x, y) = \frac{A}{(R^2 + \rho^2)^{1/2}} \exp[ik(R^2 + \rho^2)^{1/2}] \tag{10.3}$$

and

$$v = B \exp(ikR), \tag{10.4}$$

where $\rho^2 = x^2 + y^2$. The resultant complex light distribution on the plate due to these two wavefronts is thus

$$u(x, y) + v = \frac{A}{(R^2 + \rho^2)^{1/2}} \exp[ik(R^2 + \rho^2)^{1/2}] + B \exp(ikR). \tag{10.5}$$

However, if the separation R is large compared to the aperture of the recording medium, then r can be replaced by the paraxial approximation

$$r \simeq R + \frac{\rho^2}{2R} \tag{10.6}$$

in the exponent, and by R in the denominator of eq. (10.5). Thus eq. (10.5) becomes

$$u(x, y) + v = \exp(ikR)\left[\frac{A}{R} \exp\left(ik\frac{\rho^2}{2R}\right) + B\right]. \tag{10.7}$$

---

* As will be seen, a *plane* reference wave is not required; a spherical wave would serve as well. We will use a plane wave here for simplicity.

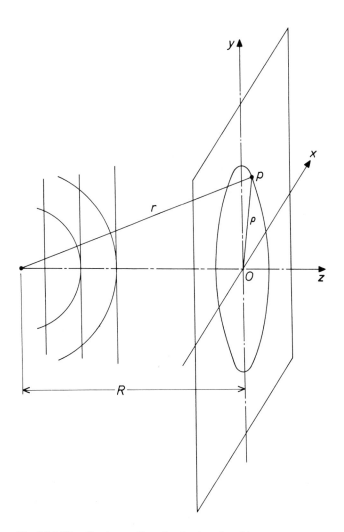

Fig. 10.1 Wavefront recording of a simple point object.

The corresponding irradiance is therefore

$$
I(x, y) = [u(x, y) + v] [u(x, y) + v]^*
$$
$$
= \left(\frac{|A|}{R}\right)^2 + |B|^2 + \frac{2|A| |B|}{R} \cos\left(\frac{k\rho^2}{2R} + \phi\right), \tag{10.8}
$$

where ϕ is the phase angle between the complex amplitudes A and B.

The exposure during recording (i.e., encoding), can be assumed to be proportional to the I of eq. (10.8). Indeed, eq. (10.8) can be recognized as describing a *Fresnel zone lens* construction (ref. 10.10). Moreover, if the wavefront construction is properly recorded in the linear region of the *T-E* characteristic of the emulsion (sec. 8.1), then the transmittance of the recorded hologram will be

$$
T(\boldsymbol{\rho}; k) = K_1 + K_2 \cos\left(\frac{k\rho^2}{2R} + \phi\right), \tag{10.9}
$$

where K_1 and K_2 are proportionality constants.

If the point-object hologram of eq. (10.9) is illuminated (i.e., decoded) by a *normally incident* monochromatic plane wave of the same wavelength λ (fig. 10.2), then by the Fresnel-Kirchhoff theory or Huygens' principle (app. B), the complex light distribution behind the hologram can be determined from the convolution theorem,

$$
E(\boldsymbol{\sigma}; k) = B \iint\limits_{S} T(\boldsymbol{\rho}; k)\, E_l^+(\boldsymbol{\sigma} - \boldsymbol{\rho}; k)\, dx\, dy, \tag{10.10}
$$

where

$$
E_l^+(\boldsymbol{\rho}; k) = -\frac{i}{\lambda l} \exp\left[ik\left(l + z + \frac{\rho^2}{2l}\right)\right]
$$

is the free-space impulse response (as derived in app. B), S denotes integration over the entire hologram surface, $\boldsymbol{\rho}(x, y, z)$ is the coordinate system at the hologram, and $\boldsymbol{\sigma}(\alpha, \beta, \gamma)$ is a separate system at a distance l from the first.

If we evaluate eq. (10.10) at a distance $l = R$ behind the hologram, then the solution is

$$
E(\boldsymbol{\sigma}; k) = C_1 + C_2 \exp\left(i\frac{k}{4R}\sigma^2\right) + C_3 \delta(\alpha, \beta), \tag{10.11}
$$

where C_1, C_2, and C_3 are the appropriate complex constants, $\sigma^2 = \alpha^2 + \beta^2$, and $\delta(\alpha, \beta)$ is the two-dimensional Dirac delta function. The three terms of eq. (10.11) may be interpreted as follows: C_1 represents the zero order (i.e., dc) diffraction, the second term is the first order virtual image diffraction, and the third term is the first order real image.

As shown in fig. 10.2, all these diffractions are overlapping; thus spurious distortions are introduced into the reconstructed image. Overlapping will occur even if the hologram is illuminated by an oblique plane wave. For such oblique illumination, the complex light distribution is given by

$$E(\sigma; k) = B \iint_S T(\rho; k) \, e^{ikx \sin\theta} E_l^+(\sigma - \rho; k) \, dx \, dy. \tag{10.12}$$

By substituting eq. (10.9) in this, the solution for $l = R$ is seen to be

$$E(\sigma; k) = C_1' + C_2' \exp\left[i \frac{k}{2R} (\sigma + R \sin\theta)^2 + \beta^2\right] + C_3' \delta(\alpha - R \sin\theta, \beta), \tag{10.13}$$

where C_1', C_2', and C_3' are the appropriate complex constants. The last two terms can be recognized again as virtual and real images, as shown in fig. 10.3. Apparently, oblique illumination is not able to separate the diffractions.

In order to make the diffractions separable, an oblique reference wave may be used during recording, as shown in fig. 10.4. The complex light amplitude distribution on the surface of the recording medium due to the reference wave is

$$v(x) = B \exp[ik(R + x \sin\theta)]. \tag{10.14}$$

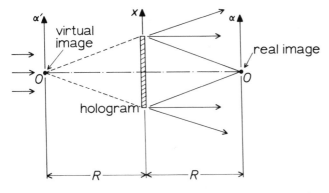

Fig. 10.2 Wavefront reconstruction of a point object. Plane monochromatic illumination.

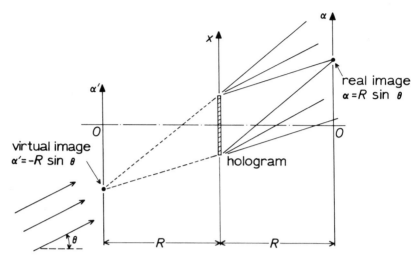

Fig. 10.3 Oblique-wavefront reconstruction of a point-object hologram. Plane monochromatic illumination.

By the usual paraxial approximation, the resultant complex light distribution on the recording medium is

$$u(x, y) + v(x) = \frac{A}{R} \exp\left[ik\left(R + \frac{\rho^2}{2R}\right)\right] + B \exp\left[ik(R + x \sin\theta)\right]. \quad (10.15)$$

The corresponding irradiance is

$$I(x, y) = [u(x, y) + v(x)]\,[u(x, y) + v(x)]^*$$
$$= |A_1|^2 + |B|^2 + 2|A_1|\,|B| \cos\left[k\left(\frac{\rho^2}{2R} - x \sin\theta\right) + \phi\right], \quad (10.16)$$

where $A_1 = A/R$, $\rho^2 = x^2 + y^2$, and ϕ is the phase angle between the complex amplitudes A and B. Again, if we assume the wavefront recording is linear, the transmittance of the off-axis hologram is

$$T(\rho; k) = K_1 + K_2 \cos\left[k\left(\frac{\rho^2}{2R} - x \sin\theta\right) + \phi\right], \quad (10.17)$$

where K_1 and K_2 are proportionality constants.

If this off-axis hologram is illuminated by a plane wave of wavelength λ and obliquity $-\theta$ (fig. 10.5), then the complex light amplitude distribution

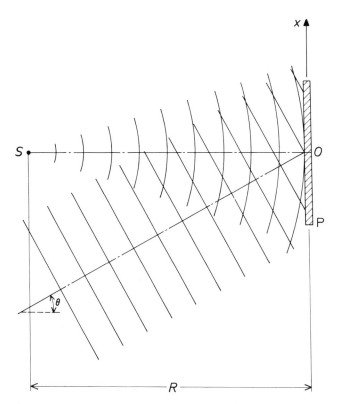

Fig. 10.4 Spatial-carrier wavefront recording of a point object. Oblique plane reference wave; S, monochromatic point source (object); P, photographic plate.

behind the hologram is

$$E(\boldsymbol{\sigma}; k) = B \iint_{S} T(\boldsymbol{\rho}; k) \exp(-kx \sin\theta)\, E_l^+(\boldsymbol{\sigma} - \boldsymbol{\rho}; k)\, dx\, dy. \tag{10.18}$$

(Note: The readout angle could be any other than $-\theta$, which we use here for simplicity.)

Again the wavefront reconstruction at distance $l = R$ behind the hologram is

$$E(\boldsymbol{\sigma}; k) = C_1 \exp(-ik\alpha \sin\theta) + C_2 \exp\left\{i\frac{k}{4R}\left[(\alpha - 2R\sin\theta)^2 + \beta^2\right]\right\} + C_3 \delta(\alpha, \beta), \tag{10.19}$$

where C_1, C_2, and C_3 are the appropriate complex constants. Figure 10.5

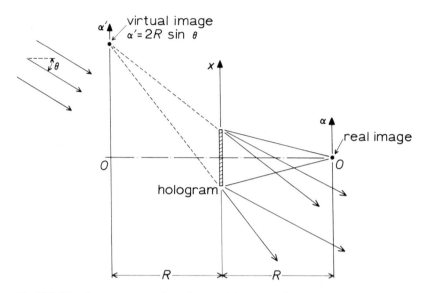

Fig. 10.5 Wavefront reconstruction of a spatial-carrier point-object hologram. Monochromatic plane wave, negative θ.

shows the real and virtual image reconstructions. It is clear that for a properly chosen angle of incidence of the reference wave, the real image may be separated from the zero order and the virtual image diffractions.

We may also let the angle of incidence of the reference wave be positive, as shown in fig. 10.6. Again, the complex light distribution at $l = R$ can be written as

$$E(\boldsymbol{\sigma}; k) = C_1' \exp(ik\alpha \sin\theta) + C_2' \exp\left(i\frac{k}{4R}\sigma^2\right) + C_3'\delta(\alpha - 2R\sin\theta, \beta),$$

$$(10.20)$$

where C_1', C_2' and C_3' are the appropriate complex constants. From fig. 10.6, we see that the virtual image diffraction is again separated from the zero order and the real image diffractions.

The complex light distribution from an extended object (fig. 10.7) may be obtained from the convolution theorem, and is

$$u(x, y) = \int\int_{S_o} O(\xi, \eta, \zeta)\, E_l^+(\boldsymbol{\rho} - \boldsymbol{\xi}; k)\, d\xi\, d\eta, \qquad (10.21)$$

where $O(\xi, \eta, \zeta)$ is the object function, and S_o denotes the surface integral over the object, viewed from the hologram aperture.

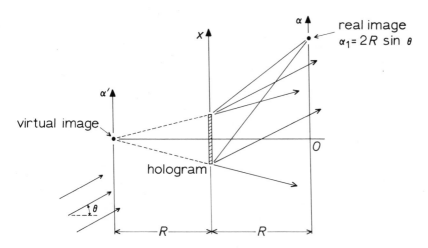

Fig. 10.6 Wavefront reconstruction of a spatial-carrier point-object hologram. Monochromatic plane wave, positive θ.

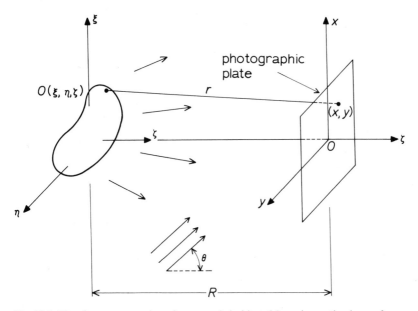

Fig. 10.7 Wavefront construction of an extended object. Monochromatic plane reference wave.

Again, under the assumption of the spatial linearity of the recording medium, the transmittance of the recorded hologram can be shown to be

$$T(\rho; k) = K[u(x, y) + v(x)] [u(x, y) + v(x)]^*$$
$$= K[|u(x, y)|^2 + B^2 + 2|u(x, y)| B^2 \cos[\phi(x, y) - kx \sin\theta], \quad (10.22)$$

where K is a proportionality constant, $v(x)$ is the oblique reference wave given in eq. (10.14), and $u(x, y) = |u(x, y)| \exp[i\phi(x, y)]$. Alternatively, eq. (10.22) could be written as

$$T(\rho; k) = K[|u(x, y)|^2 + B^2 + Bu(x, y) \exp(-ikx \sin\theta) + Bu^*(x, y)$$
$$\times \exp(ikx \sin\theta)]. \quad (10.23)$$

If the hologram is obliquely illuminated by the reference wave, as shown in fig. 10.8, then the complex light distribution behind the hologram can be written as eq. (10.18). If we consider only the term giving the real image diffraction [i.e., the last term of eq. (10.23)], then we have

$$E_r(\sigma; k) = C \exp\left(\frac{k}{2l'} \sigma^2\right) \int\int_S \left\{\int\int_{S_o} O^*(\xi, \eta, \zeta) \exp\left[-i\frac{k}{2l}(\xi^2 + \eta^2)\right]\right.$$
$$\times \exp\left[i\frac{k}{l}(\xi x + \eta y)\right] d\xi \, d\eta\left\}\left[i\frac{k}{2}\rho^2\left(\frac{1}{l'} - \frac{1}{l}\right)\right]$$
$$\times \exp\left[-i\frac{k}{l'}(\alpha x + \beta y)\right] dx \, dy, \quad (10.24)$$

where C is an appropriate complex constant, the subscript r denotes the real image diffraction, and $l' = R + \gamma$ is the separation between the holographic plate and the image coordinate $\sigma(\alpha, \beta, \gamma)$. From eq. (10.24) the real hologram image can be shown reconstructed uniquely at $l = l'$ (i.e., $R = R'$ and $\zeta = -\gamma$), as pictured in fig. 10.8. Thus,

$$E_r(\sigma; k) = CO^*(\alpha, \beta, \gamma), \text{ for } l = l'. \quad (10.25)$$

Similarly, if the hologram is illuminated at a positive oblique angle the virtual hologram image can be shown constructed uniquely at $l = l'$ (i.e., $R = -R'$, and $\zeta = \gamma'$), as shown in fig. 10.9. That is,

$$E_v(\sigma; k) = CO(\alpha', \beta', \gamma'), \text{ for } l = l. \quad (10.26)$$

Photographs of virtual and real hologram images are pictured in figs. 10.10 and 10.11, respectively. The transmittance (10.22) may be written as

$$T(\rho; k) = K[|u(x, y)|^2 + |v(x)|^2 + u(x, y) v^*(x) + u^*(x, y) v(x)], \quad (10.27)$$

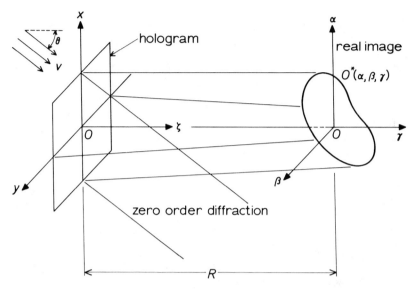

Fig. 10.8 Wavefront reconstruction of an extended object. Oblique plane monochromatic illumination, negative θ.

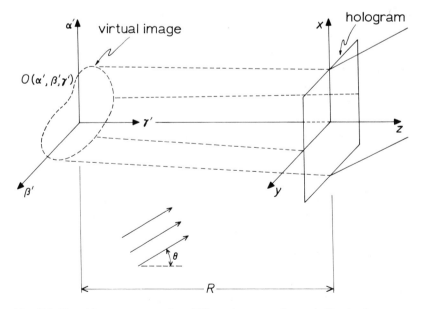

Fig. 10.9 Virtual image reconstruction. Oblique plane monochromatic illumination.

Fig. 10.10 A virtual three-dimensional hologram image.

where

$$u^*(x, y) = \int\int\limits_{S_o} O^*(\xi, \eta, \zeta) \, E_l^{+*}(\boldsymbol{\rho} - \boldsymbol{\xi}; k) \, d\xi \, d\eta$$

and

$$v^*(x) = e^{-ikx \sin\theta}.$$

The Fourier transform of $u(x, y)$ is therefore

$$U(p, q) = O(p, q) \, E_l(p, q), \tag{10.28}$$

where p and q are the spatial frequencies, and $U(p, q)$, $O(p, q)$, and $E_l(p, q)$ are the respective Fourier transforms of $u(x, y)$, $O(x, y)$, and $E_l^+(x, y)$.

Similarly, we can write the Fourier transform of $u^*(x, y)$ as

$$U^*(p, q) = O^*(p, q) \, E_l^*(p, q). \tag{10.29}$$

If the hologram of eq. (10.27) is illuminated by an oblique plane wave

Fig. 10.11 A real three-dimensional hologram image.

of negative θ, then the complex light amplitude at the immediate opposite side of the hologram can be written as

$$v^*(x) \, T(\boldsymbol{\rho}; k) = K \{ v^*(x) \, [|u(x, y)|^2 + |v(x)|^2] + u(x, y) \, [v^*(x)]^2 \\ + u^*(x, y) \, |v(x)|^2 \}, \tag{10.30}$$

where $|v(x)| = B$, a constant.

Thus the Fourier transform of eq. (10.30) may be written as

$$\mathscr{F} [v^*(x) \, T(\boldsymbol{\rho}; k)] = K\mathscr{F} \{ v^*(x) \, [|u(x, y)|^2 + |v(x)|^2] + u(x, y) \, [v^*(x)]^2 \} \\ + KB^2 O^*(p, q) \, E_l^*(p, q), \tag{10.31}$$

where $\mathscr{F}$ denotes the Fourier transform.

Since the complex light diffraction behind the hologram can be determined by the Fresnel-Kirchhoff theory, as shown in eq. (10.18), the Fourier transform of the diffraction can be written as

$$E(p, q) = \mathscr{F} [v^*(x) \, T(\boldsymbol{\rho}; k)] \, E_l(p, q), \tag{10.32}$$

where $E(p, q)$ is the Fourier transform of $E(\boldsymbol{\sigma}; k)$.

If we substitute eq. (10.31) in eq. (10.32), we have

$$E(p, q) = K\mathscr{F}\{v^*(x)\,[|u(x, y)|^2 + |v(x)|^2] + u(x, y)\,[v^*(x)]^2\}\,E_l(p, q)$$
$$+ KB^2 O^*(p, q). \tag{10.33}$$

Note that the last term of this equation is proportional to $O^*(p, q)$, the Fourier transform of the conjugate of the object function.

Similarly, if the illumination is by a wave of positive θ, then the Fourier transform of the complex light diffraction behind the hologram can be shown to be

$$E(p, q) = K\mathscr{F}\{v(x)\,[|u(x, y)|^2 + |v(x)|^2] + u^*(x, y)\,v^2(x)\} + KB^2 O(p, q). \tag{10.34}$$

The last term of this equation represents the virtual hologram image diffraction, which is proportional to the Fourier transform of the object function.

System-analog diagrams representing wavefront recording and reconstruction are given in figs. 10.12 and 10.13, respectively.

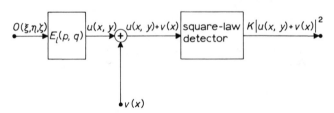

Fig. 10.12 System analog of wavefront recording.

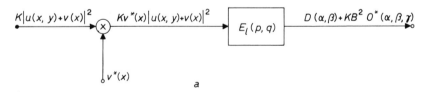

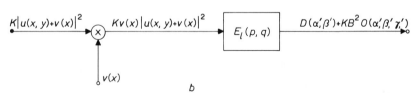

Fig. 10.13 System analogs of wavefront reconstructions: a, for real image, with $D(\alpha, \beta)$ a complex function; b, for virtual image, with $D(\alpha', \beta')$ a complex function.

10.2 Holographic Magnifications

Wavefront reconstructions are generally three-dimensional in nature. The lateral and longitudinal magnifications in wavefront reconstruction may be discussed separately.

LATERAL MAGNIFICATIONS

To obtain the lateral holographic magnifications, we may start from a wavefront recording of the three monochromatic point radiators of wavelength λ_1 (fig. 10.14). If the paraxial approximation (eq. 10.6) holds, then the complex light distributions from the three radiators are

$$u_1(\boldsymbol{\rho}; k_1) \simeq A_1 \exp\left(ik_1\left\{R_1 + \frac{1}{2R_1}\left[\left(x - \frac{h}{2}\right)^2 + y^2\right]\right\}\right),$$

$$u_2(\boldsymbol{\rho}; k_1) \simeq A_2 \exp\left(ik_1\left\{R_1 + \frac{1}{2R_1}\left[\left(x + \frac{h}{2}\right)^2 + y^2\right]\right\}\right), \qquad (10.35)$$

$$u_3(\boldsymbol{\rho}; k_1) \simeq A_3 \exp\left(ik_1\left\{L_1 + \frac{1}{2L_1}\left[(x + a)^2 + y^2\right]\right\}\right),$$

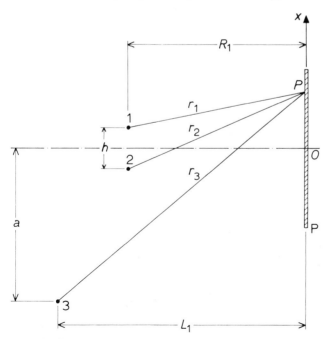

Fig. 10.14 Recording geometry for determining the lateral magnifications. 1, 2, monochromatic point sources; 3, divergent monochromatic reference source; P, photographic plate.

where $k_1 = 2\pi/\lambda_1$, and A_1, A_2, and A_3 are real constants. The corresponding irradiance is

$$I(\boldsymbol{\rho}; k_1) = (u_1 + u_2 + u_3)(u_1 + u_2 + u_3)^* = A_1^2 + A_2^2 + A_3^2 + 2A_1 A_2 \cos \frac{k_1}{R_1} hx$$

$$+ 2A_1 A_3 \cos \left\{ k_1(R_1 - L_1) + \frac{k_1}{2R_1} \left[\left(x - \frac{h}{2} \right)^2 + y^2 \right] \right.$$

$$\left. - \frac{k_1}{2L_1} [(x+a)^2 + y^2] \right\}$$

$$+ 2A_2 A_3 \cos \left\{ k_1(R_1 - L_1) + \frac{L_1}{2R_1} \left[\left(x + \frac{h}{2} \right)^2 + y^2 \right] \right.$$

$$\left. - \frac{k_1}{2L_1} [(x+a)^2 + y^2] \right\}. \tag{10.36}$$

The above equation represents two overlapping Fresnel zone lens constructions. Again, if the wavefront recording is linear, then the transmittance of the hologram is

$$T(\boldsymbol{\rho}; k_1) = K_0 + K_1 \cos \frac{k_1}{R_1} hx + K_2 [e^{i\{*\}} + e^{-i\{*\}}] + K_3 [e^{i\{**\}} + e^{-i\{**\}}], \tag{10.37}$$

where the K's are real proportionality constants, and

$$\{*\} = \left\{ k_1(R_1 - L_1) + \frac{k_1}{2R_1} \left[\left(x - \frac{h}{2} \right)^2 + y^2 \right] - \frac{k_1}{2L_1} [(x+a)^2 + y^2] \right\},$$

$$\{**\} = \left\{ k_1(R_1 - L_1) + \frac{k_1}{2R_1} \left[\left(x + \frac{h}{2} \right)^2 + y^2 \right] - \frac{k_1}{2L_1} [(x+a)^2 + y^2] \right\}.$$

If this hologram is illuminated by divergent light of wavelength λ_2, as shown in fig. 10.15,

$$u_4(\boldsymbol{\rho}; k_2) = A_4 \exp \left\{ ik \left[L_2 + \frac{1}{2L_2} [(x-b)^2 + y^2] \right] \right\}, \tag{10.38}$$

then the complex light distribution behind the hologram is

$$E(\boldsymbol{\sigma}; k_2) = \int\int_S T(\boldsymbol{\rho}; k_1) u_4(\boldsymbol{\rho}; k_2) E_l^+(\boldsymbol{\sigma} - \boldsymbol{\rho}; k_2) \, dx \, dy. \tag{10.39}$$

Since the third and fifth terms of eq. (10.37) contribute to the virtual image diffractions, and the fourth and sixth terms correspond to the real image reconstructions, the evaluation of eq. (10.39) can be performed termwise with respect to the real and virtual image reconstructions. Thus for the reconstruction of the real images we have,

$$E_r(\boldsymbol{\sigma}; k_2) = \int\int_S [K_2 e^{-i\{*\}} + K_3 e^{-i\{**\}}]\, u_4(\boldsymbol{\rho}; k_2)\, E_l^+(\boldsymbol{\rho}-\boldsymbol{\sigma}; k_2)\, dx\, dy,$$

$$(10.40)$$

where the subscript r denotes the real image.

By substitution, the above equation can be written as

$$
\begin{aligned}
E_r(\boldsymbol{\sigma}; k_2) = C_1 \int\int_S \exp\Bigg\{ &-i\frac{k_2}{2}\Bigg[\left(\frac{\lambda_2}{\lambda_1 R_1} - \frac{\lambda_2}{\lambda_1 L_1} - \frac{1}{L_2} - \frac{1}{l}\right)\rho^2 \\
&+ \frac{2}{l}\left(\alpha + l\left(\frac{b}{L_2} - \frac{\lambda_2 h}{2\lambda_1 R_1} - \frac{\lambda_2 a}{\lambda_1 L_1}\right)\right)x + \frac{2\beta}{l}\,y\Bigg]\Bigg\} dx\, dy \\
+ C_2 \int\int_S \exp\Bigg\{ &-i\frac{k_2}{2}\Bigg[\left(\frac{\lambda_2}{\lambda_1 R_1} - \frac{\lambda_2}{\lambda_1 L_1} - \frac{1}{L_2} - \frac{1}{l}\right)\rho^2 \\
&+ \frac{2}{l}\left(\alpha + l\left(\frac{b}{L_2} + \frac{\lambda_2 h}{2\lambda_1 R_1} - \frac{\lambda_2 a}{\lambda_1 L_1}\right)\right)x + \frac{2\beta}{l}\,y\Bigg]\Bigg\} dx\, dy,
\end{aligned}
$$

$$(10.41)$$

where C_1 and C_2 are the appropriate complex constants. From this equation, it is clear that the real images will be uniquely reconstructed at

$$l = \frac{\lambda_1 R_1 L_1 L_2}{\lambda_2 L_1 L_2 - \lambda_2 R_1 L_2 - \lambda_1 R_1 L_1}.$$

$$(10.42)$$

The solution of eq. (10.41), evaluated at a distance behind the hologram given by (10.42), is

$$
\begin{aligned}
E_r(\boldsymbol{\sigma}; k_2) = C_1' \delta\Bigg[&\alpha + l\left(\frac{b}{L_2} - \frac{\lambda_2 h}{2\lambda_1 R_1} - \frac{\lambda_2 a}{\lambda_1 L_1}\right), \beta\Bigg] \\
+ C_2' \delta\Bigg[&\alpha + l\left(\frac{b}{L_2} + \frac{\lambda_2 h}{2\lambda_1 R_1} - \frac{\lambda_2 a}{\lambda_1 L_1}\right), \beta\Bigg],
\end{aligned}
$$

$$(10.43)$$

where C_1' and C_2' are the appropriate complex constants, and δ is the Dirac delta function. From the construction of figs. 10.14 and 10.15, the lateral magnification of the real image can be shown to be (refs. 10.11–10.14)

$$M_{\text{lat}}^r = \frac{h_r}{h} = \left(1 - \frac{\lambda_1 R_1}{\lambda_2 L_2} - \frac{R_1}{L_1}\right)^{-1}.$$

$$(10.44)$$

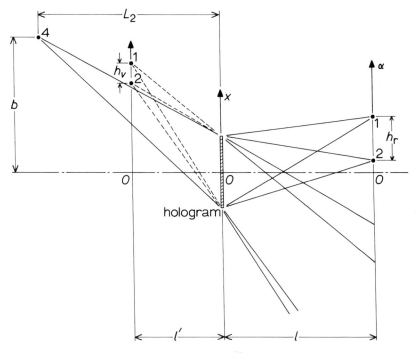

Fig. 10.15 Reconstruction geometry for determining the lateral magnifications. A monochromatic point source is at 4.

For the virtual image reconstructions, we can similarly show that

$$
\begin{aligned}
E_v(\boldsymbol{\sigma}; k_2) = C_1' \delta \left[\alpha' + l' \left(\frac{b}{L_2} + \frac{\lambda_2 h}{2\lambda_1 R_1} + \frac{\lambda_2 a}{\lambda_1 L_1} \right), \beta \right] \\
+ C_2' \delta \left[\alpha' + l' \left(\frac{b}{L_2} - \frac{\lambda_2 h}{2\lambda_1 R_1} + \frac{\lambda_2 a}{\lambda_1 L_1} \right), \beta \right],
\end{aligned}
\tag{10.45}
$$

where the subscript v denotes the virtual image diffraction, and

$$
l' = \frac{\lambda_1 R_1 L_1 L_2}{\lambda_2 R_1 L_2 - \lambda_2 L_1 L_2 - \lambda_1 L_1 R_1}.
$$

The corresponding lateral magnification is therefore (refs. 10.11–10.14)

$$
M_{\text{lat}}^v = \frac{h_v}{h} = \left(1 + \frac{\lambda_1 R_1}{\lambda_2 L_2} - \frac{R_1}{L_1} \right)^{-1}.
\tag{10.46}
$$

From eqs. (10.44) and (10.46) we can conclude that

$$M_{\text{lat}}^{\text{r}} \geq M_{\text{lat}}^{\text{v}}. \tag{10.47}$$

The equality holds where the reference and illuminating beams are both plane waves.

LONGITUDINAL MAGNIFICATIONS

The longitudinal magnifications can also be obtained for the holographic process, as shown in fig. 10.16. Again by paraxial approximation, the complex light distributions of the three monochromatic point sources of wavelength λ_1 are

$$u_1(\boldsymbol{\rho}; k_1) = A_1 \exp\left\{ik_1\left[R_1 + \frac{1}{2R_1}(x^2 + y^2)\right]\right\},$$

$$u_2(\boldsymbol{\rho}; k_1) = A_2 \exp\left\{ik_1\left[(R_1 + d) + \frac{1}{2(R_1 + d)}(x^2 + y^2)\right]\right\},$$

$$u_3(\boldsymbol{\rho}; k_1) = A_3 \exp\left\{ik_1\left[L_1 + \frac{1}{2L_1}[(x + a)^2 + y^2]\right]\right\}. \tag{10.48}$$

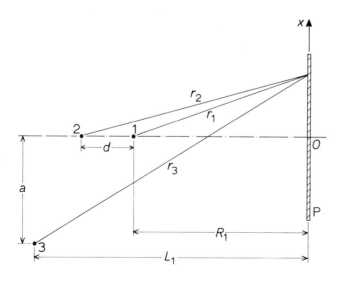

Fig. 10.16 Construction for determining the longitudinal magnifications. 1, 2, monochromatic point sources; 3, divergent reference source; P, photographic plate.

The corresponding irradiance is

$$
\begin{aligned}
I(\boldsymbol{\rho}; k_1) = {} & (u_1 + u_2 + u_3)(u_1 + u_2 + u_3)^* \\
= {} & A_1^2 + A_2^2 + A_3^2 + 2A_1 A_2 \\
& \times \cos\left\{ k_1\left[-d + \frac{1}{2}\left(\frac{1}{R_1} - \frac{1}{R_1 + d} \right)(x^2 + y^2) \right] \right\} \\
& + 2A_1 A_3 \cos\left\{ k_1\left[R_1 - L_1 + \frac{1}{2R_1}(x^2 + y^2) \right.\right. \\
& \left.\left. \qquad\qquad - \frac{1}{2L_1}[(x+a)^2 + y^2] \right] \right\} \\
& + 2A_2 A_3 \cos\left\{ k_1\left[R_1 + d - L_1 + \frac{1}{2(R_1 + d)}(x^2 + y^2) \right.\right. \\
& \left.\left. \qquad\qquad - \frac{1}{2L_1}[(x+a)^2 + y^2] \right] \right\}.
\end{aligned}
$$

(10.49)

Once again we assume the recording to be properly biased; then the transmittance is

$$
\begin{aligned}
T(\boldsymbol{\rho}; k_1) = {} & K_0 + K_1 \cos\left\{ k_1\left[-d + \frac{1}{2}\left(\frac{1}{R_1} - \frac{1}{R_1 + d} \right)(x^2 + y^2) \right] \right\} \\
& + K_2\left[e^{i\{\varDelta\}} + e^{-i\{\varDelta\}} \right] + K_3\left[e^{i\{\varDelta\varDelta\}} + e^{-i\{\varDelta\varDelta\}} \right],
\end{aligned}
$$

(10.50)

where the K's are real proportionality constants, and

$$
\begin{aligned}
\{\varDelta\} &= \left\{ k_1\left[R_1 - L_1 + \frac{1}{2R_1}(x^2 + y^2) - \frac{1}{2L_1}[(x+a)^2 + y^2] \right] \right\}, \\
\{\varDelta\varDelta\} &= \left\{ k_1\left[R_1 + d - L_1 + \frac{1}{2(R_1 + d)}(x^2 + y^2) - \frac{1}{2L_1}[(x+a)^2 + y^2] \right] \right\}.
\end{aligned}
$$

If this hologram is illuminated by a divergent beam of wavelength λ_2, (fig. 10.17), then

$$
u_4(\boldsymbol{\rho}; k_2) = A_4 \exp\left\{ ik_2\left[L_2 + \frac{1}{2L_2}[(x-b)^2 + y^2] \right] \right\},
$$

(10.51)

and the complex light distribution due to the real image diffractions (i.e., the fourth and sixth terms of eq. (10.50)) gives

$$
E_r(\boldsymbol{\sigma}; k_2) = \int\int_S \left[K_2 e^{-i\{\varDelta\}} + K_3 e^{-i\{\varDelta\varDelta\}} \right] u_4(\boldsymbol{\rho}; k_2)\, E_l^+(\boldsymbol{\rho} - \boldsymbol{\sigma}; k_2)\, dx\, dy.
$$

(10.52)

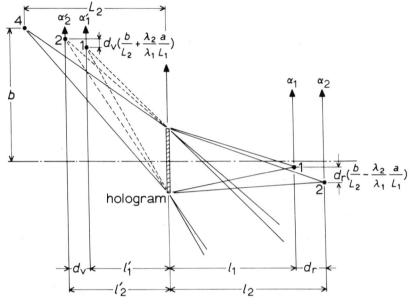

Fig. 10.17 Reconstruction geometry for determining the longitudinal magnifications. A divergent monochromatic source is at 4.

By substitution, we can put this equation into the form

$$
\begin{aligned}
E_r(\boldsymbol{\sigma}; k_2) = C_1 \iint_S \exp\Bigg\{ &-i\frac{k_2}{2}\Bigg[\left(\frac{\lambda_2}{\lambda_1 R_1} - \frac{\lambda_2}{\lambda_1 L_1} - \frac{1}{L_2} - \frac{1}{l}\right)\rho^2 \\
&+ \frac{2}{l}\left(\alpha + l\left(\frac{b}{L_2} - \frac{\lambda_2 a}{\lambda_1 L_1}\right)\right)x + \frac{2\beta}{l}y\Bigg]\Bigg\}\, dx\, dy \\
+ C_2 \iint_S \exp\Bigg\{ &-i\frac{k_2}{2}\Bigg[\left(\frac{\lambda_2}{\lambda_1 (R_1 + d)} - \frac{\lambda_2}{\lambda_1 L_1} - \frac{1}{L_2} - \frac{1}{l}\right)\rho^2 \\
&+ \frac{2}{l}\left(\alpha + l\left(\frac{b}{L_2} - \frac{\lambda_2 a}{\lambda_1 L_1}\right)\right)x + \frac{2\beta}{l}y\Bigg]\Bigg\}\, dx\, dy,
\end{aligned}
$$

(10.53)

where C_1 and C_2 are the appropriate complex constants. From the above equation, it is clear that the real images will be uniquely reconstructed at $l = l_1$,

$$
l_1 = \frac{\lambda_1 R_1 L_1 L_2}{\lambda_2 L_1 L_2 - \lambda_2 R_1 L_2 - \lambda_1 R_1 L_1},
$$

(10.54)

and at $l = l_2$,

$$l_2 = \frac{\lambda_1 L_1 L_2 (R_1 + d)}{\lambda_2 L_1 L_2 - \lambda_2 L_2 (R_1 + d) - \lambda_1 L_1 (R_1 + d)}. \tag{10.55}$$

Thus the solution of eq. (10.53) may be expressed termwise,

$$E_r(\boldsymbol{\sigma}; k_2) = C_1' \delta \left[\alpha + l_1 \left(\frac{b}{L_1} - \frac{\lambda_2 a}{\lambda_1 L_1} \right), \beta \right]_{l = l_1}$$
$$+ C_2' \delta \left[\alpha + l_2 \left(\frac{b}{L_2} - \frac{\lambda_2 a}{\lambda_1 L_1} \right), \beta \right]_{l = l_2}, \tag{10.56}$$

where C_1' and C_2' are the appropriate complex constants.

From eqs. (10.54) and (10.55), the longitudinal separation of the real images can be shown to be

$$d_r = l_2 - l_1$$
$$= \frac{\lambda_1 \lambda_2 (L_1 L_2)^2 d}{[\lambda_2 L_1 L_2 - \lambda_2 (R_1 + d) L_2 - \lambda_1 (R_1 + d) L_1][\lambda_2 L_1 L_2 - \lambda_2 R_1 L_2 - \lambda_1 R_1 L_1]}. \tag{10.57}$$

If the separation d is small compared with R_1, then the longitudinal magnification can be written (refs. 10.11–10.14) as

$$M_{\text{long}}^r = \frac{d_r}{d} \simeq \frac{\lambda_1 \lambda_2 (L_1 L_2)^2}{[\lambda_2 L_1 L_2 - \lambda_2 R_1 L_2 - \lambda_1 R_1 L_1]^2}, \quad \text{for} \quad d \ll R_1. \tag{10.58}$$

Furthermore, if we recall the lateral magnification for the real image reconstructions of eq. (10.44), then we can write down the following relation (refs. 10.11–10.14):

$$M_{\text{long}}^r \simeq \frac{\lambda_1}{\lambda_2} (M_{\text{lat}}^r)^2, \quad \text{for} \quad d \ll R_1. \tag{10.59}$$

Similarly, for the virtual image reconstructions, we can show that

$$E_v(\boldsymbol{\sigma}; k_2) = C_1' \delta \left[\alpha' + l_1' \left(\frac{b}{L_1} + \frac{\lambda_2 a}{\lambda_1 L_1} \right), \beta \right]_{l = l_1'}$$
$$+ C_2' \delta \left[\alpha' + l_2' \left(\frac{b}{L_2} + \frac{\lambda_2 a}{\lambda_1 L_1} \right), \beta \right]_{l = l_2'}, \tag{10.60}$$

where

$$l_1' = \frac{\lambda_1 L_1 L_2 R_1}{\lambda_2 R_1 L_2 - \lambda_2 L_1 L_2 - \lambda_1 L_1 R_1},$$

$$l_2' = \frac{\lambda_1 L_1 L_2 (R_1 + d)}{\lambda_2 L_2 (R_1 + d) - \lambda_2 L_1 L_2 - \lambda_1 L_1 (R_1 + d)}.$$

The corresponding longitudinal magnification is

$$M_{\text{long}}^{\text{v}} = \frac{d_{\text{v}}}{d} = \frac{\lambda_1 \lambda_2 (L_1 L_2)^2}{[\lambda_2 L_2 R_1 - \lambda_2 L_1 L_2 - \lambda_1 L_1 R_1]^2}, \quad \text{for} \quad d \ll R_1. \tag{10.61}$$

From eq. (10.46), once again we can write (refs. 10.11–10.14)

$$M_{\text{long}}^{\text{v}} \simeq \frac{\lambda_1}{\lambda_2} (M_{\text{lat}}^{\text{v}})^2, \text{ for } d \ll R_1. \tag{10.62}$$

Furthermore, from eqs. (10.58) and (10.61), we have

$$M_{\text{long}}^{\text{v}} \leq M_{\text{long}}^{\text{r}}, \tag{10.63}$$

where the equality holds for plane reference and illuminating beams.

It is interesting to note that, as can be seen from eqs. (10.56) and (10.60) or from fig. 10.17, there is a translational distortion (i.e., the image is twisted for a three-dimensional object). In practice it is possible to remove these translational distortions of the real and virtual images by setting the illuminating beam at

$$b = \pm \frac{\lambda_2 L_2}{\lambda_1 L_1} a, \tag{10.64}$$

where the $+$ and $-$ signs are for the removal of the real and virtual image translational distortions, respectively. The translational distortion for both real and virtual images may be removed by setting $a = b = 0$. However, by doing this, we lose the separation of the real, virtual, and zero order diffractions.

From eqs. (10.59) and (10.62), a distortion due to different lateral and longitudinal magnifications may be expected for a three-dimensional object. However, this distortion will be minimized if we set $M_{\text{lat}} = \lambda_2/\lambda_1$. In this case, eqs. (10.59) and (10.62) give

$$M_{\text{lat}} = M_{\text{long}}, \quad \text{for} \quad d \ll R_1. \tag{10.65}$$

10.3 Resolution Limits

In general, the lateral holographic resolutions are limited by the size of the hologram aperture, the spatial frequency limit of the recording medium, and aberrations in the wavefront reconstruction. Only the first two effects will be discussed here. The longitudinal resolution is limited by the frequency bandwidth of the illuminating beam, which will also be demonstrated.

LATERAL RESOLUTION LIMITS

Let us consider first the resolution limit imposed by the size of the hologram aperture. Recall eq. (10.41), where the surface integral over the hologram aperture S is not assumed to be of infinite extent, but finite within the hologram aperture. Thus,

$$T(\mathbf{\rho}; k)=0 \quad \text{for} \quad |x|>\frac{Lx}{2}, |y|>\frac{Ly}{2}. \tag{10.66}$$

Then at the distance where the real images are reconstructed, the solution of eq. (10.41) is

$$
\begin{aligned}
E(\mathbf{\sigma}; k_2)=C_1 L_x L_y & \frac{\sin\left[\dfrac{\pi L_x}{l\lambda_2}(\alpha+\alpha_1)\right]}{\dfrac{\pi L_x}{l\lambda_2}(\alpha+\alpha_1)} \frac{\sin\left(\dfrac{\pi L_y}{l\lambda_2}\beta\right)}{\dfrac{\pi L_y}{l\lambda_2}\beta} \\
+ C_2 L_x L_y & \frac{\sin\left[\dfrac{\pi L_x}{l\lambda_2}(\alpha+\alpha_2)\right]}{\dfrac{\pi L_x}{l\lambda_2}(\alpha+\alpha_2)} \frac{\sin\left(\dfrac{\pi L_y}{l\lambda_2}\beta\right)}{\dfrac{\pi L_y}{l\lambda_2}\beta},
\end{aligned}
\tag{10.67}
$$

where

$$\alpha_1=l\left(\frac{b}{L_2}-\frac{\lambda_2 h}{2\lambda_1 R_1}-\frac{\lambda_2 a}{\lambda_1 L_1}\right),$$

$$\alpha_2=l\left(\frac{b}{L_2}+\frac{\lambda_2 h}{2\lambda_1 R_1}-\frac{\lambda_2 a}{\lambda_1 L_1}\right),$$

C_1 and C_2 are the appropriate complex constants, and l is given by eq. (10.42).

If we use the Rayleigh criterion (sec. 4.6) for the lateral resolution limit, then the minimum resolvable distance for the real images (fig. 10.18) can be shown to be

$$h_{\text{r, min}}=\frac{l\lambda_2}{L_x}. \tag{10.68}$$

Accordingly, from the real-image lateral magnification, we may conclude that

$$h\geq\frac{l\lambda_2}{L_x}(M_{\text{lat}}^{\text{r}})^{-1}=\frac{\lambda_1 R_1}{L_x}. \tag{10.69}$$

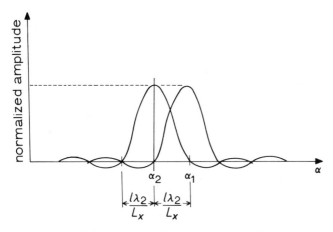

Fig. 10.18 Rayleigh criterion for the lateral resolution limit of a hologram.

Here equality holds for the minimum resolvable distance of h (refs. 10.13–10.14), i.e.,

$$h_{r, \, min} = \frac{\lambda_1 R_1}{L_x}. \tag{10.70}$$

Similarly, for the lateral resolution limit of the virtual images, we can show that

$$h \geq -\frac{l'\lambda_2}{L_x} (M_{lat}^v)^{-1} = \frac{\lambda_1 R_1}{L_x}, \tag{10.71}$$

where l' is defined by eq. (10.45). The corresponding minimum resolvable distance is therefore

$$h_{v, \, min} = \frac{\lambda_1 R_1}{L_x}, \tag{10.72}$$

which is identical to eq. (10.70), as expected. Accordingly, the minimum resolvable distance in holography is proportional to the wavelength of the coherent source used for the wavefront construction and to the distance between the object and the recording medium, and it is inversely proportional to the size of the hologram aperture. In other words, the lateral resolution limit is set during the recording (encoding), not during the reconstruction (decoding).

Now let us consider the resolution limit set by the spatial frequency limit of the film. In this case, the limits of the surface integral of eq. (10.41) do not extend over the entire hologram aperture, but only from $x = x_2$

to $x = x_1$ and $y = -\dfrac{x_1 - x_2}{2}$ to $y = \dfrac{x_1 - x_2}{2}$ for the first integral, and $x = x_2'$

to $x = x_1'$ and $y = -\dfrac{x_1' - x_2'}{2}$ to $y = \dfrac{x_1' - x_2'}{2}$ for the second integral. The limiting parameters are defined by

$$x_1 = \frac{\lambda_1 v_2 R_1 L_1 + \frac{1}{2} h L_1 + R_1 a}{L_1 - R_1},$$

$$x_2 = \frac{-\lambda_1 v_2 R_1 L_1 + \frac{1}{2} h L_1 + R_1 a}{L_1 - R_1},$$

$$x_1' = \frac{\lambda_1 v_2 R_1 L_1 - \frac{1}{2} h L_1 + R_1 a}{L_1 - R_1}, \qquad (10.73)$$

$$x_2' = \frac{-\lambda_1 v_2 R_1 L_1 - \frac{1}{2} h L_1 + R_1 a}{L_1 - R_1},$$

where v_2 is the high spatial frequency limit of the recording medium. Then at the distance where the real image is reconstructed, the solution of eq. (10.41) is

$$E(\boldsymbol{\sigma}; k_2) = C_1 (\Delta x)^2 \frac{J_1 \left\{ \dfrac{\pi \Delta x}{l \lambda_2} [(\alpha + \alpha_1)^2 + \beta^2]^{1/2} \right\}}{\dfrac{\pi \Delta x}{l \lambda_2} [(\alpha + \alpha_1)^2 + \beta^2]^{1/2}}$$

$$+ C_2 (\Delta x)^2 \frac{J_1 \left\{ \dfrac{\pi \Delta x}{l \lambda_2} [(\alpha + \alpha_1)^2 + \beta^2]^{1/2} \right\}}{\dfrac{\pi \Delta x}{l \lambda_2} [(\alpha + \alpha_2)^2 + \beta^2]^{1/2}}, \qquad (10.74)$$

where C_1 and C_2 are the appropriate complex constants, α_1 and α_2 are defined in eq. (10.67), J_1 is the first order Bessel function, and

$$\Delta x = x_1 - x_2 = x_1' - x_2' = \frac{2 \lambda_1 v_2 R_1 L_1}{L_1 - R_1}.$$

Accordingly, we can show that the minimum resolvable distance is

$$h_{r, \min} = 1.22 \frac{\lambda_1 R_1}{\Delta x}, \qquad (10.75)$$

and similarly,

$$h_{v, \min} = 1.22 \frac{\lambda_1 R_1}{\Delta x}. \qquad (10.76)$$

Again, eqs. (10.75) and (10.76) are identical, and the minimum resolvable distance is inversely proportional to the spatial frequency limit v_2 of the recording medium.

LONGITUDINAL RESOLUTION LIMIT

Let us now consider the holographic longitudinal resolution limit, which is set by the finite frequency bandwidth (i.e., the quasi-monochromaticity) of the illuminating beam. We recall the two-point-object hologram of eq. (10.50). If it is assumed that the hologram is illuminated by a quasi-monochromatic divergent source with a finite bandwidth Δv, then the minimum resolvable longitudinal distance of the real image reconstruction may be shown to be

$$d_{r,\,min} \simeq \Delta l_r, \quad \text{for} \quad d \ll R_1, \tag{10.77}$$

where

$$\Delta l_r = l_r' - l_r'',$$
$$l_r' = \frac{\lambda_1 R_1 L_1 L_2}{\lambda' L_1 L_2 - \lambda' R_1 L_2 - \lambda_1 R_1 L_1},$$
$$l_r'' = \frac{\lambda_1 R_1 L_1 L_2}{\lambda'' L_1 L_2 - \lambda'' R_1 L_2 - \lambda_1 R_1 L_1},$$

and λ' and λ'' are the respective low and high cutoff wavelengths of the source. From eq. (10.58) we can conclude that

$$d \geq \Delta l_r (M_{long}^r)^{-1}, \tag{10.78}$$

where

$$M_{long}^r = \frac{\lambda_1 \lambda_2 (L_1 L_2)^2}{[\lambda_2 L_1 L_2 - \lambda_2 R_1 L_2 - \lambda_1 R_1 L_1]^2},$$

and $\lambda_2 = (\lambda' \lambda'')^{1/2}$ is the mean wavelength of the source. Therefore the minimum resolvable longitudinal distance may be shown to be

$$d_{r,\,min} = \Delta l_r (M_{long}^r)^{-1}. \tag{10.79}$$

Similarly, for virtual image reconstruction, we can show that the minimum resolvable longitudinal distance is

$$d_{v,\,min} = \Delta l_v (M_{long}^v)^{-1}, \tag{10.80}$$

where

$$\Delta l_v = l_v' - l_v'',$$

$$l'_v = \frac{\lambda_1 L_1 R_1 L_2}{\lambda' R_1 L_2 - \lambda' L_1 L_2 - \lambda_1 L_1 R_1},$$

$$l''_v = \frac{\lambda_1 L_1 R_1 L_2}{\lambda'' R_1 L_2 - \lambda'' L_1 L_2 - \lambda_1 L_1 R_1},$$

$$M^v_{\text{long}} = \frac{\lambda_1 \lambda_2 (L_1 L_2)^2}{[\lambda_2 L_2 R_1 - \lambda_2 L_1 L_2 - \lambda_1 L_1 R_1]^2},$$

$$\lambda_2 = (\lambda' \lambda'')^{1/2}.$$

10.4 Bandwidth Requirements

From the Fresnel-Kirchhoff theory (app. B), the complex light distribution on the recording medium from a diffuse object is

$$u(x, y) = \iint_{s_0} O(\xi, \eta)\, E_l^+ (\rho - \xi;\, k)\, d\xi\, d\eta, \tag{10.81}$$

where $O(\xi, \eta)$ is the object function projected onto the (ξ, η) plane of the $\xi(\xi, \eta, \zeta)$ coordinates, and s_0 denotes integration over the object surface that gives rise to O.

If we take the Fourier transform of eq. (10.81), then it becomes

$$U(p, q) = O(p, q)\, E_l(p, q), \tag{10.82}$$

where

$$E_l(p, q) = -\frac{i}{\lambda l} \exp\left\{ -i\frac{l}{2k}(p^2 + q^2) \right\},$$

and p and q are the corresponding spatial frequencies.

Since $|E_l(p, q)| = 1/\lambda l$, we have

$$|U(p, q)|^2 = \left(\frac{1}{\lambda l}\right)^2 |O(p, q)|^2. \tag{10.83}$$

Equation (10.83) implies that the spatial power spectrum of the light distribution is proportional to the spatial power spectrum of the object function $O(\xi, \eta)$. Therefore the preservation of the spatial frequency spectrum of the object depends on the frequency spectrum response of the recording medium. The spatial frequency bandwidth is most often limited by the size of the hologram aperture but not by the film spatial frequency limit. However, as we have seen in the last section, the resolution limits in holography are set both by the size of the hologram aperture and by the spatial frequency limit of the film. It is clear that whichever of these resolution limits comes first will set the resolution limit of the hologram.

As shown in fig. 10.19, the resolution requirements may be determined from the field angles as follows. We let the tips of the object act like secondary point radiators; then the complex light distributions on the recording medium due to these radiators are

$$u_1(\boldsymbol{\rho}; k) = A_1 \exp\left\{ik\left[R_1 + \frac{1}{2R_1}\left[(x-h_1)^2 + y^2\right]\right]\right\},$$

$$u_2(\boldsymbol{\rho}; k) = A_2 \exp\left\{ik\left[R_1 + \frac{1}{2R_1}\left[(x+h_2)^2 + y^2\right]\right]\right\}, \qquad (10.84)$$

$$u_3(\boldsymbol{\rho}; k) = A_3 \exp\left\{ik\left[L_1 + \frac{1}{2L_1}\left[(x+a)^2 + y^2\right]\right]\right\}.$$

For simplicity, the A's are assumed to be positive real constants. Then

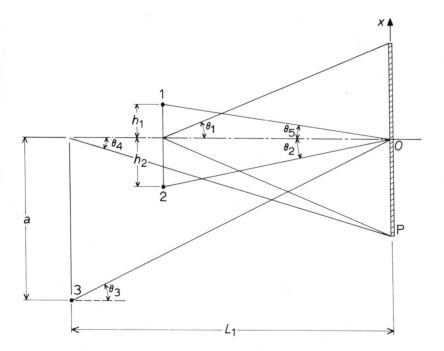

Fig. 10.19 Geometry for determining the spatial frequency bandwidth of a hologram. 3, divergent reference source; P, photographic plate.

the light intensity distribution on the recording medium is

$$I(\boldsymbol{\rho}; k) = (u_1 + u_2 + u_3)(u_1 + u_2 + u_3)^*$$

$$= A_1^2 + A_2^2 + A_3^2 + 2A_1 A_2 \cos \frac{k}{R_1} (h_1 + h_2) x$$

$$+ 2A_1 A_3 \cos \left\{ k(R_1 - L_1) + \frac{k}{2R_1} [(x - h_1)^2 + y^2] \right.$$

$$\left. - \frac{k}{2L_1} [(x + a)^2 + y^2] \right\}$$

$$+ 2A_2 A_3 \cos \left\{ k(R_1 - L_1) + \frac{k}{2R_1} [(x + h_2)^2 + y^2] \right.$$

$$\left. - \frac{k}{2L_1} [(x + a)^2 + y^2] \right\}.$$

(10.85)

From eq. (10.85), the spatial phase shift due to u_1 and u_3 is

$$\phi_{13}(x, y) = k \left\{ R_1 - L_1 + \frac{1}{2R_1} [(x - h_1)^2 + y^2] - \frac{1}{2L_1} [(x + a)^2 + y^2] \right\}, \quad (10.86)$$

and that due to u_2 and u_3 is

$$\phi_{23}(x, y) = k \left\{ R_1 - L_1 + \frac{1}{2R_1} [(x + h_2)^2 + y^2] - \frac{1}{2L_1} [(x + a)^2 + y^2] \right\}. \quad (10.87)$$

The corresponding spatial frequencies in the x coordinate can be determined to be

$$p_{13}(x) = \frac{\partial \phi_{13}(x, y)}{\partial x} = k \left[\left(\frac{1}{R_1} - \frac{1}{L_1} \right) x - \frac{h_1}{R_1} - \frac{a_1}{L_1} \right] \quad (10.88)$$

and

$$p_{23}(x) = \frac{\partial \phi_{23}(x, y)}{\partial x} = k \left[\left(\frac{1}{R_1} - \frac{1}{L_1} \right) x + \frac{h_2}{R_1} - \frac{a_1}{L_1} \right]. \quad (10.89)$$

Thus the highest positive spatial frequency allowed by the size of the hologram aperture is

$$\nu_{\text{high}} = \frac{1}{2\pi} p_{23}(x) \bigg|_{x = L_x/2} = \frac{1}{\lambda} \left[\left(\frac{1}{R_1} - \frac{1}{L_1} \right) \frac{L_x}{2} + \frac{h_2}{R_1} - \frac{a}{L_1} \right]. \quad (10.90)$$

This can be written in terms of the field angles,

$$v_{\text{high}} = \frac{1}{\lambda} [\tan\theta_1 - \tan\theta_4 + \tan\theta_2 - \tan\theta_3]. \tag{10.91}$$

Similarly, the negative spatial frequency limit is

$$v_{\text{low}} = \frac{1}{2\pi} p_{13}(x)\Big|_{x=-L_x/2} = \frac{1}{\lambda}\left[\left(-\frac{1}{R_1} + \frac{1}{L_1}\right)\frac{L_x}{2} - \frac{h_1}{R_1} - \frac{a}{L_1}\right], \tag{10.92}$$

or equivalently,

$$v_{\text{low}} = \frac{1}{\lambda} [-\tan\theta_1 + \tan\theta_4 - \tan\theta_5 - \tan\theta_3]. \tag{10.93}$$

Therefore, the spatial frequency bandwidth limited by the size of the hologram aperture is

$$\Delta v = v_{\text{high}} - v_{\text{low}} = \frac{1}{\lambda}\left[\left(\frac{1}{R_1} - \frac{1}{L_1}\right)L_x + \frac{1}{R_1}(h_1 + h_2)\right], \tag{10.94}$$

or equivalently,

$$\Delta v = \frac{1}{\lambda} [2\tan\theta_1 - 2\tan\theta_4 + \tan\theta_2 + \tan\theta_5]. \tag{10.95}$$

It is interesting to note that, if the size of the hologram aperture is sufficiently small (i.e., as it approaches a point), the spatial frequency bandwidth reduces to

$$\Delta v_1 \simeq \frac{1}{\lambda R_1}(h_1 + h_2) = \frac{1}{\lambda}[\tan\theta_2 + \tan\theta_5]. \tag{10.96}$$

On the other hand, if the object is sufficiently small, the spatial frequency bandwidth becomes

$$\Delta v_2 \simeq \frac{L_x}{\lambda}\left(\frac{1}{R_1} - \frac{1}{L_1}\right) = \frac{2}{\lambda}[\tan\theta_1 - \tan\theta_4]. \tag{10.97}$$

Thus we conclude that

$$\Delta v \leq \Delta v_1 + \Delta v_2. \tag{10.98}$$

It may be emphasized that the spatial frequency bandwidth in wavefront recording depends upon two factors: The field angles from the object to the recording medium, and the angles of the beams received from the tips of the object (i.e., s_1 and s_2).

Obviously, if the object is large compared to the hologram aperture, then eq. (10.98) reduces to

$$\Delta v \simeq \Delta v_1 = \frac{1}{\lambda} \left[\tan \theta_2 + \tan \theta_5 \right]. \tag{10.99}$$

On the other hand, if the size of the hologram aperture is large compared with the object, then

$$\Delta v \simeq \Delta v_2 = \frac{2}{\lambda} \left[\tan \theta_1 - \tan \theta_4 \right]. \tag{10.100}$$

Furthermore, from eqs. (10.94) or (10.95) it can be seen that a reduction of spatial frequency bandwidth is possible, if we place the divergent reference beam (i.e., s_3), on the same plane as the object (ref. 10.15). The spatial frequency bandwidth is then

$$\Delta v = \frac{1}{\lambda} (\tan \theta_2 + \tan \theta_5) = \Delta v_1, \tag{10.101}$$

which is independent of the size of the object relative to the hologram aperture. If the size of the object is large compared with the hologram aperture, then the reduction in spatial frequency is not significant. However if the size of the hologram aperture is large compared with the size of the object, then the spatial frequency bandwidth reduction would be considerable. This second condition may be of some importance in the application of wavefront reconstruction to microscopy.

Similarly, if the divergent reference beam in fig. 10.19 is replaced by an oblique plane wave, then it can be shown that the positive spatial frequency limit is

$$v'_{\text{high}} = \frac{1}{\lambda} \left[\tan \theta_1 + \tan \theta_2 - \sin \theta \right], \tag{10.102}$$

and the negative spatial frequency limit is

$$v'_{\text{low}} = \frac{1}{\lambda} \left[-\tan \theta_1 - \tan \theta_5 - \sin \theta \right], \tag{10.103}$$

where θ is the oblique angle of incidence of the reference wave. Therefore, the spatial frequency bandwidth allowed by the size of the hologram aperture is

$$\Delta v' = \Delta v'_{\text{high}} - \Delta v'_{\text{low}} = \frac{1}{\lambda} \left[2 \tan \theta_1 + \tan \theta_2 + \tan \theta_5 \right]. \tag{10.104}$$

Again, if the size of the hologram aperture is sufficiently small, then the spatial frequency bandwidth is

$$\Delta v_1' = \Delta v_1 \simeq \frac{1}{\lambda} [\tan \theta_2 + \tan \theta_5]. \tag{10.105}$$

On the contrary, if the size of object is very small, then the spatial frequency bandwidth is

$$\Delta v_2' \simeq \frac{2}{\lambda} \tan \theta_2. \tag{10.106}$$

Thus we can conclude that

$$\Delta v' \leq \Delta v_1' + \Delta v_2'. \tag{10.107}$$

Obviously, the reduction of spatial frequency bandwidth for a plane reference wave is impossible.

Applying the preceding considerations to a general three-dimensional object (fig. 10.20), we obtain the spatial frequency bandwidths

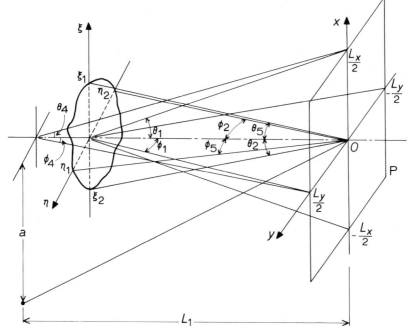

Fig. 10.20 Geometry for determining the spatial frequency bandwidth of a hologram of an extended object. 3, divergent reference source; P, photographic plate.

$$\Delta v_x = \frac{1}{\lambda} \left[2 \tan \theta_1 - 2 \tan \theta_4 + \tan \theta_2 + \tan \theta_5 \right] \tag{10.108}$$

and

$$\Delta v_y = \frac{1}{\lambda} \left[2 \tan \phi_1 - 2 \tan \phi_4 + \tan \phi_2 + \tan \phi_5 \right], \tag{10.109}$$

where Δv_x and Δv_y are the respective bandwidths in the x and y coordinate axes. If the divergent reference beam is replaced by a plane reference wave, then the corresponding x and y spatial frequency bandwidths are

$$\Delta v_x' = \frac{1}{\lambda} \left[2 \tan \theta_1 + \tan \theta_2 + \tan \theta_5 \right] \tag{10.110}$$

and

$$\Delta v_y' = \frac{1}{\lambda} \left[2 \tan \phi_1 + \tan \phi_2 + \tan \phi_5 \right]. \tag{10.111}$$

10.5 Holographic Aberrations

In sec. 10.2 we developed a general and simple procedure for calculating the holographic image magnification or demagnification. Those calculations, however, were based on the paraxial approximation, for which the reconstructed image exhibits no aberrations. In practice, however, magnified (or demagnified) hologram images frequently suffer from aberrations. It is the purpose of this section to evaluate the five primary aberrations (ref. 10.16) of holographic images, namely: spherical aberration, coma, astigmatism, curvature of field, and distortion. We will also discuss the conditions under which these aberrations may be minimized or eliminated.

Let us now identify the essential geometrical parameters of wavefront construction and reconstruction, as shown in fig. 10.21. In this figure a plane wavefront is used for recording, and a spherical wavefront is used for reconstruction. The explicit form of the complex light field of the hologram image on the σ coordinate system, for a two-dimensional object, is

$$E(\sigma; k_2) = C \int\int_{S_2} \left\{ \int\int_{S_1} O(\xi, \eta) \exp\left[ik_1 (x \sin \theta_1 - r_1) \right] d\xi \, d\eta \right\}$$

$$\times \exp\left[ik_2 \left(\frac{\rho^2}{2R} - x \sin \theta_2 \right) \right] \exp(ik_2 r_2) \, dx \, dy, \tag{10.112}$$

where C is a complex constant, $O(\xi, \eta)$ is the two-dimensional object function, $k_1 = 2\pi/\lambda_1$, with λ_1 the recording wavelength, $k_2 = 2\pi/\lambda_2$, with λ_2 the

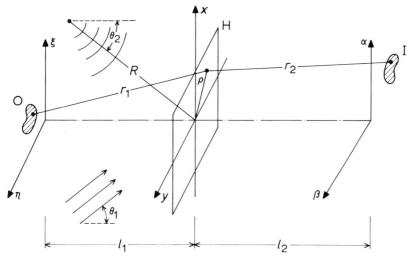

Fig. 10.21 Geometry for wavefront construction and reconstruction. O, object; H, hologram; I, image. Divergent reconstruction beam incident at angle θ_2, and plane reference wave incident at θ_1.

reconstructing wavelength, $\rho^2 = x^2 + y^2$, and S_1 and S_2 denote the surface integral of the object function and the transmission function of the hologram respectively. It is also clear that the first surface integral represents the wavefront construction, the second exponential represents the hologram illumination, and the last exponential represents the diffraction from the hologram.

Referring to fig. 10.21, we can write the distances r_1 and r_2 as

$$r_1 = l_1 \left[1 + \frac{(x-\xi)^2 + (y-\eta)^2}{l_1^2} \right]^{1/2} \tag{10.113}$$

and

$$r_2 = l_2 \left[1 + \frac{(\alpha-x)^2 + (\beta-y)^2}{l_2^2} \right]^{1/2}. \tag{10.114}$$

Then by binomial expansion (ref. 10.17), eqs. (10.113) and (10.114) can be written as

$$r_1 = l_1 + \frac{1}{2l_1} \left[(x-\xi)^2 + (y-\eta)^2 \right]$$
$$- \frac{1}{8l_1^3} \left[(x-\xi)^2 + (y-\eta)^2 \right]^2 + \cdots, \tag{10.115}$$

$$r_2 = l_2 + \frac{1}{2l_2}[(\alpha - x)^2 + (\beta - y)^2]$$
$$- \frac{1}{8l_2^3}[(\alpha - x)^2 + (\beta - y)^2]^2 + \cdots . \tag{10.116}$$

If the first two terms of eqs. (10.115) and (10.116) are retained, we have the paraxial approximations, and the explicit form of eq. (10.112) will be

$$E(\boldsymbol{\sigma}; k_2) = C' \iint_{S_2} \left\{ \iint_{S_1} O(\xi, \eta) \exp\left[ik_1 \left(x \sin\theta_1 - \frac{(x-\xi)^2 + (y-\eta)^2}{2l_1} \right) \right] d\xi \, d\eta \right\}$$
$$\times \exp\left[ik_2 \left(\frac{\rho^2}{2R} - x \sin\theta_2 \right) \right] \exp\left[ik \frac{(\alpha - x)^2 + (\beta - y)^2}{2l_2} \right] dx \, dy, \tag{10.117}$$

where C is an appropriate complex constant. Equation (10.117) is the form we have used in the previous sections. However, if we retain the first three terms of eqs. (10.115) and (10.116), then the third order aberrations may be calculated.

It may be emphasized that aberrations of the wavefront construction and reconstruction process depend on the exponential argument containing r_1 and r_2. In the usual paraxial approximation, the quadratic exponential factor ρ^2 [for example, in eq. (10.53)], is eliminated by imposing the lens condition

$$\frac{1}{R} + \frac{1}{l_2} = \frac{\lambda_2}{\lambda_1} \frac{1}{l_1}. \tag{10.118}$$

However, in the nonparaxial case it is no longer sufficient to impose the condition expressed by eq. (10.118) in order to eliminate the higher order exponential term. These nonvanishing terms in the exponent constitute the aberrations in hologram images. To investigate the aberrations, we can begin with the evaluation of the phase factor $\Delta\phi = k_2 r_2 - k_1 r_1$ of the construction-reconstruction process. After a long but straightforward calculation, $\Delta\phi$ is seen to be

$$\Delta\phi = -\frac{1}{8}\left(\frac{k_2}{l_2^3} - \frac{k_1}{l_1^3} \right)\rho^4 + \frac{1}{2}\left(\frac{Mk_2}{l_2^3} - \frac{k_1}{l_1^3} \right)\rho^2 K^2$$
$$-\frac{1}{2}\left(\frac{M^2 k_2}{l_2^3} - \frac{k_1}{l_1^3} \right)K^4 - \frac{1}{4}\left(\frac{M^2 k_2}{l_2^3} - \frac{k_1}{l_1^3} \right)\rho^2\tau^2 + \frac{1}{2}\left(\frac{M^3 k_2}{l_2^3} - \frac{k_1}{l_1^3} \right)\tau^2 K^2, \tag{10.119}$$

where $\rho^2 = x^2 + y^2$, $\tau^2 = \xi^2 + \eta^2$, $K^2 = \xi x + \eta y$, and $M = \lambda_2 l_2 / \lambda_1 l_1$. The lateral magnification M can be derived from eq. (10.44). By comparing eq. (10.119) with the general treatment of lens aberrations given in sec. 5.3 of ref. 10.16, we can see that the first term of eq. (10.119) is the spherical aberration; the second term is the coma; the third term is the astigmatism; the fourth term is the curvature of field; and the last term is the distortion.

We see that the five primary aberrations which occur for physical lenses also occur in holography. It may also be noted that the higher order (beyond the third order) aberrations in holography may be calculated; however, we will not attempt that here.

Now let us consider the conditions under which these primary holographic aberrations may be corrected. In order to do so, we set each term of eq. (10.119) equal to zero. The results are tabulated in table 10.1 (taken from ref. 10.13).

From table 10.1, it is clear that if we correct one of the aberrations, the others generally cannot be corrected. However, there are two exceptions, namely: for unity lateral magnification $(M = 1)$, all the aberrations will vanish; and in any case, the astigmatism and curvature of field can be corrected together.

We should note in conclusion that for a more general holographic process the reference and the reconstructing processes will both employ spherical wavefronts. The third order aberrations in this general case can be determined by a procedure similar to the foregoing. For such an investigation the reader can refer to the article by Meier (ref. 10.11). According to this paper, all five primary aberrations can be made to dis-

Table 10.1 *Aberrations and conditions for their correction*

Aberration	Condition
Spherical aberration	$\dfrac{\lambda_2}{\lambda_1}\left(\dfrac{l_2}{l_1}\right)^3 = 1$
Coma	$\dfrac{\lambda_2}{\lambda_1}\left(\dfrac{l_2}{l_1}\right)^3 = M \, ; \, l_1 = l_2$
Astigmatism and curvature of field	$\dfrac{\lambda_2}{\lambda_1}\left(\dfrac{l_2}{l_1}\right)^3 = M^2 \, ; \, \dfrac{\lambda_2}{\lambda_1} = \dfrac{l_2}{l_1}$
Distortion	$\dfrac{\lambda_2}{\lambda_1}\left(\dfrac{l_2}{l_1}\right)^3 = M^3 \, ; \, \lambda_1 = \lambda_2$

appear simultaneously if, and only if, the condition of unity magnification is met. It is also clear that to achieve unity magnification the construction reference beam and the reconstruction beam are both required to be plane waves and to have the same wavelength.

10.6 Spatially Incoherent Holography

The holographic process was originally conceived to be a coherent imaging process. However, several techniques are available in which a spatially incoherent source may be used. Such a technique was first suggested by Mertz and Young (ref. 10.18), and later the theory and experiment were extended by Lohmann (ref. 10.19), Stroke and Restrick (ref. 10.20), and Cochran (ref. 10.21). In this section we will adopt the technique suggested by Cochran.

It is well known that the light scattered by a point on an incoherently illuminated object will not interfere with the light scattered by any other point of the object. However, with special arrangement of the optical setup, it is possible to split the light field from each point of the illuminated object and then rejoin the parts in such a way that an interference pattern is formed. Thus each object point will be recorded in a suitable interference pattern. If the resulting hologram is illuminated by a coherent source, then each of the fringe patterns will be reconstructed into a unique image point. In other words, in making an incoherent hologram, each object point is made to form its own reference beam.

Figure 10.22 is a diagram of the triangular interferometer devised by Cochran. This device consists of two lenses (L_1 and L_2) of different focal lengths (f_1 and f_2). The lenses are separated by a path length $f_1 + f_2$ and their focal points coincide at P, as shown in the figure. The object plane O and the recording plane H are each located at a path length f_1 from lens L_1, and a path length f_2 from lens L_2. Light may travel from plane O to plane H by two different paths, namely the clockwise and counterclockwise paths around the interferometer. For example, for the clockwise path, light travels a distance f_1 from plane O to lens L_1 by means of the reflection from the beam splitter BS. From lens L_1 to lens L_2, the light ray travels a distance of $f_1 + f_2$. From lens L_2 to plane H the light travels a distance of f_2 by means of BS. Because of this arrangement of the optical system, it is clear that an illuminated object located at plane O will be imaged onto plane H. It is also clear that, with this particular arrangement of the lenses, the image produced on plane H is magnified by an amount

$$M_1 = -\frac{f_2}{f_1}, \quad \text{for the clockwise path.} \tag{10.120}$$

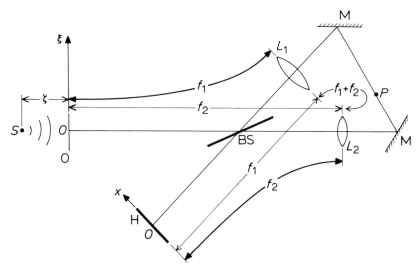

Fig. 10.22 Triangular interferometer for producing incoherent holograms. H, hologram; M, M, mirrors; BS, beam splitter.

In a similar manner, it can be seen that the illuminated object at plane O will also be imaged onto plane H on a counterclockwise path. However, the image magnification is

$$M_2 = -\frac{f_1}{f_2} = \frac{1}{M_1}, \quad \text{for the counterclockwise path.} \tag{10.121}$$

For simplicity in illustration, let us consider a single point object S, located at distance ζ behind O. The complex light distribution on O due to S is

$$O(\xi, \eta) = C \exp\left[i \frac{k}{2\zeta} (\xi^2 + \eta^2) \right], \tag{10.122}$$

where C is a complex constant and $k = 2\pi/\lambda$.

From the discussion given previously it is clear that, due to the light distribution given by eq. (10.122), two spherical wavefronts will be formed at plane H, with magnifications of M_1 and M_2. The resultant complex light field at plane H is the combination of these two wavefronts,

$$u(x, y) = C_1 \exp\left\{ i \frac{k}{2\zeta} \left[\left(\frac{x}{M_1} \right)^2 + \left(\frac{y}{M_1} \right)^2 \right] \right\}$$
$$+ C_2 \exp\left\{ i \frac{k}{2\zeta} \left[\left(\frac{x}{M_2} \right)^2 + \left(\frac{y}{M_2} \right)^2 \right] \right\}, \tag{10.123}$$

where C_1 and C_2 are the appropriate complex constants. The corresponding irradiance is

$$I(x, y) = |C_1|^2 + |C_2|^2 + 2 |C_1| |C_2| \cos \left[\frac{k}{2\zeta} (M_2^2 - M_1^2) \rho^2 + \phi \right], \quad (10.124)$$

where $\rho^2 = x^2 + y^2$, $\phi =$ phase angle between C_1 and C_2, and $M_1 = M_2^{-1}$.

If we insert a photographic plate at plane H to record the interference pattern given by eq. (10.124), the resultant transmittance function of the recorded plate is

$$T(x, y) = K_1 + K_2 \cos \left[\frac{k}{2\zeta} (M_2^2 - M_1^2) \rho^2 + \phi \right], \quad (10.125)$$

where K_1 and K_2 are the proportionality constants.

It can be seen that eq. (10.125) describes a Fresnel zone lens construction with a focal length of

$$f = \zeta \left(\frac{1}{M_2^2 - M_1^2} \right). \quad (10.126)$$

By the substitution of eqs. (10.120) and (10.121), eq. (10.126) can be written as

$$f = \zeta \left(\frac{f_1^2 f_2^2}{f_1^4 - f_2^4} \right). \quad (10.127)$$

If the transparency of eq. (10.125) is illuminated by a coherent source, it is clear that virtual and real hologram images will be reconstructed.

Let us now extend this concept to a more complicated case; that of a multitude of mutually incoherent point sources. Each of these sources will generate an interference pattern of its own; however, the patterns are incoherent with respect to each other. The total irradiance is the sum of the individual irradiances due to each of the point sources. The resulting transmittance function of the recorded transparency will therefore be the summation of these interference patterns. Each point source determines the center and focal length of a Fresnel zone lens, and thus a three-dimensional holograph image will be formed. A photograph of an incoherent hologram image reconstruction is given in fig. 10.23.

Although incoherent holography is an attractive technique, there are several drawbacks in practice at present. One of the main disadvantages is that each of the elementary Fresnel zone lenses is formed by the interference of two extremely small portions of the total light incident on the recording medium, whereas in coherent holography the light field from each object point interferes with the whole reference beam incident on

Fig. 10.23 An image produced by a spatially incoherent hologram. (Permission by P. J. Peters.)

the film. Another disadvantage is the accumulation of the bias level of the wavefront recording; thus in practice incoherent holography has only been successfully applied to a relatively small number of resolvable point objects. These major drawbacks may become insignificant, depending on the future research and newly developed techniques in incoherent holography.

10.7 Reflection Holography

By a simple rearrangement of the optical setup for the coherent wavefront construction process, it is possible to obtain holographic image reconstruction by means of incoherent "white" light illumination. This image reconstruction process is entirely dependent upon reflection from the recorded hologram, rather than on transmission through the hologram. Since this new technique mainly utilizes the thick emulsion of the photographic plate, reflection (or white light) holography is also known as thick emulsion holography.

The wavefront construction of a reflection hologram, which we will illustrate in a moment, is very similar to the basic concept underlying Lippmann's color photography (ref. 10.23). Thus reflection holography is also known as color holography. The basic theory of reflection holography was first described by Denisyuk (ref. 10.24) in 1962. However the concept was not fully appreciated here until 1966, when a sequence of papers was published by Stroke and Labeyrie (ref. 10.25), by Lin et al. (ref. 10.26), and by Leith et al. (ref. 10.27).

In this section we will first study the reflection holography of a simple point object and then extend the technique to holography of a three-dimensional object. To record a reflection hologram, the coherent object light field and the reference waves are introduced from opposite sides of the recording medium, as shown in fig. 10.24. From this figure, we may write the complex light distribution in the photographic emulsion as

$$u(\mathbf{\rho}; k) = A \exp\left\{ik\left[(R+z) + \frac{\rho^2}{2(R+z)}\right]\right\}, \quad \text{for} \quad -\Delta z \le z \le 0, \qquad (10.128)$$

and the complex light field due to the reference plane wave as

$$v(\mathbf{\rho}; k) = B \exp\left[-ik(R+z+x\sin\theta)\right], \quad \text{for} \quad -\Delta z \le z \le 0, \qquad (10.129)$$

where A and B are complex constants, $\rho^2 = x^2 + y^2$, and $k = 2\pi/\lambda$. The

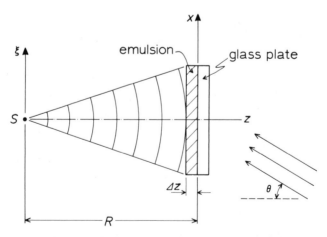

Fig. 10.24 Coherent wavefront construction of a point-object reflection hologram. S, monochromatic point source. An oblique monochromatic reference wave is incident from behind the photographic plate.

corresponding irradiance is therefore

$$I(\mathbf{\rho}; k) = (u + v)(u + v)^*$$

$$= |A|^2 + |B|^2 + 2|A| |B| \cos\left\{k\left[2(R+z) + \frac{\rho^2}{2(R+z)} + x \sin\theta\right] + \phi\right\},$$

$$\text{for } -\Delta z \leq z \leq 0, \qquad (10.130)$$

where ϕ is the phase angle between A and B. For $\Delta z \gg \lambda$, it may be seen from eq. (10.130) that the irradiance is sinusoidally varying along the z direction within the emulsion. If this wavefront recording is linear in the developed photographic grain density, then the density function of the recorded reflection hologram may be written as

$$D(\mathbf{\rho}; k) = K_1 + K_2 \cos\left\{k\left[2(R+z) + \frac{\rho^2}{2(R+z)} + x \sin\theta\right] + \phi\right\},$$

$$\text{for } -\Delta z \leq z \leq 0, \qquad (10.131)$$

where K_1 and K_2 are the appropriate positive constants.

Indeed it can be seen that there is a sequence of very thin holograms arranged in parallel in the photographic emulsion, which act as reflecting planes. Since in practice the reference angle θ is very small, the spacing of these reflecting planes is about $\lambda/2$. If we assume that the reflectance of these thin holograms is proportional to the density of the developed photographic grains, then the reflectance function of the overall hologram may be written as

$$r(\mathbf{\rho}; k) = K_1' + K_2' \cos\left\{k\left[2(R+z) + \frac{\rho^2}{2(R+z)} + x \sin\theta\right] + \phi\right\},$$

$$\text{for } -\Delta z \leq z \leq 0, \qquad (10.132)$$

where K_1' and K_2' are the appropriate positive constants.

Now, if the hologram described by eq. (10.132) is illuminated by incoherent white light (fig. 10.25), then only a single reconstructing wavelength which satisfies Bragg's law (ref. 10.28) will be strongly reflected. Thus by means of the Fresnel-Kirchhoff theory, the reflected complex light field of this selected wavelength can be calculated to be

$$E(\mathbf{\sigma}; k) = C_1(z) \exp(-ik\alpha \sin\theta)$$

$$+ C_2(z) \exp\left\{\frac{k}{4(R+z)}[(\alpha - 2(R+z)\sin\theta)^2 + \beta^2]\right\} \qquad (10.133)$$

$$+ C_3(z)\,\delta(\alpha, \beta), \quad \text{for } -\Delta z \leq z \leq 0,$$

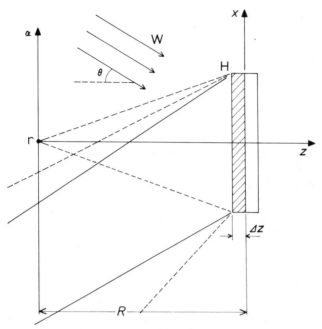

Fig. 10.25 Real image reconstruction, reflection hologram. H, hologram; W, white light illumination; r, real image.

where $C_1(z)$, $C_2(z)$, and $C_3(z)$ are complex functions of z, and $\delta(\alpha, \beta)$ is the Dirac delta function. Needless to say, the first term of eq. (10.133) is the zero order diffraction, the second term is the first order divergent term, and the last term obviously is the real hologram image term.

To reconstruct the virtual image, we illuminate the reflection hologram from behind with an oblique white light (fig. 10.26). The geometry shown in this figure for virtual image reconstruction is similar to that shown previously for the real image.

The above considerations can be extended to multiwavelength recording. To do so, let the object field and the reference wave of fig. 10.24 be derived from two coherent light sources of wavelength λ_1 and λ_2. Then the complex light field in the emulsion, due to the two-wavelength point source, is

$$u(\boldsymbol{\rho}; k_1, k_2) = u_1(\boldsymbol{\rho}; k_1) + u_2(\boldsymbol{\rho}; k_2)$$
$$= A_1 \exp\left\{ik_1\left[z + \frac{\rho^2}{2(R+z)}\right]\right\} + A_2 \exp\left\{ik_2\left[z + \frac{\rho^2}{2(R+z)}\right]\right\},$$

$$(10.134)$$

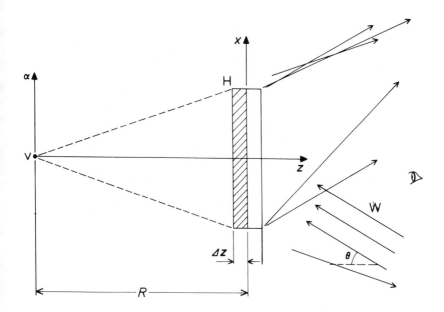

Fig. 10.26 Virtual image reconstruction, reflection hologram. H, hologram; W, white light illumination; v, virtual image.

and that due to the two-wavelength reference field is

$$
\begin{aligned}
v(\boldsymbol{\rho}; k_1, k_2) &= v_1(\boldsymbol{\rho}; k_1) + v_2(\boldsymbol{\rho}; k_2) \\
&= B_1 \exp\left[-ik_1(z + x \sin\theta)\right] + B_2 \exp\left[-ik_2(z + x \sin\theta)\right],
\end{aligned}
\tag{10.135}
$$

where the A's and B's are complex constants and $\rho^2 = x^2 + y^2$. Since u and v are derived from two independent coherent sources, the time average over the exposure of $u_1 u_2^*$ and $u_1^* u_2$ may be assumed negligible. Thus the reflectance of the hologram may be written as

$$
\begin{aligned}
r(\boldsymbol{\rho}; k_1, k_2) &= K \langle I \rangle \\
&= K\Delta t \left\{ |A_1|^2 + |A_2|^2 + |B_1|^2 + |B_2|^2 \right. \\
&\quad + 2|A_1|\,|B_1| \cos\left[k_1\left(2z + \frac{\rho^2}{2(R+z)} + x \sin\theta\right) + \phi_1\right] \\
&\quad \left. + 2|A_2|\,|B_2| \cos\left[k_2\left(2z + \frac{\rho^2}{2(R+z)} + x \sin\theta\right) + \phi_2\right] \right\},
\end{aligned}
\tag{10.136}
$$

where K is a proportionality constant, Δt is the exposure time, $\langle I \rangle$ denotes the time average of the irradiance over Δt, and ϕ_1 and ϕ_2 are the constant phase angles.

It may be seen from eq. (10.136) that this two-wavelength reflection hologram essentially consists of two sets of thin holograms within the emulsion, such that the spacing in one set is $\lambda_1/2$, and in the other set, $\lambda_2/2$. For real image reconstruction, a multiwavelength reflection hologram can be illuminated by an oblique white light, as shown in Fig. 10.25. Then by Bragg's law and the Fresnel-Kirchhoff theory, the complex reflected field can be evaluated. The solution is

$$
\begin{aligned}
E(\boldsymbol{\sigma}; k_1, k_2) = {} & C_{11}(z)\exp(-ik_1\alpha\sin\theta) + C_{12}(z)\exp(-ik_2\alpha\sin\theta) \\
& + C_{21}(z)\exp\left\{\frac{k_1}{4(R+z)}\left[(\alpha-2(R+z)\sin\theta)^2+\beta^2\right]\right\} \\
& + C_{22}(z)\exp\left\{\frac{k_2}{4(R+z)}\left[(\alpha-2(R+z)\sin\theta)^2+\beta^2\right]\right\} \\
& + C_{31}(z)\,\delta(k_1;\alpha,\beta) + C_{32}(z)\,\delta(k_2;\alpha,\beta),
\end{aligned}
\tag{10.137}
$$

where the $C(z)$'s are complex functions of z.

It can be seen from this equation that the real image is constructed by the two wavelengths λ_1 and λ_2. The image irradiance due to Δz is therefore

$$
I_r(\boldsymbol{\sigma}; k_1, k_2) =
\begin{cases}
\left|\displaystyle\int_{-\Delta z}^{0} C_{31}(z)\,dz\right|^2 + \left|\displaystyle\int_{-\Delta z}^{0} C_{32}(z)\,dz\right|^2, & \alpha=\beta=0 \\[6pt]
0, & \text{otherwise,}
\end{cases}
\tag{10.138}
$$

which is the sum of the image irradiances contributed by λ_1 and λ_2,

$$
I_r(\boldsymbol{\sigma}; k_1, k_2) = I_{r,1}(\boldsymbol{\sigma}; k_1) + I_{r,2}(\boldsymbol{\sigma}; k_2),
\tag{10.139}
$$

where the subscript r refers to the real image reconstruction. For virtual image reconstruction, the hologram can be illuminated by an oblique white light, as shown in fig. 10.26.

From this simple point-object reflection hologram, it is easy to extend the concept to the case of a three-dimensional object. Let us replace the point object S of fig. 10.24 by a three-dimensional object function $O(\xi, \eta, \zeta)$, and let the object and reference fields be of wavelengths λ_1 and λ_2. Then the complex light field within the photographic emulsion due to the object field is

$$
\begin{aligned}
u(\boldsymbol{\rho}; k_1, k_2) &= u_1(\boldsymbol{\rho}; k_1) + u_2(\boldsymbol{\rho}; k_2) \\
&= \iint_{S_o} O(\xi, \eta, \zeta)\left[E_l^+(\boldsymbol{\rho}-\boldsymbol{\xi}; k_1) + E_l^+(\boldsymbol{\rho}-\boldsymbol{\xi}; k_2)\right] d\xi\, d\eta,
\end{aligned}
\tag{10.140}
$$

where S_o denotes the surface integral over the object function, and E_l^+ denotes the spatial impulse response.

Retaining the assumptions we have made previously, we may write the reflectance of a full-color (i.e., one constructed from light of wavelengths $\lambda_1, \lambda_2, ..., \lambda_N$) hologram as

$$r(\rho; k_n) = K \left\{ \sum_{n=1}^{N} |u_n(\rho; k_n)|^2 + |B_n|^2 \right.$$

$$\left. + 2|u_n(\rho; k_n)| |B_n| \cos[\phi_n(\rho; k_n) + k_n x \sin\theta] \right\}, \tag{10.141}$$

where

$$u_n(\rho; k_n) = |u_n(\rho; k_n)| \exp[i\phi_n(\rho; k_n)], \text{ for } n = 1, 2, ..., n.$$

If this full-color hologram is illuminated by an incoherent white light, in the manner of fig. 10.25, then by means of the Fresnel-Kirchhoff theory the real image light field may be obtained:

$$E_r(\sigma; k_n) = \sum_{n=1}^{N} C_n(z) O(k_n; \alpha, \beta, \gamma), \tag{10.142}$$

where the subscript r denotes the real image term, and the $C_n(z)$'s are complex functions of z.

The corresponding irradiance is

$$I_r(\rho; k_n) = \sum_{n=1}^{N} K_n |O(k_n; \alpha, \beta, \gamma)|^2, \tag{10.143}$$

which is the sum of the image irradiance due to $\lambda_1, \lambda_2, ..., \lambda_N$. By a similar procedure, one can obtain the virtual hologram image.

One might expect the colors of the hologram image to be the same as those of the object. However, in practice, the wavelengths reflected are shorter than those of the recordings. This is due to emulsion shrinkage after the development and fixing process. In order to maintain the original wavelengths, we must prevent this shrinkage. There are certain techniques available for reswelling the emulsion after the fixing process. By careful application of these techniques, it is possible to prevent the emulsion shrinkage enough to preserve the original wavelengths over a broad spectral range.

Figure 10.27 is a photograph of a virtual image reconstruction produced from a single-wavelength reflection hologram under white light illumination.

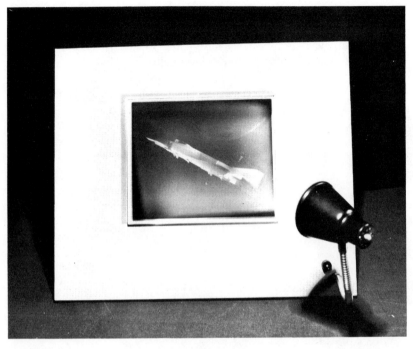

Fig. 10.27 A virtual image produced by white light illumination of a single-wavelength reflection hologram. (Courtesy of C. Charnetski, Conductron Corp.)

The analysis of linear holography given in this chapter has been based upon the point-source concept and linear system theory. Although some of the expressions seem rather long, the calculations were made in a straightforward fashion. The main use of this approach, of course, is for those problems that are simple enough to be directly evaluated. In a manner similar to that of geometrical optics, the holographic magnifications, resolutions, bandwidth requirements, etc., can in such cases be obtained in this simple way.

Problems

10.1 Referring to eqs. (10.44), (10.46), (10.58) and (10.61):

(a) derive the lateral and longitudinal magnifications for the case where a hologram is illuminated by an oblique monochromatic *plane* (rather than divergent) wave of wavelength λ_2.

(b) derive the lateral and longitudinal magnifications if the hologram is made and reconstructed by *convergent* reference and illumination beams. Show the relationship between the virtual and real image magnifications.

10.2 In a certain wavefront recording, the coherent light is derived from an argon laser of wavelength 4880 Å. The reference wave is from a divergent point source, which is located at 1 m away from the x axis of the photographic plate and 0.5 m below the optical axis of the recording system, as shown in fig. 10.28, and the object to be recorded is located at about 0.5 m away from the recording aperture. If the recorded transparency (i.e., hologram) is illuminated by a normally incident divergent point source, using light from a helium-neon laser of wavelength 6328 Å, and located at 0.5 m away from the hologram, then
(a) determine the corresponding lateral and longitudinal magnifications for the real and virtual images.
(b) compute the locations of the images.
(c) calculate the position of the divergent point source so that the translational distortions described in sec. 10.2 may be eliminated.

10.3 In a certain holographic process, the hologram is made by a divergent reference beam and the hologram images are reconstructed by a convergent source of the same wavelength. Calculate the locations and relationship between the divergent and convergent beams in order to have unity lateral magnification.

10.4 Suppose a hologram transparency is recorded in the manner shown in fig. 10.14. If this hologram is uniformly enlarged by a factor m, and this

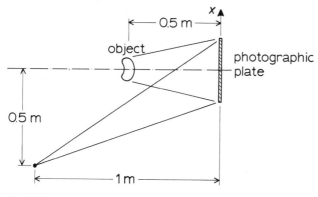

Fig. 10.28

enlarged hologram is illuminated by a divergent monochromatic source of different wavelength, as shown in fig. 10.29, then determine the real and virtual hologram image positions and their image magnifications.

10.5 Given a 10×12 cm, high resolution photographic plate. If holographic recording takes place with the illuminated object located on the optical axis of the recording system, at a distance of 24 cm in front of the plate, determine the lateral resolution limits of the hologram images.

10.6 In a certain holographic process, the recording aperture (i.e., the photographic plate) is assumed to be an 8×10 cm rectangle. The center of a circular object transparency of 6 cm diameter is placed on the optical axis, parallel to the plate, at a distance of 20 cm. If a plane reference wave oriented at 45° with respect to the major axis of the recording plate is used, determine the spatial frequency bandwidths of the major and minor axes of the hologram aperture.

10.7 With reference to fig. 10.21, by interchanging the reference plane wave and the reconstruction divergent beam for the holographic process, determine the third order primary aberrations, as well as the conditions for their corrections. Compare the results with eq. (10.119) and table 10.1.

10.8 If the separation between the object plane and the recording aperture of a holographic process is large enough to satisfy the Fraunhofer diffraction condition, show that the recorded wavefront is essentially a Fourier transform hologram (i.e., a hologram of the corresponding Fourier transform).

10.9 With reference to prob. 10.8, if the reference wave is not used during the recording, show that the resulting transparency is the power spectrum

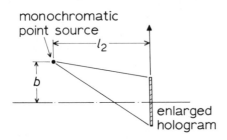

Fig. 10.29

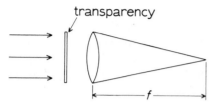

Fig. 10.30

of the object function. If this transparency is illuminated by a normally incident monochromatic plane wave, and the transmitted light field is imaged at the back focal length of a positive lens (fig. 10.30), show that the image irradiance is proportional to the square of the autocorrelation function of the object.

References

10.1 D. Gabor, A New Microscope Principle, *Nature*, **161** (1948), 777.

10.2 D. Gabor, Microscopy by Reconstructed Wavefronts, *Proc. Royal Soc.*, ser. A, **197** (1949), 454.

10.3 D. Gabor, Microscopy by Reconstructed Wavefronts, II, *Proc. Phys. Soc.*, ser B, **64** (1951), 449.

10.4 G. L. Rogers, Gabor Diffraction Microscopy: The Hologram as a Generalized Zone Plate, *Nature*, **166** (1950), 237.

10.5 H. M. A. El-Sum, Reconstructed Wavefront Microscopy, doctoral dissertation, Stanford University, 1952 (available from University Microfilms, Ann Arbor, Michigan).

10.6 A. Lohmann, Optical Single-Sideband Transmission Applied to the Gabor Microscope, *Opt. Acta*, **3** (1956), 97.

10.7 E. N. Leith and J. Upatnieks, Reconstructed Wavefront and Communication Theory, *J. Opt. Soc. Am.*, **52** (1962), 1123.

10.8 E. N. Leith and J. Upatnieks, Wavefront Reconstruction with Continuous-tone Objects, *J. Opt. Soc. Am.*, **53** (1963), 1377.

10.9 E. N. Leith and J. Upatnieks, Wavefront Reconstruction with Diffused Illumination and Three-dimensional Objects, *J. Opt. Soc. Am.*, **54** (1964), 1295.

10.10 J. B. DeVelis and G. O. Reynolds, *Theory and Applications of Holography*, Addison-Wesley, Reading, Mass., 1967.

10.11 R. W. Meier, Magnification and Third-order Aberrations in Holography, *J. Opt. Soc. Am.*, **55** (1965), 987.

10.12 E. N. Leith, J. Upatnieks, and K. A. Haines, Microscopy by Wavefront Reconstruction, *J. Opt. Soc. Am.*, **55** (1965), 981.

10.13 J. A. Armstrong, Fresnel Holograms: Their Imaging Properties and Aberrations, *IBM J. Develop.*, **9** (1965), 171.

10.14 F. I. Diamond, Magnification and Resolution in Wavefront Reconstruction, *J. Opt. Soc. Am.*, **57** (1967), 503.

10.15 J. T. Winthrop and C. R. Worthington, X-Ray Microscopy by Successive Fourier Transformation, *Phys. Letters*, **15** (1965), 124.

10.16 M. Born and E. Wolf, *Principles of Optics*, 2nd rev. ed., Pergamon Press, New York, 1964, pp. 211 ff.

10.17 H. B. Dwight, *Table of Integrals and Other Mathematical Data*, 3rd ed., Macmillan, New York, 1957.

10.18 L. Mertz and N. O. Young, Fresnel Transformations of Optics, in K. J. Habell (ed.), *Proceedings of the Conference on Optical Instruments and Techniques*, Wiley, New York, 1963, pp. 305 ff.

10.19 A. W. Lohmann, Wavefront Reconstruction for Incoherent Objects, *J. Opt. Soc. Am.*, **55** (1965), 1555.

10.20 G. W. Stroke and R. C. Restrick III, Holography with Spatially Noncoherent Light, *Appl. Phys. Letters*, **7** (1965), 229.

10.21 G. Cochran, New Method of Making Fresnel Transforms with Incoherent Light, *J. Opt. Soc. Am.*, **56** (1966), 1513.

10.22 P. J. Peters, Incoherent Holograms with Mercury Light Source, *Appl. Phys. Letters*, **8** (1966), 209.

10.23 M. G. Lippmann, La Photographie des Couleurs, *Compt. Rend.*, **112** (1891), 274.

10.24 Y. N. Denisyuk, Photographic Reconstruction of the Optical Properties of an Object in Its Own Scattered Radiation Field, *Soviet Physics-Doklady*, **7** (1962), 1275.

10.25 G. W. Stroke and A. E. Labeyrie, White-light Reconstruction of Holographic Images Using the Lippmann-Bragg Diffraction Effect, *Phys. Letters*, **20** (1966), 368.

10.26 L. H. Lin et al., Multicolor Holographic Image Reconstruction with White-light Illumination, *Bell System Tech. J.*, **45** (1966), 659.

10.27 E. N. Leith et al., Holographic Data Storage in Three-dimensional Media, *Appl. Opt.*, **5** (1966), 1303.

10.28 F. W. Sears, *Optics*, Addison-Wesley, Cambridge, Mass., 1949, p. 243.

Chapter 11 Analysis of Nonlinear Holograms

The detailed mechanisms of the transfer modulation functions and non-linear effects in wavefront reconstruction processes have been discussed by Kozma (ref. 11.1), Freisem and Zelenka (ref. 11.2), Goodman and Knight (ref. 11.3), and Bryngdahl and Lohmann (ref. 11.4). However, the purpose of this chapter (which is based upon refs. 11.5–11.7) is to study the nonlinear effects in wavefront reconstructions from an elementary system theory point of view. As we have seen in the previous chapters, a hologram may be conveniently represented by a black-box input-output system analog. Therefore the study of nonlinear effects in holography may be approached as the study of a nonlinear system. Since the wavefront reconstruction of an extended three-dimensional object may be regarded as composed of a large number of resolvable point-object constructions, we may begin with an elementary point concept, and then extend to the more general case.

11.1 Finite-Point Analysis

In the on-axis wavefront construction depicted in fig. 10.1, a monochromatic point source is located a distance R away from the recording medium, and a plane reference wave of the same wavelength is traveling from left to right toward the photographic plate. The exposure due to the combination of these two wavefronts on the recording medium may be approximated by [eq. (10.8)]

$$E(\rho; k) = I(\rho; k)\, t \simeq \left[\left(\frac{|A|}{R} \right)^2 + |B|^2 + \frac{2|A|\,|B|}{R} \cos\left(\frac{k\rho^2}{2R} + \phi \right) \right] t, \qquad (11.1)$$

where A and B are the complex amplitudes of the point source and reference waves, respectively, ϕ is the phase angle between A and B, $\rho^2 = x^2 + y^2$, $k = 2\pi/\lambda$, with λ the wavelength, and t is the exposure time.

From the amplitude transmittance characteristic of a physical photographic plate, as shown in fig. 11.1, we note that the linear region is very narrow. From eq. (11.1), the quiescent point (the bias) depends on the amplitudes of the two waves arriving at the recording plate, i.e.,

$$E_Q = \left[\left(\frac{|A|}{R} \right)^2 + |B|^2 \right] t. \qquad (11.2)$$

If the amplitude of the reference wave is much greater than that of the spherical wave distributed on the photographic film, i.e., if $|B| \gg |A|/R$,

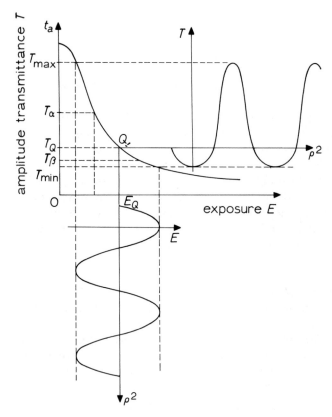

Fig. 11.1 Five-point analysis of a nonlinear hologram.

then eq. (11.1) becomes

$$E(\rho; k) \simeq \left[|B|^2 + \frac{2|A|\,|B|}{R} \cos\left(\frac{k\rho^2}{2R} + \phi\right) \right] t. \tag{11.3}$$

The quiescent value of the exposure for this case is

$$E_Q \simeq |B|^2\, t, \quad \text{for} \quad |B| \gg \frac{|A|}{R}, \tag{11.4}$$

which is independent of $|A|$. Thus, the linear operating region of the transmission amplitude versus exposure intensity can be achieved by properly adjusting the magnitude of B, and hence E_Q, of the reference beam.

On the other hand, if $|A|/R$ is either comparable to or greater than $|B|$, then it is difficult to maintain the holographic process within the bounds of the linear region of the amplitude transmittance characteristic. Thus nonlinear distortion of the amplitude transmission will result, as shown in fig. 11.1.

Either from eq. (11.1) or from fig. 11.1, it is apparent that amplitude transmittance of the point-object hologram is periodic in the ρ^2 axis. From the elementary Fourier theorem, the amplitude transmittance of this hologram may be expanded in a Fourier series on the ρ^2 axis, or

$$T(\rho; k) = \frac{a_0}{2} + \sum_{n=1}^{\infty} a_n \cos \frac{nk}{2R} \rho^2, \tag{11.5}$$

where $\rho(x, y, z)$ denotes the given coordinate system, and a_n is the corresponding Fourier coefficient:

$$a_n = \frac{2}{\lambda R} \int_0^{\lambda R} T(\rho; k) \cos \frac{nk}{2R} \rho^2 \, d\rho^2, \quad n = 0, 1, 2, 3, \dots. \tag{11.6}$$

Note: Without loss of generality, we can drop the phase angle ϕ from the calculation. From eq. (11.5), the amplitude transmittance $T(\rho; k)$ is that of the sum of an infinite number of Fresnel zone lenses (sec. 4.5):

$$T(\rho; k) = T_0 + \sum_{n=1}^{\infty} T_n(\rho; k), \tag{11.7}$$

where

$$T_0 = \frac{a_0}{2},$$

$$T_n(\rho; k) = a_n \cos \frac{nk}{2R} \rho^2, \quad n = 1, 2, 3, \dots.$$

The focal lengths of the zone lenses are therefore

$$f_n = \frac{R}{n}, \quad n = 1, 2, 3, \dots. \tag{11.8}$$

If a recorded photographic plate (hologram) having an amplitude transmittance given by eq. (11.7) is illuminated by a monochromatic plane wave of wavelength λ (fig. 11.2), the complex light field behind the hologram may be determined from the convolution theorem,

$$E(\sigma; k) = \iint_s T(\rho; k) \, E_l^+(\sigma - \rho; k) \, dx \, dy, \tag{11.9}$$

where

$$E_l^+ (\boldsymbol{\rho}; k) = -\frac{i}{\lambda l} \exp \left[ik \left(l + \gamma - z + \frac{\rho^2}{2l} \right) \right]$$

is the free space impulse response, S denotes the integration over the entire surface of $T(\boldsymbol{\rho}; k)$, and $\boldsymbol{\sigma}(\alpha, \beta, \gamma)$ and $\boldsymbol{\rho}(x, y, z)$ denote the coordinate systems.

By substituting eq. (11.7) in eq. (11.9), we obtain

$$E(\boldsymbol{\sigma}; k) = \int\int_S \left[T_0 + \sum_{n=1}^{\infty} T_n(\boldsymbol{\rho}; k) \right] E_l^+ (\boldsymbol{\sigma} - \boldsymbol{\rho}; k) \, dx \, dy. \tag{11.10}$$

Without loss of generality, we may take the surface S to be of infinite extent. Since we are interested in the hologram image reconstruction due to each of the Fresnel zone lenses, we will evaluate the above integral term by term with respect to the focal length of the corresponding lens. Thus

$$
\begin{aligned}
E(\boldsymbol{\sigma}_n; k) = & -i \frac{na_n}{2\lambda R} \exp \left(i \frac{kR}{n} \right) \left\{ \frac{\lambda R}{2n} \exp \left[i \left(\frac{nk}{4R} \sigma_n^2 + \frac{\pi}{2} \right) \right] \right. \\
& \left. + \exp \left(i \frac{nk}{2R} \sigma_n^2 \right) \delta(\alpha_n, \beta_n) \right\}, \quad n = 1, 2, 3, \ldots,
\end{aligned}
\tag{11.11}
$$

where $E(\boldsymbol{\sigma}_n; k)$ is the complex light field due to the zone lens $T_n(\boldsymbol{\rho}; k)$ alone at its focal length f_n and where $\sigma_n^2 = \alpha_n^2 + \beta_n^2$.

Obviously the first term of eq. (11.11) represents the virtual image (i.e., divergent) term, and the second term is the real image (convergent) term. The higher order reconstructed images lie between the first order real and virtual images, as shown in fig. 11.2. These higher order images are due to the nonlinearity of the hologram. The degree of the nonlinearity of the hologram may be defined as

$$\text{nonlinearity (percent)} = \frac{\left(\sum_{n=2}^{\infty} a_n^2 \right)^{1/2}}{a_1} \times 100. \tag{11.12}$$

The analysis of nonlinear holograms is similar to that used in nonlinear amplifier theory; and to determine the Fourier coefficients a five-point

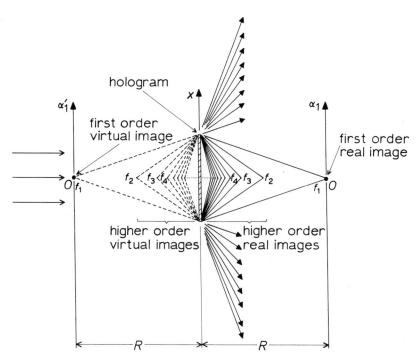

Fig. 11.2 Wavefront reconstruction of an on-axis point-object nonlinear hologram. Monochromatic plane wave illumination.

analysis (ref. 11.8) as shown in fig. 11.1 may be adequate. Thus,

$$\frac{a_0}{2} = \frac{T_{\max} + T_{\min}}{6} + \frac{T_\alpha + T_\beta}{3},$$

$$a_1 = \frac{T_{\max} - T_{\min}}{3} + \frac{T_\alpha - T_\beta}{3},$$

$$a_2 = \frac{T_{\max} + T_{\min}}{4} - \frac{T_Q}{2}, \tag{11.13}$$

$$a_3 = \frac{T_{\max} - T_{\min}}{6} - \frac{T_\alpha - T_\beta}{3},$$

$$a_4 = \frac{T_{\max} + T_{\min}}{12} - \frac{T_\alpha + T_\beta}{3} + \frac{T_Q}{2}.$$

It is clear that for a more accurate result, a higher order analysis can be applied. Furthermore, for extreme distortion, the hologram is composed of totally transparent and opaque Fresnel zones. The corresponding amplitude transmittance may be written as

$$T(\boldsymbol{\rho}; k) = \frac{1}{2} + \frac{2}{\pi} \sum_{n=1}^{\infty} \frac{-(-1)^n}{2n-1} \cos \frac{k}{2R} \rho^2. \tag{11.14}$$

Equation (11.14) differs from eq. (11.7) only in the Fourier coefficients. The image reconstructions can be shown to be located at f_n, which is identical to the case of eq. (11.8). However, the degree of nonlinearity may be shown to be somewhat higher than all the unsaturated cases. On the other hand, for extreme linearity in the holographic recording, it can be seen that the irradiances of the higher order reconstructions vanish.

Let us suppose that the recorded photographic plate of eq. (11.5) is illuminated by an oblique monochromatic plane wave of wavelength λ, as shown in fig. 11.3. Again by the Fresnel-Kirchhoff theory, the complex light field behind the hologram is

$$E(\boldsymbol{\sigma}; k) = \iint_S T(\boldsymbol{\rho}; k)\, e^{ikx\sin\theta}\, E_l^+(\boldsymbol{\sigma} - \boldsymbol{\rho}; k)\, dx\, dy. \tag{11.15}$$

By substituting eq. (11.5) in eq. (11.15), we see that the complex light field

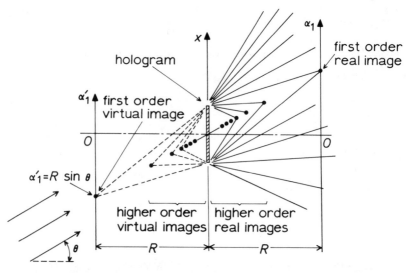

Fig. 11.3 Oblique plane monochromatic illumination of an on-axis nonlinear hologram.

$E(\boldsymbol{\sigma}_n; k)$ due to $T_n(\boldsymbol{\rho}; k)$ at f_n is

$$E(\boldsymbol{\sigma}_n; k) = -i\frac{a_n}{2}\exp\left(i\frac{kR}{n}\right)\cdot\left\{\frac{1}{2}\exp\left[i\frac{\pi}{2} - \frac{kR}{2n}\sin^2\theta\right]\right.$$
$$\times\exp\left[i\frac{nk}{2R}\left(\sigma_n + \frac{R}{n}\sin\theta\right)^2\right]$$
$$\left. + \exp\left(i\frac{nk}{2R}\sigma_n^2\right)\delta\left(\alpha_n - \frac{R}{n}\sin\theta, \beta_n\right)\right\}, \quad n = 1, 2, \ldots. \quad (11.16)$$

The two terms of eq. (11.16) can be recognized again as virtual and real hologram images, as shown in fig. 11.3. Apparently, oblique illumination is not able to separate the first order hologram image from the higher order image reconstructions. In order to do so, an oblique reference wave should be used during the wavefront construction, as will be seen in the next section.

11.2 Off-Axis Nonlinear Hologram
In order to separate the zero order, virtual, and real hologram image diffractions, an oblique reference wave may be used for recording, as shown in fig. 11.4. The resulting exposure may be approximated by

$$E(\boldsymbol{\rho}; k) = \left\{|A_1|^2 + |B|^2 + 2|A_1|\,|B|\cos\left[k\left(\frac{\rho^2}{2R} - x\sin\theta\right) + \phi\right]\right\}t, \quad (11.17)$$

where A_1 and B are the complex light amplitude distributions from the point source and the reference beam, respectively, ϕ is the phase angle between them, and t is the exposure time.

It is a simple procedure to show that the exposure described by eq. (11.17) consists of concentric circles with a common center at

$$x = R\sin\theta, \quad y = 0;$$

the corresponding maximum and minimum exposures occur at

$$\rho = \left[R\left(n\lambda - \frac{\lambda}{\pi}\phi\right) + R\sin^2\theta\right]^{1/2}, \quad n = 0, 1, 2, \ldots.$$

If we let $x = x_1 + R\sin\theta$, then we may write eq. (11.17) as

$$E(\boldsymbol{\rho}; k) = \left\{|A_1|^2 + |B|^2 + 2|A_1|\,|B|\cos\left[\frac{k}{2R}\rho_1^2 - \frac{kR}{2}\sin^2\theta + \phi\right]\right\}t. \quad (11.18)$$

Except for a constant phase factor, eq. (11.18) is identical to eq. (11.1).

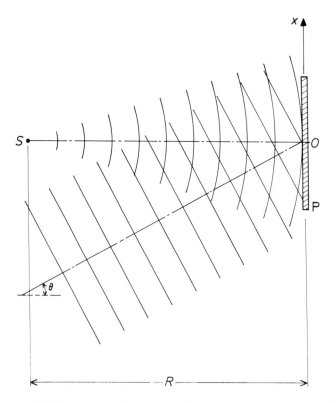

Fig. 11.4 Off-axis wavefront construction of a nonlinear hologram. Oblique plane monochromatic reference wave. P, photographic plate.

Therefore, by the same argument as used in the previous section, the amplitude transmittance of the hologram is

$$T(\boldsymbol{\rho}; k) = \frac{a_0}{2} + \sum_{n=1}^{\infty} a_n \cos\left[nk\left(\frac{\rho^2}{2R} - x\sin\theta\right) + n\phi \right], \qquad (11.19)$$

where

$$a_n = \frac{1}{\lambda R} \int_{-\lambda R}^{\lambda R} T(\boldsymbol{\rho}; k) \cos\left[nk\left(\frac{\rho^2}{2R} - x\sin\theta\right) + n\phi \right] d\rho^2, \quad n = 0, 1, 2, \ldots$$

Then by proper translation of the x axis, we can show that the Fourier components in eq. (11.19) are identical to those in eq. (11.5). Hence, the amplitude transmittance $T(\boldsymbol{\rho}; k)$ can be expressed as the sum of an infinite number of Fresnel zone lenses, that is

$$T(\boldsymbol{\rho}; k) = T_0 + \sum_{n=1}^{\infty} T_n(\boldsymbol{\rho}; k), \tag{11.20}$$

where $T_0 = a_0/2$, and

$$T_n(\boldsymbol{\rho}; k) = a_n \cos\left[nk\left(\frac{\rho^2}{2R} - x \sin\theta\right) + n\phi\right], \quad n = 1, 2, \ldots.$$

If this hologram is illuminated by an oblique monochromatic plane wave of wavelength λ (fig. 11.5), then the complex light amplitude distribution may be evaluated by the convolution equation,

$$E(\boldsymbol{\rho}; k) = \int\int_S T(\boldsymbol{\rho}; k)\, e^{-ikx \sin\theta} E_l^+(\boldsymbol{\sigma} - \boldsymbol{\rho}; k)\, dx\, dy. \tag{11.21}$$

Note that the readout angle could be any other than θ, but for simplicity we have used the same θ. Again, the wavefront reconstructions due to the individual Fresnel zone lenses are exhibited by the termwise evaluation

$$E(\boldsymbol{\sigma}_n; k) = K_1 \exp\left\{i\frac{nk}{4R}\left[\left(\alpha_n - R\left(1 + \frac{1}{n}\right)\sin\theta\right)^2 + \beta_n^2\right]\right\}$$
$$+ K_2\, \delta\left[\alpha_n - R\left(1 - \frac{1}{n}\right)\sin\theta, \beta_n\right], \quad n = 1, 2, 3, \ldots, \tag{11.22}$$

where K_1 and K_2 are the appropriate complex constants.

As shown in fig. 11.5, the sequence of higher order images lies between the primary front and back focal points of the hologram. The same figure shows that by a proper choice of the angle θ of the reference wave, the first order real image can be separated from all of the unwanted images.

On the other hand, if the point-object hologram is illuminated by a monochromatic plane wave with the same wavelength but positive obliquity as shown in fig. 11.6, then the complex light amplitude distribution at the focal plane of the associated zone lens is

$$E(\boldsymbol{\sigma}_n; k) = K_1' \exp\left\{i\frac{nk}{4R}\left[\left(\alpha_n - R\left(1 - \frac{1}{n}\right)\sin\theta\right)^2 + \beta_n^2\right]\right\}$$
$$+ K_2'\, \delta\left[\alpha_n - R\left(1 + \frac{1}{n}\right)\sin\theta, \beta_n\right], \quad n = 1, 2, 3, \ldots, \tag{11.23}$$

where K_1' and K_2' are the appropriate complex constants. From eq. (11.23),

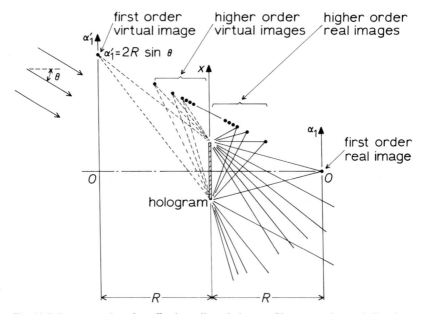

Fig. 11.5 Reconstruction of an off-axis nonlinear hologram. Plane monochromatic illumination, negative θ.

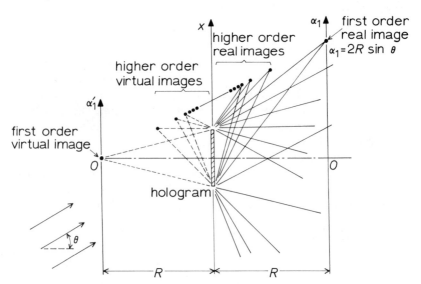

Fig. 11.6 Reconstruction of an off-axis nonlinear hologram. Plane monochromatic illumination, positive θ.

once again a sequence of higher order diffracted images is determined. As before, it is possible to separate the first order virtual image from the higher order diffractions.

Since in most holographic applications the light amplitude distribution on the recording medium from the object is approximately uniform, the transmittance of a nonlinear hologram can be written as

$$T(\boldsymbol{\rho}; k) = \sum_{n=0}^{\infty} T_n(\boldsymbol{\rho}; k), \tag{11.24}$$

where

$$T_n(\boldsymbol{\rho}; k) = a_n \cos\{n[kx \sin\theta + \phi(x, y)]\},$$

$$a_n = \frac{1}{d} \int_d T(\boldsymbol{\rho}; k) \cos\{n[kx \sin\theta + \phi(x, y)]\} \, d\rho,$$

and d is the corresponding spatial period. Of course the Fourier components can be determined by graphical analysis, as in the previous section.

First order image reconstruction can therefore be separated from the unwanted higher order diffractions by off-axis wavefront recording. An illustration of the image reconstructions from an off-axis nonlinear hologram is given in fig. 11.7. It can be seen from this figure that the minimum oblique angle of the reference wave can be determined from the size of the object and the hologram aperture. A photograph of a nonlinear hologram image, reconstructed by means of a beam of laser light, is given in fig. 11.8. It can be seen from this figure that, besides the two first order image reconstructions near the zero order diffraction in the center, there are two sets of the higher order reconstructions.

11.3 Spurious Distortion

At this point it is possible to calculate the distortion due to all the unwanted signals (i.e., the zero and higher order, as well as first order virtual or real hologram images), which will be defined as spurious distortion. To do this, we may recall the case of fig. 11.5. Equation (11.21) is evaluated at $l = R$,

$$E(\boldsymbol{\sigma}; k)\bigg|_{l=R} = \int_S \int T(\boldsymbol{\rho}; k) \, e^{-ikx\sin\theta} E_l^+(\boldsymbol{\sigma} - \boldsymbol{\rho}; k)\bigg|_{l=R} dx \, dy; \tag{11.25}$$

then the complex light amplitude distribution at $l = R$ is

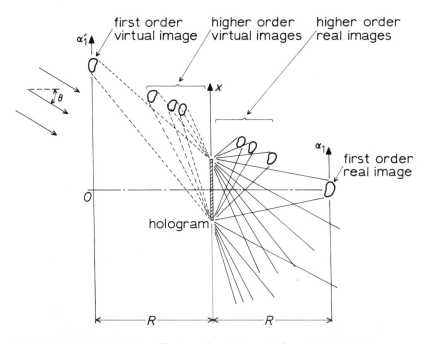

Fig. 11.7 Reconstruction of an off-axis nonlinear hologram of an extended object.

Fig. 11.8 Nonlinear hologram images reconstructed by means of a laser beam.

$$E(\boldsymbol{\sigma};k)\bigg|_{l=R} = \frac{a_0}{2}\exp\left[ik\left(R - \frac{R}{2}\sin^2\theta - \alpha\sin\theta\right)\right]$$

$$+\frac{1}{2}e^{ik(R-\alpha\sin\theta)}\sum_{n=1}^{\infty}\frac{a_n}{n+1}$$

$$\times\exp\left[i\left(n\phi - \frac{n+1}{2}kR\sin\theta + \frac{kn}{2R(n+1)}\sigma^2\right)\right]$$

$$+\frac{a_1}{2}\exp\left[i\left(kR + \frac{k}{2R}\sigma^2 - \phi - \frac{\pi}{2}\right)\right]\delta(\alpha,\beta)$$

$$-\frac{1}{2}e^{ik(R-\alpha\sin\theta)}\sum_{n=2}^{\infty}\frac{a_n}{n-1}$$

$$\times\exp\left[i\left(\frac{n-1}{2}kR\sin^2\theta - n\phi + \frac{kn}{2R(n-1)}\sigma^2\right)\right]. \quad (11.26)$$

The degree of spurious distortion D_s may be defined as the total un-wanted irradiance in the neighborhood of a reconstructed image divided by the irradiance of that image, i.e.,

$$D_s = \frac{\displaystyle\sum_{n=2}^{\infty}\left(\frac{a_n}{n-1}\right)^2 + \sum_{n=1}^{\infty}\left(\frac{a_n}{n+1}\right)^2 + a_0^2}{a_1^2}, \quad (11.27)$$

for $\theta = 0$. If there is no nonlinear distortion, then for $\theta = 0$ the distortion is

$$D_s = \frac{\left(\dfrac{a_1}{2}\right)^2 + a_0^2}{a_1^2}.$$

If the unwanted signals be called noise,* then the signal-to-noise ratio may be defined by

$$\text{signal-to-noise ratio} = \frac{1}{D_s}. \quad (11.28)$$

It is a simple matter to show that the degree of spurious distortion for the case shown in fig. 11.6 is given by eq. (11.27).

Furthermore, as implied in figs. 11.5, 11.6, and 11.7, for a given finite hologram aperture, it is possible to reduce the spurious distortion to a minimum by an appropriate choice of the reference-beam angle.

* Reluctantly called "noise," since the unwanted signals are deterministic. Strictly speaking, noise should be probabilistic in nature.

We should note that Fourier-decomposition analysis of nonlinear holograms may be difficult to apply when the irradiance of the recording field is nonuniform. However, the method not only provides a direct picture of what causes the distortion, but also provides a simple graphical procedure for obtaining the behavior of the hologram. In any event, the location of the reconstructed higher order images can be determined by means of this method; qualitatively, the degree of nonlinear and spurious distortion can be easily determined. Of course, an analysis of nonlinear holograms can be obtained by utilizing the exact mathematical relation between the exposure and the amplitude transmittance of the recording emulsion. This may be accomplished by expanding the transfer characteristic of the amplitude transmittance in a power series (refs. 11.1–11.4), or in a suitable set of orthogonal functions. However, this method is more complicated, and is difficult to apply in practice.

11.4 Effect of Emulsion Thickness Variation on Wavefront Reconstruction

Precise holographic-image reconstruction is required in certain engineering applications, e.g., in contouring, holographic interferometry, microscopic wavefront reconstruction, etc. We will now investigate the effects of photographic emulsion thickness variations on the precision of image reconstruction (ref. 11.9). In the following, a general mathematical approach will be developed for wavefront recording and reconstruction on a holographic plate of nonuniform emulsion thickness. We will see that the emulsion thickness variation does not affect wavefront recording in any essential way. However, the emulsion thickness variation will appreciably affect the precision of the hologram image. A simplified example of the effect will be illustrated.

EFFECT ON HOLOGRAPHIC RECORDING

Given a photographic plate in which the thickness of the thin emulsion* is nonuniform over the plate (fig. 11.9). Then the phase delay due to the emulsion thickness at points in the coordinate system (x, y) may be written as

$$\phi(x, y) = k \left[z_0 + (\eta - 1) \, z(x, y) \right], \tag{11.29}$$

where $z(x, y)$ is the thickness variation of the emulsion, z_0 is the maximum thickness of the emulsion, η is the index of refraction of the emulsion, and k is the wave number.

* A thin emulsion is defined in the same way as a thin lens: a light ray entering a point of the coordinate system (x, y) from one side of the emulsion emerges at approximately the same point on the other side of the coordinate system.

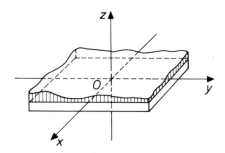

Fig. 11.9 Cross section of a photographic plate, showing emulsion thickness variation.

If the thickness variation $z(x, y)$ varies smoothly over the surface of the photographic plate, then the corresponding phase transmittance may be approximated by a Taylor expansion,

$$\phi(x, y) = \phi(0, 0) + \frac{\partial \phi(0, 0)}{\partial x} x + \frac{\partial \phi(0, 0)}{\partial y} y + \frac{1}{2} \frac{\partial^2 \phi(0, 0)}{\partial x^2} x^2$$
$$+ \frac{1}{2} \frac{\partial^2 \phi(0, 0)}{\partial y^2} y^2 + \frac{\partial^2 \phi(0, 0)}{\partial x \, \partial y} xy + \cdots . \tag{11.30}$$

If this photographic plate is used as the recording medium in the geometry shown in fig. 11.10, then the effect of the emulsion thickness variation on the scattered light is given by

$$u'(x, y) = u(x, y) \exp[i\phi(x, y)] \tag{11.31}$$

and

$$v'(x, y) = v \exp\{i[kx \sin\theta + \phi(x, y)]\}, \tag{11.32}$$

where $u(x, y)$ and $v \exp(ikx \sin\theta)$ are the complex light amplitude distributions scattered from the object and from the reference wave, respectively, without the emulsion thickness effect, with v a positive real constant. The resulting irradiance of the photographic plate is

$$\begin{aligned} I(\rho; k) &= [u'(x, y) + v'(x, y)][u'(x, y) + v'(x, y)]^* \\ &= |u(x, y)|^2 + v^2 + vu(x, y) \exp(-ikx \sin\theta) \\ &\quad + vu^*(x, y) \exp(ikx \sin\theta). \end{aligned} \tag{11.33}$$

If the wavefront construction is biased so that the transmittance of the plate remains always linear in irradiance, and if the lateral emulsion shrinkage does not occur during the photographic process, then the transmittance of the recorded hologram is

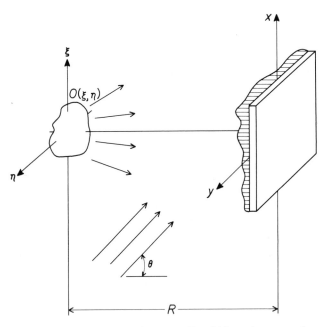

Fig. 11.10 Geometry of wavefront recording. Oblique plane monochromatic reference wave.

$$T(\boldsymbol{\rho}; k) = K\left[|u(x, y)|^2 + v^2 + u(x, y)\, v\, \exp(-ikx\sin\theta)\right.$$
$$\left. + vu^*(x, y)\, \exp(ikx\sin\theta)\right]. \tag{11.34}$$

From either of eqs. (11.33) or (11.34), it is clear that the emulsion thickness variation has not affected the wavefront construction.

EFFECT ON IMAGE RECONSTRUCTION

If the recorded hologram of eq. (11.34) is illuminated by a plane wavefront of the same wavelength, with the geometry as shown in fig. 11.11, then the light amplitude distribution behind the hologram is

$$E(\boldsymbol{\sigma}; k) = \int\int_s T(\boldsymbol{\rho}; k)\, v\, \exp(-ikx\sin\theta)\, \exp[i\phi(x, y)]\, E_l^+(\boldsymbol{\sigma}-\boldsymbol{\rho}; k)\, dx\, dy,$$
$$\tag{11.35}$$

where

$$E_l^+(\boldsymbol{\rho}; k) = -\frac{i}{\lambda l}\exp[ik(l-z+\gamma)]\exp\left[i\frac{k}{2l}(x^2+y^2)\right]$$

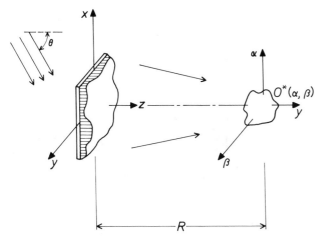

Fig. 11.11 Reconstruction of the real image. Plane monochromatic illumination, negative θ.

is the spatial impulse response, l is the separation between the two co-ordinate systems $\rho(x, y, z)$ and $\sigma(\alpha, \beta, \gamma)$, and S denotes the integration over the hologram surface.

Equation (11.35) can be evaluated termwise with respect to the order of image diffraction. Thus for real image reconstruction, the complex light amplitude at $l = R$ can be written as

$$E_r(\sigma; k) = v^2 \iint\limits_{S} u^*(x, y) \exp\left[i\phi(x, y)\right] E_l^+ (\sigma - \rho; k) \, dx \, dy, \tag{11.36}$$

where the subscript r denotes the real image.

Let the conjugate light amplitude distribution be

$$u^*(x, y) = \iint\limits_{S_1} O^*(\xi, \eta) E_l^{+*}(\rho - \xi; k) \, d\xi \, d\eta, \tag{11.37}$$

where $O(\xi, \eta)$ is the object function; S_1 denotes the surface integral of the object, viewed from the hologram aperture. By substitution of eq. (11.37), eq. (11.36) becomes

$$E_r(\sigma; k) = K_1 \iint\limits_{S} \left\{ \iint\limits_{S_1} O^*(\xi, \eta) \exp\left[-i\frac{k}{2R}(\xi^2 + \eta^2)\right]\right.$$

$$\times \exp\left[i\frac{k}{R}(\xi x + \eta y)\right] d\xi\, d\eta\Bigg\} \exp[i\phi(x, y)]$$

$$\times \exp\left[i\frac{k}{2R}(\alpha^2 + \beta^2)\right] \exp\left[-i\frac{k}{R}(\alpha x + \beta y)\right] dx\, dy, \qquad (11.38)$$

where K_1 is an appropriate complex constant.

Alternatively, eq. (11.38) can be written as

$$E_r(\boldsymbol{\sigma}; k) = K_1 \exp\left[i\frac{k}{2R}(\alpha^2 + \beta^2)\right] \int\int_S \left\{ G^*(x_0, y_0) * \exp\left[i\frac{R}{2k}(x_0^2 + y_0^2)\right] \right\}$$

$$\times \exp[i\phi(x, y)] \exp\left[-i\frac{k}{R}(\alpha x + \beta y)\right] dx\, dy, \qquad (11.39)$$

where $G^*(x_0, y_0)$ is the Fourier transform of the conjugate object function $O^*(\xi, \eta)$.

If there is no emulsion thickness variation, i.e., if $\phi(x, y) = 0$, the image reconstruction will be unique:

$$E_r(\boldsymbol{\sigma}; k) = K_1 O^*(\alpha, \beta). \qquad (11.40)$$

Furthermore, if the phase delay is a first-degree equation (i.e., a perfect wedge),

$$\phi(x, y) = b_0 + b_1 x + b_2 y, \qquad (11.41)$$

then the precision of the image reconstruction will also be unique, with a lateral translation

$$E_r(\boldsymbol{\sigma}; k) = K_1 O^*\left(\alpha - \frac{l}{k}b_1, \beta - \frac{l}{k}b_2\right). \qquad (11.42)$$

On the other hand, if the higher order terms of the phase delay due to the emulsion thickness variation are considered, then exact precision of the image reconstruction would not be expected. In the following, a simplified example will be given. We will see that a very small amount of emulsion thickness variation could severely affect the precision of image reconstruction, and would have to be restricted in some engineering applications.

SIMPLIFIED EXAMPLE

We have seen that impulse-response techniques are applicable to linear holography, since superposition holds. Therefore, in the following example the effect due to the emulsion thickness variation will be derived for a discrete-point source. Let us consider the wavefront recording of two point objects separated by a distance h (fig. 11.12). If these monochromatic point sources are assumed to have identical wavelength λ, then the light amplitude distribution on the (x, y) coordinate system may be approximated by

$$u_1(\rho; k) = A_1 \exp\left\{i\left[kR + \frac{k}{2R}\left(\left(x - \frac{h}{2}\right)^2 + y^2\right) + \phi(x, y)\right]\right\} \tag{11.43}$$

and

$$u_2(\rho; k) = A_2 \exp\left\{i\left[kR + \frac{k}{2R}\left(\left(x + \frac{h}{2}\right)^2 + y^2\right) + \phi(x, y)\right]\right\}, \tag{11.44}$$

where $\phi(x, y)$ is the phase delay due to the emulsion thickness variation, and A_1 and A_2 are arbitrary positive constants.

If a plane reference wave (of the same wavelength as that of the object wave) is used [i.e., the effect of the emulsion is simply that given by eq. (11.32)], then the transmittance of the recorded plate is

$$T(\rho; k) = \left\{a_0 + a_1 \cos\frac{k}{R} hx + \frac{a_2}{2}\left[\exp(i\wedge) + \exp(-i\wedge)\right]\right.$$
$$\left. + \frac{a_3}{2}\left[\exp(i\wedge \wedge) + \exp(-i\wedge \wedge)\right]\right\} \exp[i\phi(x, y)], \tag{11.45}$$

where

$$\wedge = kR + \frac{k}{2R}\left[\left(x - \frac{h}{2}\right)^2 + y^2\right]$$

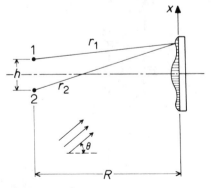

Fig. 11.12 Geometry for recording two point objects. Plane monochromatic reference wave, positive θ.

and

$$\wedge \wedge = kR + \frac{k}{2R}\left[\left(x+\frac{h}{2}\right)^2 + y^2\right].$$

If this hologram is illuminated by a plane wave of the same wavelength (fig. 11.13), then the light amplitude distribution behind the hologram may be determined by the Fresnel-Kirchhoff theory, i.e., by eq. (11.35). However, the evaluation of eq. (11.35) is in general very complicated, since the phase delay $\phi(x, y)$ is not generally specified. In order to see some of the effect of the emulsion thickness variation, we will carry out the computation for a simplified case. Let the phase delay be approximated by a second order polynomial,

$$\phi(x, y) \simeq b_0 + b_1 x + b_2 y + b_3 x^2 + b_4 y^2, \tag{11.46}$$

where the b's are arbitrary real constants.

It can be recognized that the third and fifth terms of eq. (11.45) contribute to the virtual images, and the fourth and sixth terms to the real images. Thus if we separate the Fresnel-Kirchhoff integral for the real image terms we obtain

$$E_r(\sigma; k) = K_1 \int_{S_x} \exp\left[-i\frac{k}{2}\left(\frac{1}{R} - \frac{1}{l} - \frac{2}{k}b_3\right)x^2\right]$$

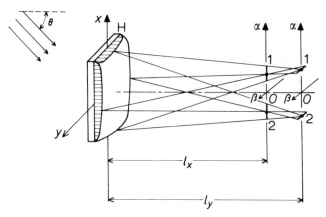

Fig. 11.13 Reconstruction of the real two-point image. Plane monochromatic illumination, negative θ.

$$\times \exp\left[-i\frac{k}{l}\left(\alpha-\frac{lh}{2R}-\frac{l}{k}b_1\right)\right]dx$$

$$\times \int_{S_y} \exp\left[-i\frac{k}{2}\left(\frac{1}{R}-\frac{1}{l}-\frac{2}{k}b_4\right)y^2\right]\exp\left[-i\frac{k}{l}\left(\beta-\frac{l}{k}b_2\right)\right]dy$$

$$+K_2 \int_{S_x} \exp\left[-i\frac{k}{2}\left(\frac{1}{R}-\frac{1}{l}-\frac{2}{k}b_3\right)x^2\right]$$

$$\times \exp\left[-i\frac{k}{l}\left(\alpha+\frac{lh}{2R}-\frac{l}{k}b_1\right)\right]dx$$

$$\times \int_{S_y} \exp\left[-i\frac{k}{2}\left(\frac{1}{R}-\frac{1}{l}-\frac{2}{R}b_4\right)y^2\right]\exp\left[-i\frac{k}{l}\left(\beta-\frac{l}{k}b_2\right)\right]dy,$$

$$(11.47)$$

where K_1 and K_2 are the appropriate complex constants, and S_x and S_y are integrations in the x and y directions, respectively. Without loss of generality, in the following the surface integral S will be assumed to be of infinitely large extent.

From eq. (11.47) the unique image reconstructions with respect to the x and y coordinates will occur at

$$l_x = \frac{kR}{k-2Rb_3} \tag{11.48}$$

and

$$l_y = \frac{kR}{k-2Rb_4}, \tag{11.49}$$

where l_x and l_y are the separations between the hologram coordinate system $\rho(x, y)$ and the image reconstruction coordinate system $\sigma(\alpha, \beta)$. Thus at $l=l_x$ for the x integral, and $l=l_y$ for the y integral, eq. (11.47) becomes

$$E_r(\sigma; k) = K_1'\delta\left(\alpha-\frac{lh}{2R}-\frac{l}{k}b_1\right)\bigg|_{l=l_x}\delta\left(\beta-\frac{l}{k}b_2\right)\bigg|_{l=l_y}$$
$$+ K_2'\delta\left(\alpha+\frac{lh}{2R}-\frac{l}{k}b_1\right)\bigg|_{l=l_x}\delta\left(\beta-\frac{l}{k}b_2\right)\bigg|_{l=l_y}, \tag{11.50}$$

where K_1' and K_2' are the appropriate complex constants.

From eq. (11.50) or fig. 11.13, there is an astigmatic effect in the image reconstructions. It is clear that this astigmatism will vanish for the case

$l_x = l_y$ (i.e., $b_3 = b_4$). The quadratic nature of the phase delay characteristic will also cause some degree of lateral magnification,

$$M_{\text{lat}, x}^{\text{r}} = \frac{k}{k - 2Rb_3} \tag{11.51}$$

and

$$M_{\text{lat}, y}^{\text{r}} = \frac{k}{k - 2Rb_4}, \tag{11.52}$$

where the superscript r denotes the real images, and the subscripts lat, x and lat, y are for the lateral magnifications with respect to the x and y axes.

The longitudinal magnifications can be determined with respect to the lateral magnifications due to the quadratic phase delay. Thus

$$M_{\text{long}, x}^{\text{r}} \simeq (M_{\text{lat}, x}^{\text{r}})^2 \tag{11.53}$$

and

$$M_{\text{long}, y}^{\text{r}} \simeq (M_{\text{lat}, y}^{\text{r}})^2. \tag{11.54}$$

From eqs. (11.53) and (11.54), the astigmatic effect can be seen. It is also clear that, as we have stated, the astigmatism disappears for $b_3 = b_4$.

Further, the angular magnification of the image reconstruction can be shown to be independent of the quadratic variation of the emulsion thickness. If higher order variations of the phase delay are considered (which is beyond the scope of the present work), more complicated image distortions may be determined.

In a similar manner as the foregoing, the virtual image lateral magnifications can be shown to be

$$M_{\text{lat}, x}^{\text{v}} = \frac{k}{k + 2Rb_3} \tag{11.55}$$

and

$$M_{\text{lat}, y}^{\text{v}} = \frac{k}{k + 2Rb_4}. \tag{11.56}$$

The corresponding longitudinal magnifications are

$$M_{\text{long}, x}^{\text{v}} \simeq (M_{\text{lat}, x}^{\text{v}})^2 \tag{11.57}$$

and

$$M_{\text{long}, y}^{\text{v}} \simeq (M_{\text{lat}, y}^{\text{v}})^2. \tag{11.58}$$

Again, the astigmatic effect disappears for $b_3 = b_4$.

In order to have some practical insight into the effect of emulsion thickness variation on hologram images, we may assume that the emulsion thickness variation of the holographic plate duplicates that of a positive lens. The phase delay is then

$$\phi(x, y) = k \left[\eta z_0 - (\eta - 1) \frac{x^2 + y^2}{2R_1} \right], \tag{11.59}$$

where η is the refractive index of the emulsion, z_0 is the maximum thickness of the emulsion, and R_1 is the radius of curvature of the lens. Then the lateral magnifications for the corresponding real and virtual images are

$$M_{lat}^r = \frac{R_1}{R_1 + R(\eta - 1)}, \tag{11.60}$$

$$M_{lat}^v = \frac{R_1}{R_1 - R(\eta - 1)}. \tag{11.61}$$

Suppose we let an object of lateral dimension $h = 30$ cm be holographed at a distance $R = 30$ cm away from the photographic plate. If the reconstructed real image is measured to be 0.2 cm shorter, then (assuming $\eta = 1.5$), the radius of curvature of the emulsion is seen to be $R_1 = 2240$ cm. Thus, from this very simple example, it is clear that a very small variation of the emulsion thickness does affect the accuracy of the image reconstruction.

Problems

11.1 Given the T-E transfer characteristic of Fig. 8.19, and a sinusoidal grating input exposure

$$E(x, y) = 1000 + 1000 \cos px \text{ erg/cm}^2, \quad \text{for every } y.$$

Determine, by means of a five-point analysis, the corresponding higher order amplitude transmittances and the degree of nonlinearity.

11.2 Consider a holographic recording process in which the complex light field incident on the recording aperture from a coherently illuminated object is

$$u(x, y) = A(x, y) \exp[i\phi(x, y)],$$

where $A(x, y)$ is uniformly distributed over the hologram aperture, and the complex light field from the reference beam is

$$v(x, y) = B \exp(ikx \sin\theta),$$

where B is an appropriate constant. Using the T-E transfer characteristic of fig. 11.14, with an object-to-reference beam ratio of 75 percent (i.e., $A/B = 0.75$) and a quiescent bias exposure $E_Q = 37.5$ μJoule/cm^2, determine the degree of nonlinearity for this holographic process, and the first, second, and third order diffraction efficiencies.

11.3 Repeat prob. 11.2, for unity object-to-reference beam ratio.

11.4 For the holographic process described in prob. 11.2, by means of a five-point analysis:

(a) Make a rough plot of the degree of nonlinearity as a function of exposure for the following object-to-reference beam ratios: $A/B = 1$, $A/B = 3/4$, $A/B = 1/2$, and $A/B = 1/4$.

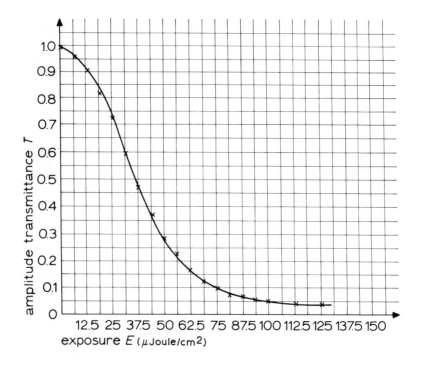

(b) Plot the corresponding first order diffraction efficiency as a function of exposure.

(c) Discuss the significance of the graphical results obtained in (a) and (b).

11.5 In certain holographic processes (e.g., for some Fourier holograms), the light amplitude field $A(x, y)$ may not be uniformly distributed over the recording aperture.

(a) May the finite-point analysis, as described in sec. 11.1, be applied for a qualitative analysis? Show what a qualitative finite-point analysis would or would not be able to uncover.

(b) Illustrate that, for a more accurate analysis, the nonlinear holographic analysis can be performed on a cycle-to-cycle basis.

11.6 Figure 11.15, shows the geometrical parameters in a certain non-linear holographic recording. The object O is of height h, the hologram aperture is of length L_x, and the separation between the object and the hologram is R. Determine the minimum angle θ of the plane reference wave for which the first order real image can be entirely separated from the higher order diffractions as well as from the zero order diffraction.

11.7 Let us assume that the phase delay of a certain photographic emulsion may be approximated by a second order polynomial,

$$\phi(x, y) \simeq b_0 + b_1 x + b_2 y + b_3 x^2 + b_4 y^2.$$

If the holographic image is constructed and reconstructed by plane wavefronts (i.e., both the reference and illuminating beams are plane), then by means of a technique similar to that described in sec. 10.5, determine the five primary holographic aberrations.

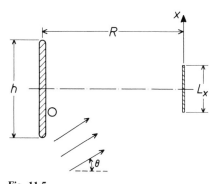

Fig. 11.5

References

11.1 A. Kozma, Photographic Recording of Spatially Modulated Coherent Light, *J. Opt. Soc. Am.*, **56** (1966), 428.

11.2 A. A. Friesem and J. S. Zelenka, Effects of Film Nonlinearities in Holography, *Appl. Opt.*, **6** (1967), 1755.

11.3 J. W. Goodman and G. R. Knight, Effects of Film Nonlinearities on Wavefront Reconstruction Images of Diffuse Objects, *J. Opt. Soc. Am.*, **58** (1968), 1276.

11.4 O. Bryngdahl and A. Lohmann, Nonlinear Effects in Holography, *J. Opt. Soc. Am.*, **58** (1968), 1325.

11.5 F. T. S. Yu, Analysis of Nonlinear Holograms, *J. Opt. Soc. Am.*, **58** (1968), 1550.

11.6 F. T. S. Yu, Five-Point Analysis of Nonlinear Holograms, *J. Opt. Soc. Am.*, **59** (1969), 360.

11.7 F. T. S. Yu, Off-axis Nonlinear Hologram and Spurious Distortion, *Opt. Commun.*, **1** (1970), 427.

11.8 F. E. Terman, *Electronic and Radio Engineering*, 4th ed., McGraw-Hill, New York, 1955, p. 326.

11.9 F. T. S. Yu and A. D. Gara, Effect of Emulsion Thickness Variations on Wavefront Reconstruction, *J. Opt. Soc. Am.*, **59** (1969), 1530, and *Appl. Opt.* **10** (1971), 1324.

Chapter 12   Linear Optimization in Holography

The analysis of nonlinear holograms has been treated in the previous chapter. In this chapter we will study the linearization of nonlinear holograms from a nonlinear systems viewpoint (refs. 12.1–12.4).

Wavefront reconstruction has already been widely used in modern optics. We can expect it to be extensively applied in the research and development of optical computers, optical memories, etc., as well as in optical communications. Therefore, it may be vitally important to learn how to apply linear optimization techniques to holographic processes.

We will first discuss the application of a conventional optimization technique to holography (ref. 12.5). However, we will demonstrate thereafter that this conventional optimization may not, in fact, be generalizable to most holographic processes (refs. 12.6–12.7). Therefore, following our investigation of this conventional optimization process, a general optimal linearization method for the photographic emulsion will be discussed. The application of this linearization technique to a simple point-object hologram will be demonstrated, and its extension to more complicated objects will be illustrated. Finally, an application of this linear optimization technique to complex spatial filtering will also be given (ref. 12.8).

12.1 Conventional Optimization

For the conventional photographic process (sec. 8.1), there is only one linear optimum condition of the H and D curve. In this section we will apply this conventional photographic optimization to holography; we will see that, in fact, there is only one linear optimum condition for wavefront recording. However, this conventional optimization process may not be required or even obtainable in most holographic processes.

Assume the conventional wavefront recording geometry shown in fig. 12.1. The complex light field scattered from the illuminated object onto the emulsion is

$$u(x, y) = A(x, y) \exp[i\psi(x, y)], \tag{12.1}$$

and the reference wave field on the emulsion is

$$v(x, y) = B \exp(ikx \sin \theta). \tag{12.2}$$

Then the corresponding irradiance of the wavefront recording is

$$I(x, y) = A^2(x, y) + B^2 + 2A(x, y) B \cos[kx \sin \theta - \psi(x, y)].$$

If the recording is on the linear portion of the H and D curve (sec. 8.1), then the amplitude transmittance of the hologram may be written as

$$T(x, y) = K [I(x, y)]^{\gamma/2}, \tag{12.3}$$

where K is a proportionality constant, and γ is the slope of the linear region of the H and D curve.

If the hologram described by eq. (12.3) is illuminated by a monochromatic plane wave, then the transmitted light field immediately behind the hologram is

$$E(x, y) = K_1 [A^2(x, y) + B^2 + 2A(x, y) B \cos \phi(x, y)]^{\gamma/2}, \tag{12.4}$$

where K_1 is an appropriate positive constant, and $\phi = k \, x \sin \theta - \psi(x, y)$. Alternatively, eq. (12.4) may be expressed as

$$E(x, y) = K_2 \left\{ 1 + \left[\frac{A(x, y)}{B} \right]^2 + 2 \frac{A(x, y)}{B} \cos \phi(x, y) \right\}^{\gamma/2}, \tag{12.5}$$

where $K_2 = K_1 B^\gamma$.

If we restrict A and B by the inequality

$$\left[\frac{A(x, y)}{B} \right]^2 + 2 \frac{A(x, y)}{B} \cos \phi(x, y) < 1, \quad \text{for every} \quad (x, y), \tag{12.6}$$

that is,

$$\left| \frac{A(x, y)}{B} \right| < 0.414,$$

then eq. (12.5) may be written as a binomial expansion. Thus it becomes

$$\begin{aligned}
E(x, y) = K \Bigg\{ 1 &+ \frac{\gamma}{2} \left[\left(\frac{A(x, y)}{B} \right)^2 + 2 \left(\frac{A(x, y)}{B} \right) \cos \phi(x, y) \right] \\
&+ \frac{1}{2!} \frac{\gamma}{2} \left(\frac{\gamma}{2} - 1 \right) \left[\left(\frac{A(x, y)}{B} \right)^2 + 2 \left(\frac{A(x, y)}{B} \right)^2 \cos \phi(x, y) \right]^2 \\
&+ \frac{1}{3!} \frac{\gamma}{2} \left(\frac{\gamma}{2} - 1 \right) \left(\frac{\gamma}{2} - 2 \right) \left[\left(\frac{A(x, y)}{B} \right)^2 \right. \\
&\left. + 2 \left(\frac{A(x, y)}{B} \right)^2 \cos \phi(x, y) \right]^3 + \cdots \Bigg\}.
\end{aligned} \tag{12.7}$$

After expanding the bracket [ ] terms of eq. (12.7), it can be written in the form

$$
\begin{aligned}
E(x, y) = K \Bigg\{ &\left[1 + \left(\frac{\gamma}{2}\right)^2 \left(\frac{A(x, y)}{B}\right)^2 + \frac{1}{2!}\frac{\gamma}{2}\left(\frac{\gamma}{2}-1\right)\left(\frac{\gamma}{2}-3\right)\left(\frac{A(x, y)}{B}\right)^4 \right. \\
&\left. + \frac{1}{3!}\frac{\gamma}{2}\left(\frac{\gamma}{2}-1\right)\left(\frac{\gamma}{2}-2\right)\left(\frac{A(x, y)}{B}\right)^6 + \cdots \right] \\
&+ 2\left[\frac{\gamma}{2}\left(\frac{A(x, y)}{B}\right) + \frac{1}{2!}\left(\frac{\gamma}{2}\right)^2\left(\frac{\gamma}{2}-1\right)\left(\frac{A(x, y)}{B}\right)^3 + \cdots \right] \\
&\times \cos\phi(x, y) + 2\left[\frac{1}{2!}\frac{\gamma}{2}\left(\frac{\gamma}{2}-1\right)\left(\frac{A(x, y)}{B}\right)^2 \right. \\
&\left. + \frac{3}{3!}\frac{\gamma}{2}\left(\frac{\gamma}{2}-2\right)\left(\frac{A(x, y)}{B}\right)^4 + \cdots \right]\cos[2\phi(x, y)] \\
&+ 2\left[\frac{1}{3!}\frac{\gamma}{2}\left(\frac{\gamma}{2}-1\right)\left(\frac{\gamma}{2}-2\right)\left(\frac{A(x, y)}{B}\right)^3 + \cdots \right] \\
&\times \cos[3\phi(x, y)] + \cdots \Bigg\}.
\end{aligned}
\tag{12.8}
$$

We see that eq. (12.8) may be written as

$$
E(x, y) = K \sum_{n=1}^{\infty} a_n(\gamma) \cos[n\phi(x, y)],
\tag{12.9}
$$

where the $a_n(\gamma)$ are the corresponding coefficients. Then the degree of nonlinearity of the hologram (eq. 11.12) is

$$
\text{nonlinearity (percent)} = \frac{\left[\displaystyle\sum_{n=2}^{\infty} a_n^2(\gamma)\right]^{1/2}}{a_1(\gamma)} \times 100.
\tag{12.10}
$$

It can be seen from eq. (12.8) that the degree of nonlinearity approaches zero whenever γ approaches 2. Thus $\gamma=2$ is the optimum condition in the sense of minimum nonlinear distortion. To achieve the value $\gamma=2$ in practice, a two-step contact printing process as described in sec. 8.1 may be applied.

The conventional linear optimization process just described optimizes in the sense of minimizing the nonlinear distortion. However, the linear optimization technique which we will describe in the following sections optimizes in the sense of maximizing the reconstruction of the first order linear hologram image. In other words, in the holographic process we could allow a certain degree of nonlinear wavefront recording, and still have the best first order hologram image reconstruction.

12.2 Linear Optimization Technique

The *T-E* curve for a photographic emulsion is a monotonically decreasing function (sec. 8.1), which may be expanded into a finite power series

$$T(E) = \sum_{n=0}^{N} a_n E^n, \quad \text{for} \quad E \geq 0,$$
(12.11)

where T is the amplitude transmittance, E is the exposure, and the a_n's are the real coefficients. Clearly, from the boundary conditions on the amplitude transmittance $[T(0) = 1$ and $T(\infty) = 0]$, we can conclude:

$$a_0 = 1$$

and

$$\lim_{E \to \infty} \sum_{n=1}^{N} a_n E^n = -1.$$

Let us now choose to replace eq. (12.11) with a linear approximated transmittance,

$$T^{\dagger}(E) = \lambda_0 + \lambda_1 E.$$
(12.12)

Then we have to choose the parameters λ_0 and λ_1 to approximate the nonlinear transmittance of eq. (12.11) as closely as possible. Let us now consider the difference of eqs. (12.11) and (12.12),

$$T - T^{\dagger} = (a_0 - \lambda_0) + (a_1 - \lambda_1) E + a_2 E^2 + a_3 E^3 + \cdots + a_N E^N.$$
(12.13)

We seek to minimize the mean square integral

$$\iint_S (T - T^{\dagger})^2 \, dx \, dy,$$
(12.14)

where S denotes integration over the surface of the photographic plate. Let us define the spatial ensemble average,

$$\langle E^n \rangle \triangleq \iint_S E^n(x, y) \, dx \, dy, \quad n = 1, 2, 3, \ldots, N.$$
(12.15)

Then the optimal choice of the parameters λ_0 and λ_1 can be determined by

$$\frac{\partial}{\partial \lambda_n} \iint_S (T - T^{\dagger})^2 \, dx \, dy = 0, \quad n = 0, 1.$$
(12.16)

The solution of eq. (12.16) is

$$\lambda_0 = \frac{\sum_{n=0}^{N} a_n(\langle E^n \rangle \langle E^2 \rangle - \langle E^{n+1} \rangle \langle E \rangle)}{\langle E^2 \rangle - \langle E \rangle^2} \tag{12.17}$$

and

$$\lambda_1 = \frac{\sum_{n=0}^{N} a_n(\langle E^{n+1} \rangle - \langle E^n \rangle \langle E \rangle)}{\langle E^2 \rangle - \langle E \rangle^2}. \tag{12.18}$$

The corresponding matrix representation is

$$\begin{bmatrix} 1 & \langle E \rangle \\ \\ \langle E \rangle & \langle E^2 \rangle \end{bmatrix} \begin{bmatrix} \lambda_0 \\ \\ \lambda_1 \end{bmatrix} = \begin{bmatrix} \sum_{n=0}^{N} a_n \langle E^n \rangle \\ \\ \sum_{n=0}^{N} a_n \langle E^{n+1} \rangle \end{bmatrix} \tag{12.19}$$

Equations (12.17) and (12.18), or eq. (12.19) are the best linear approximations with respect to eq. (12.12), as determined by the least-mean-square-error criterion of eq. (12.14). It may be emphasized that eq. (12.18) can be considered to be a generalized first order transmittance (i.e., first order transfer function) for the photographic emulsion. In the following section, the application of this optimal linearization method in holography will be discussed.

12.3 Optimal Linearization

The object recorded in a hologram may be considered to be composed of a large number of infinitesimal point objects. Therefore, in the optimal linearization, we could start from a point object, and then extend the results by superposition.

Suppose a monochromatic point source and an oblique plane reference wave are of the same wavelength; and suppose that the recording geometry is as shown in fig. 11.4. Then the irradiance contributed by these two wavefronts on the recording medium can be approximated by

$$I(x, y) = |A|^2 + |B|^2 + 2|A| |B| \cos\left[k\left(\frac{\rho^2}{2R} - x \sin\theta \right) + \psi \right], \tag{12.20}$$

where A and B are the complex amplitudes of the spherical and plane wave fronts, respectively, ψ is the phase angle between A and B, $\rho^2 = x^2 + y^2$, and k is the wave number. By a translation of the spatial coordinate system, eq. (12.20) can be put into the form

$$I(x, y) = |A|^2 + |B|^2 + 2|A| |B| \cos\left[\frac{k}{2R} \rho_1^2 - \frac{kR}{2} \sin^2\theta + \psi \right], \tag{12.21}$$

where $x_1 = x - R \sin\theta$, and $\rho_1^2 = x_1^2 + y^2$. For convenience in notation, in the following A and B will be used to represent $|A|$ and $|B|$, respectively, unless otherwise specified. Rewrite eq. (12.21),

$$I = (A^2 + B^2)\left[1 + \frac{2AB}{A^2 + B^2}\cos\left(\frac{k}{2R}\rho_1^2 + \phi\right)\right],\tag{12.22}$$

where $\phi = -\frac{1}{2}kR\sin^2\theta + \psi$. Thus,

$$E^n = (It)^n = t^n(A^2 + B^2)^n\left[1 + \frac{2AB}{A^2 + B^2}\cos\left(\frac{k}{2R}\rho_1^2 + \phi\right)\right]^n, n = 1, 2, \dots, N+1,\tag{12.23}$$

where t is the exposure time. By performing the integration of eq. (12.23) over a period of the ρ_1^2 axis (ref. 12.6), the ensemble average can be obtained:

$$\langle E^n\rangle = \langle I^n\rangle\, t^n = t^n \sum_{2r=0}^{n} \frac{n!}{(n-2r)\,!(r!)^2}(A^2 + B^2)^{n-2r}(AB)^{2r}.\tag{12.24}$$

The parameters λ_0 and λ_1 can be obtained by the substitution of eq. (12.24) in eqs. (12.17) and (12.18) respectively. The optimal linear amplitude transmittance is therefore

$$T^\dagger(x, y) = \sum_{n=0}^{N} a_n\langle E^n\rangle + 2\lambda_1 tAB\cos\left(\frac{k}{2R}\rho_1^2 + \phi\right).\tag{12.25}$$

Equation (12.25) is a linear approximation for the given A and B, and it is also the best approximation with respect to eq. (12.12) and to the mean-square-error criterion of eq. (12.16).

Furthermore, from Eqs. (12.18) and (12.24), λ_1 is a function of three variables, i.e., $\lambda_1 = \lambda_1(A, B, t)$. Therefore, for a fixed amplitude of the reference wavefront B, the optimum value of λ_1 may be determined by the partial derivatives

$$\frac{\partial\lambda_1}{\partial t} = 0,\tag{12.26}$$

$$\frac{\partial\lambda_1}{\partial A} = 0.\tag{12.27}$$

On the other hand, if the values of A and B are given, then the optimum value of λ_1 can be determined from eq. (12.26). The variable λ_1 can be considered to be a generalized first order transmittance or the first order diffraction for this point-object hologram. Therefore, the optimum value of λ_1 is also the best first order diffraction of this wavefront recording, with respect to the optimal linearization of eqs. (12.12) and (12.16).

If the irradiance of the reference wave is much larger than that of the spherical wave (resulting in only weak nonlinear distortion), then eq. (12.25) may be approximated by

$$T^\dagger(x, y) \simeq \sum_{n=0}^{N} a_n t^n B^{2n} + 2 \sum_{n=1}^{N} n a_n t^n B^{2n-1} A \cos\left(\frac{k}{2R} \rho_1^2 + \phi\right), \quad \text{for } B \gg A.$$

(12.28)

It must be emphasized that, in this weak distortion, it is possible to define a linearization such that the distortion of the hologram will be negligible.

As an example, for the Kodak 649F (D $-$ 19, 5 min) photographic emulsion, which is commonly used in holography, the T-E curve may be approximated by a third order polynomial (refs. 12.9–12.10),

$$T = \sum_{n=0}^{3} (-1)^n a_n E^n, \quad \text{for} \quad E \geq 0,$$

(12.29)

where the a_n's are the positive real coefficients. If the input signal (the irradiance) is that of eq. (12.21), then we have

$$\langle E \rangle = t(A^2 + B^2),$$

$$\langle E^2 \rangle = t^2 \left[(A^2 + B^2)^2 + 2(AB)^2\right],$$

$$\langle E^3 \rangle = t^3 \left[(A^2 + B^2)^3 + 6(AB)^2 (A^2 + B^2)\right],$$

$$\langle E^4 \rangle = t^4 \left[(A^2 + B^2)^4 + 12(AB)^2 (A^2 + B^2)^2 + 6(AB)^4\right].$$

(12.30)

Hence,

$$\lambda_0 = a_0 + a_2 t^2 \left[2(AB)^2 - (A^2 + B^2)^2\right]$$
$$+ a_3 t^3 (A^2 + B^2) \left[2(A^2 + B^2)^2 - 3(AB)^2\right] \qquad (12.31)$$

and

$$\lambda_1 = -a_1 + 2a_2 t(A^2 + B^2) - 3a_3 t^2 \left[(A^2 + B^2)^2 + (AB)^2\right],$$

(12.32)

which depend here on amplitudes A and B of the input irradiance, but not on the spatial frequency. (In some cases they may depend on both the amplitude and the spatial frequency.) The corresponding optimal linear transmittance is

$$T^\dagger(x, y) = a_0 - a_1 t(A^2 + B^2) + a_2 t^2 [(A^2 + B^2)^2 + 2(AB)^2] - a_3 t^3 [(A^2 + B^2)^3$$
$$+ 6(AB)^2 (A + B)^2] + 2ABt\{-a_1 + 2a_2 t(A^2 + B^2)$$
$$- 3a_3 t^2 [(A^2 + B^2)^2 + (AB)^2]\} \cos\left(\frac{k}{2R} \rho_1^2 + \phi\right), \qquad (12.33)$$

where $a_0 = 1$.

For a given amplitude B, the optimum value of λ_1 can be obtained by

$$\frac{\partial \lambda_1}{\partial t} = a_2(A^2 + B^2) - 3a_3 t[(A^2 + B^2)^2 + (AB)^2] = 0 \qquad (12.34)$$

and

$$\frac{\partial \lambda_1}{\partial A} = 2a_2 - 3a_3 t(2A^2 + 3B^2) = 0. \qquad (12.35)$$

The solutions for eqs. (12.34) and (12.35) are

$$A = B \qquad (12.36)$$

and

$$t = \frac{2a_2}{15a_3 B^2}. \qquad (12.37)$$

The optimum value of λ_1 is therefore

$$(\lambda_1)_{op} = -a_1 + \frac{4a_2^2}{15a_3}, \qquad (12.38)$$

and the corresponding optimal linear transmittance (for $A = B$) is

$$T^\dagger(x, y) = 1 - \frac{4a_1 a_2}{15a_3} + \frac{32a_2^3}{675a_3^2} + 2\left(-\frac{2a_1 a_2}{15a_3} + \frac{8a_2^3}{225a_3^2}\right) \cos\left(\frac{k}{2R} \rho_1^2 + \phi\right). \qquad (12.39)$$

On the other hand, if A and B are given, then the optimum value of λ_1 occurs at

$$t = \frac{a_2}{3a_3} \frac{A^2 + B^2}{(A^2 + B^2)^2 + (AB)^2}. \qquad (12.40)$$

For $B \gg A$, i.e., weak nonlinear distortion, eq. (12.40) may be approximated by

$$t \simeq \frac{a_2}{3a_3 B^2}.$$
(12.41)

The corresponding optimum value of λ_1 is

$$(\lambda_1)_{\text{op}} \simeq -a_1 + \frac{a_2^2}{3a_3},$$
(12.42)

and the optimum linear transmittance is therefore

$$T^\dagger(x, y) = 1 - \frac{a_1 a_2}{3a_3} + \frac{2a_2^3}{27a_3^2} + 2\frac{A}{B}\left[-\frac{a_1 a_2}{3a_3} + \frac{a_2^3}{9a_3^2}\right]$$

$$\times \cos\left(\frac{k}{2R}\rho_1^2 + \phi\right), \quad \text{for} \quad B \gg A.$$
(12.43)

In this case, the nonlinear distortion in the hologram is negligible.

The application of this optimal linearization method to the extended object shown in fig. 12.1 follows from superposition. As usual, the light amplitude distribution from the extended object on the recording medium can be determined by the Fresnel-Kirchhoff theory,

$$A(x, y) = \iint_S O(\xi, \eta)\, E_l^+(\boldsymbol{\rho} - \boldsymbol{\xi}; k)\, d\xi\, d\eta,$$
(12.44)

where

$$E_l^+(\boldsymbol{\rho} - \boldsymbol{\xi}; k) = \left(-\frac{i}{\lambda l}\right) \exp\left\{ik\left[l - \zeta + \frac{(x - \xi)^2}{2l} + \frac{(y - \eta)^2}{2l}\right]\right\},$$

and where $\boldsymbol{\xi}(\xi, \eta, \zeta)$ and $\boldsymbol{\rho}(x, y, z)$ are the coordinate systems, l is the separation of the coordinate systems, $O(\xi, \eta)$ is the object function, S denotes the integration over the effective surface of the object, and $A(x, y) = |A(x, y)|\, e^{i\psi(x, y)}$ is the complex light amplitude. Again for convenience in notation, in the following $A(x, y)$ and B will be used to denote $|A(x, y)|$ and $|B|$, respectively.

Thus, the irradiance of this wavefront recording is

$$I(x, y) = A^2(x, y) + B^2 + 2A(x, y)\, B \cos[xk \sin\theta - \psi(x, y)],$$
(12.45)

and the corresponding nth power exposure can be written as

$$E^n(x, y) = t^n[A^2(x, y) + B^2]^n \left[1 + \frac{2A(x, y)\, B}{A^2(x, y) + B^2} \cos\phi(x, y)\right]^n,$$
(12.46)

where $\phi(x, y) = xk \sin\theta - \psi(x, y)$.

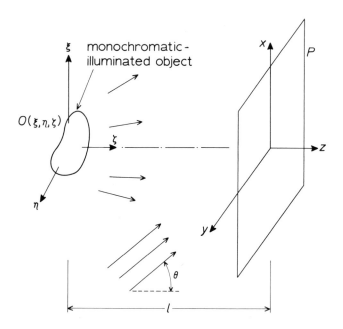

Fig. 12.1 Conventional wavefront recording. The object emits monochromatic light, and the reference wave is plane monochromatic.

Calculating the ensemble average over the surface of the photographic plate (the hologram), we have

$$\langle E^n \rangle = t^n \sum_{r=0}^{N} \frac{n!}{(n-r)!\,r!} \frac{2^r}{S} \iint_S [A^2(x, y) + B^2]^{n-r}$$

$$\times A^r(x, y) \, B^r \cos^r[\phi(x, y)] \, dx \, dy. \qquad (12.47)$$

The parameters λ_0 and λ_1 can be obtained from eqs. (12.17) and (12.18) respectively. The optimum value of λ_1 may be determined by solving $\partial \lambda_1 / \partial t = 0$, therefore determining the optimal hologram linearization.

Obtaining a unique solution of eq. (12.47) may be difficult. However, in most holographic processes the light amplitude distribution $A(x, y)$ is approximately uniform over the recording aperture. Therefore the exposure ensemble average may be approximated by

$$\langle E^n \rangle \simeq t^n \sum_{2r=0}^{N} \frac{n!}{(n-2r)!\,(r!)^2} [A^2(x_0, y_0) + B^2]^{n-2r} [A(x_0, y_0) \, B]^{2r}, \quad (12.48)$$

where $A(x_0, y_0)$ is the maximum value of $A(x, y)$ on the recording aperture.

It may be noted that eq. (12.48) is essentially identical to eq. (12.24), the point object case.

Furthermore, if the transmittance of the recording medium is approximated by a third order polynomial, as given in eq. (12.29), then the parameter λ_1 may be obtained:

$$\lambda_1 = -a_1 - 2a_2 t \left[A^2(x_0, y_0) + B^2 \right]$$
$$- 3a_3 t^2 \left\{ \left[A^2(x_0, y_0) + B^2 \right]^2 + A^2(x_0, y_0) \, B^2 \right\}. \qquad (12.49)$$

The parameter λ_1 again depends on the amplitudes $A(x_0, y_0)$ and B of the recording irradiance, but is independent of the spatial frequency. The corresponding optimal linear transmittance is

$$T^\dagger(x, y) = \sum_{n=0}^{3} (-1)^n a_n \langle E^n \rangle + 2\lambda_1 A(x_0, y_0) \, B \cos\left[\phi(x, y) \right]. \qquad (12.50)$$

The optimum value of λ_1 may be obtained by solving

$$\frac{\partial \lambda_1}{\partial A(x_0, y_0)} = 2a_2 - 3a_3 t \left[2A^2(x_0, y_0) + 3B^2 \right] = 0 \qquad (12.51)$$

and

$$\frac{\partial \lambda_1}{\partial t} = a_2 \left[A^2(x_0, y_0) + B^2 \right] - 3a_3 t \left\{ \left[A^2(x_0, y_0) + B^2 \right]^2 + A^2(x_0, y_0) \, B^2 \right\} = 0. \qquad (12.52)$$

Thus we have

$$B = A(x_0, y_0) \qquad (12.53)$$

and

$$t = \frac{2a_2}{15a_3 B^2}. \qquad (12.54)$$

The optimum value of λ_1 is therefore

$$(\lambda_1)_{\text{op}} = -a_1 + \frac{4a_2^2}{15a_3}, \quad \text{for} \quad B = A(x_0, y_0), \qquad (12.55)$$

and the corresponding optimal linear transmittance is

$$T^\dagger(x, y) = 1 - \frac{4a_1 a_2}{15a_3} + \frac{32a_2^3}{675a_3^2} + 2\left(-\frac{2a_1 a_2}{15a_3} + \frac{8a_2^3}{225a_3^2} \right) \cos\left(\frac{k}{2R} \rho_1^2 + \phi \right),$$
$$\text{for} \quad B = A(x_0, y_0). \qquad (12.56)$$

On the other hand, if $A(x_0, y_0) \neq B$, then the optimum value of λ_1 occurs at

$$t = \frac{a_2}{3a_3} \frac{A^2(x_0, y_0) + B^2}{[A^2(x_0, y_0) + B^2]^2 + A^2(x_0, y_0) B^2}. \tag{12.57}$$

For $B \gg A(x_0, y_0)$, i.e., for weak nonlinear distortion, eq. (12.57) may be approximated by

$$t \simeq \frac{a_2}{3a_3 B^2}. \tag{12.58}$$

The corresponding optimum value of λ_1 is

$$(\lambda_1)_{op} \simeq -a_1 + \frac{a_2^2}{3a_3}, \quad \text{for} \quad B \gg A(x_0, y_0), \tag{12.59}$$

and the linear optimum transmittance of the nonlinear hologram is therefore

$$T^\dagger(x, y) = 1 - \frac{a_1 a_2}{3a_3} + \frac{2a_2^3}{27a_3^2} + 2\frac{A(x_0, y_0)}{B}\left[-\frac{a_1 a_2}{3a_3} + \frac{a_2^3}{9a_3^2}\right]\cos[\phi(x, y)]. \tag{12.60}$$

In this case, the nonlinear distortion in the hologram is negligible.

Equations (12.53) – (12.60) are essentially identical to the equations for the point object case. Nevertheless, it may be concluded that in most holographic linear optimization processes, a one-to-one object-reference beam ratio and an appropriate optimum exposure is required.

Figure 12.2 gives the experimentally determined first order diffraction efficiency* as a function of exposure for various object-reference beam ratios. The recording medium used was a Kodak 649F (D−19, 5 min) photographic plate. It can be seen from this figure that the optimum first order hologram image occurs for the unity object-reference beam ratio. Figure 12.3 is a series of photographs of three-dimensional first order optimum hologram images for various object-reference beam ratios. Again it can be seen that the optimum image reconstruction occurs for unity object-reference beam ratio.

To recapitulate: In nonlinear holographic processes, the separation of the first order hologram image from the higher order images is possible

---

* The diffraction efficiency is defined as

$$\text{Diffraction Efficiency} \triangleq \frac{\text{Output (Hologram Image Irradiance)}}{\text{Input (Incident Irradiance)}}.$$

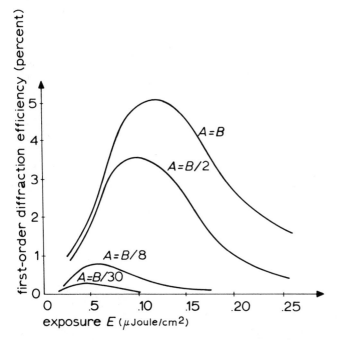

Fig. 12.2 First order diffraction efficiency as a function of exposure.

(chap. 11). Therefore, instead of confining the recording to the linear region of the photographic emulsion, it is possible to obtain optimum hologram image reconstruction from a nonlinear holographic recording. It may also be emphasized that in most holographic processes the phase information is far more important than the amplitude information. Therefore, as long as the phase information is preserved in the wavefront construction, an amplitude distortion will not cause any defect in the hologram image reconstruction, but instead generates higher order diffractions. These higher order diffractions can be spatially separated from the first order hologram image diffraction.

12.4 Syntheses of Optimum Nonlinear Spatial Filters
In sec. 7.3 we learned that a complex matched filter can be synthesized by means of coherent optical processes. An example of such a filter has already been given (see fig. 7.10). It can be readily seen that the filter recording is indeed a Fourier transform hologram. Since the signal in complex spatial filter synthesis can be considered to be the summation of signals

$A = B$

$A = B/2$

$A = B/8$

$A = B/30$

Fig. 12.3 Hologram images for various object/reference beam ratios.

from a large number of resolvable point sources, the Fourier transform of the signal can be treated as the superposition of a large number of monochromatic plane waves. By reasoning along the lines developed in the previous section, we may first optimize for a simple plane wave (i.e., the Fourier transform of a point signal), and then extend this concept to more complicated signals.

Suppose the signal transparency $s(x, y)$ is replaced by a monochromatic point source which is located at the origin of the (x, y) coordinate system. Then the irradiance at the Fourier transform plane due to the point signal and the reference point source is

$$I(p, q) = S^2 + R^2 + 2SR \cos(x_0 p + \phi), \tag{12.61}$$

where S and R are the light amplitudes of the two plane waves from the signal and the reference sources, ϕ is the phase angle, $p = k\alpha/f$ and $q = k\beta/f$ are the corresponding spatial frequency coordinates, with k the wave number and f the focal length of the lens, and (α, β) the coordinate system of the photographic plate.

By the technique of the previous section, we see that the linear optimum filter is

$$H^\dagger(p, q) = T^\dagger(p, q) = \sum_{n=0}^{N} a_n \langle E^n(p, q) \rangle + 2\lambda_1 t SR \cos(x_0 p + \phi). \tag{12.62}$$

It may also be noted that λ_1 is a function of S, R, and t; its optimum value can be obtain from

$$\frac{\partial \lambda_1(S, R, t)}{\partial S} = \frac{\partial \lambda_1(S, R, t)}{\partial R} = \frac{\partial \lambda_1(S, R, t)}{\partial t} = 0. \tag{12.63}$$

As an illustration, if the transmittance of the recording medium is approximated by a third order polynomial, as in eq. (12.29), then the linear optimization condition occurs at

$$S = R, \qquad t = \frac{2a_2}{15a_3 R^2}. \tag{12.64}$$

The corresponding optimal linear spatial filter is

$$H^\dagger(p, q) = T^\dagger(p, q) = a_0 - \frac{4a_1 a_2}{15a_3} + \frac{32a_2^3}{675a_3^2} + 2\left(-\frac{2a_1 a_2}{15a_3} + \frac{8a_2^3}{225a_3^2}\right)$$
$$\times \cos(x_0 p + \phi). \tag{12.65}$$

If $R \gg S$ (i.e., weak nonlinear distortion), then the optimum linear filter is

$$H^{\dagger}(p, q) = 1 - \frac{a_1 a_2}{3a_3} + \frac{2a_2^3}{27a_3^2} + 2\frac{S}{R}\left(-\frac{a_1 a_2}{3a_3} + \frac{a_2^3}{9a_3^2}\right)\cos(x_0 p + \phi), \text{ for } R \gg S.$$
$$(12.66)$$

The application of this linear optimization to a more complicated non-linear spatial filter may be accomplished. The complex light field from the signal is the Fourier transform of the signal; i.e.,

$$S(p, q) = \int\int s(x, y) \exp[-i(px + qy)] \, dx \, dy,$$

where $S(p, q) = |S(p, q)| \, e^{i\phi(p, q)}$. For convenience in notation, in the following $S(p, q)$ will mean $|S(p, q)|$. The irradiance of the filter recording can be written as

$$I(p, q) = S^2(p, q) + R^2 + 2RS(p, q)\cos[x_0 p + \phi(p, q)]. \tag{12.68}$$

It can be seen from eq. (12.68) that the Fourier transform of the signal $S(p, q)$ is carried by a one-dimensional spatial frequency x_0 along the p axis. The corresponding ensemble average is

$$\langle E^n(p, q)\rangle = t^n \sum_{r=0}^{N} \frac{n!}{(n-r)!\,r!} \frac{2^r}{\Sigma} \int\int_{\Sigma} [S^2(p, q) + R^2]^{n-r}$$
$$\times S^r(p, q)\, R^r \cdot \cos[x_0 p + \phi(p, q)]\, dp\, dq, \tag{12.69}$$

where Σ denotes the integration over the spatial frequency plane. In practice, a unique solution of eq. (12.69) is difficult to find. However for a small spatially bounded signal $s(x, y)$, the light amplitude distribution $S(p, q)$ may be assumed to be approximately uniform within the spatial frequency of interest; i.e.,

$$S(p, q) \simeq S(0, 0), \tag{12.70}$$

where $S(0, 0)$ is the maximum value of $S(p, q)$. Then eq. (12.69) may be approximated by

$$\langle E^n(p, q)\rangle \simeq t^n \sum_{2r=0}^{n} \frac{n!}{(n-2r)!\,(r!)^2} [S^2(0, 0) + R^2]^{n-2r}[RS(0, 0)]^{2r}. \tag{12.71}$$

As an example, if the transfer characteristic of the photographic film is approximated by the third order power series given in eq. (12.29), then the optimum value of λ_1 may be calculated by means of

$$\frac{\partial \lambda_1}{\partial S(0, 0)} = 0, \quad \frac{\partial \lambda_1}{\partial R} = 0, \quad \frac{\partial \lambda_1}{\partial t} = 0. \tag{12.72}$$

The solutions are:

$$R \simeq S(0, 0), \quad t \simeq \frac{2a_2}{15a_3 R^2}.$$

(12.73)

Again we see that a one-to-one signal-reference beam ratio and an appropriate optimum exposure is required.

12.5 Analysis of Signal Detection by Optimum Nonlinear Spatial Filtering

In the analysis of optimum nonlinear spatial filtering, we may again start from the simple point signal, and then extend to a more complicated case. If we recall eq. (12.61), we see that the nonlinear spatial filter is a periodic function. Therefore nonlinear filtering may be expressed by

$$H(p, q) = H_0 + \sum_{n=1}^{\infty} H_n(p, q),$$

(12.74)

where $H_0 = B_0/2$ is the zero order bias level, and $H_n(p, q) = B_n \cos [n(x_0 p + \phi)]$. The Fourier coefficients B_n may be determined by

$$B_n = \frac{x_0}{2\pi} \int_{-\pi/x_0}^{\pi/x_0} H(p, q) \cos(n x_0 p) \, dp, \quad n = 0, 1, 2, \ldots.$$

(12.75)

If the nonlinear spatial filter is inserted in the spatial frequency plane of a coherent processing system (fig. 12.4), the complex light distribution at

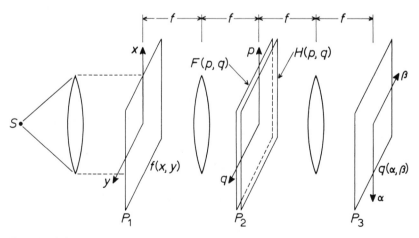

Fig. 12.4 Coherent complex spatial filtering.

the output plane will be

$$g(\alpha, \beta) = \int\int_{-\infty}^{\infty} F(p, q) H(p, q) \exp[-i(p\alpha + q\beta)] \, dp \, dq, \tag{12.76}$$

where $F(p, q)$ is the Fourier transform of the input signal $f(x, y)$. By substituting eq. (12.74) into eq. (12.76), and interchanging the summation and integration, we have

$$g(\alpha, \beta) = \sum_{n=0}^{\infty} \int\int_{-\infty}^{\infty} B_n F(p, q) \cos[n(x_0 p + \phi)] \exp[-i(p\alpha + q\beta)] \, dp \, dq. \tag{12.77}$$

If the input signal is assumed to be a Dirac delta function,

$$f(x, y) = \delta(x, y), \tag{12.78}$$

then the output signal may be written as

$$
\begin{aligned}
g(\alpha, \beta) = \frac{B_0}{2} \, \delta(\alpha, \beta) &+ \sum_{n=1}^{\infty} \frac{B_n}{2} \, e^{in\phi} \, \delta(\alpha + nx_0, \beta) \\
&+ \sum_{n=1}^{\infty} \frac{B_n}{2} \, e^{-in\phi} \, \delta(-\alpha - nx_0, -\beta).
\end{aligned}
\tag{12.79}
$$

The first term of eq. (12.79) is the zero order diffraction, which appears at the origin of the output plane, the first summation includes the convolutions in different orders, and the last summation covers the crosscorrelation of different orders. The zero order diffraction and the convolutions are of no particular interest here, and the first order crosscorrelation will serve for signal detection.

From eq. (12.79), it is clear that the higher order crosscorrelations are spatially separated in the output plane. The value of the nth order crosscorrelation is proportional to the corresponding Fourier component B_n. Thus for an optimum first order crosscorrelation, an optimum value of B_1 should be used. In fact, the linear optimization in the synthesis of the nonlinear spatial filter in the previous section was accomplished by the optimization of B_1.

A general analysis of nonlinear spatial filtering for a more complicated signal is difficult to achieve in practice. However, from eq. (12.68), it is clear that the complex quantity $S(p, q)$ is spatially modulated by a spatial carrier frequency x_0. In order to insure the separation of the crosscorrelation from the zero order diffraction, the spatial carrier frequency must be sufficiently high, and may be approximated by

$$x_0 > l_f + \tfrac{3}{2} l_s, \tag{12.80}$$

where l_f and l_s are the spatial lengths in the x direction of the input signal $f(x, y)$ and the detecting signal $s(x, y)$ respectively. Moreover, it is clear from eq. (12.68) that the nonlinear distortion due to the film transfer characteristic can occur only in the amplitude modulation and not in the phase modulation. Therefore, the nonlinear spatial filter yielding the signal $s(x, y)$ may be approximated, by means of the Fourier decomposition theorem, by

$$H(p, q) \simeq H_0(p, q) + \sum_{n=1}^{\infty} H_n(p, q), \tag{12.81}$$

where $H_0(p, q) = \tfrac{1}{2} B_0(p, q)$ and $H_n(p, q) = B_n(p, q) \cos\{n[x_0 p + \phi(p, q)]\}$. The spatial Fourier components $B_n(p, q)$ may be found by finite-point analysis (sec. 11.1) on a cycle by cycle basis with respect to the spatial carrier frequency x_0.

It may be emphasized that, although $B_1(p, q)$ is not quite identical to $S(p, q)$, there is a great deal of similarity between them, particularly between their corresponding inverse Fourier transforms,

$$\mathscr{F}^{-1}\{B_1(p, q) \exp[i\phi(p, q)]\} \sim K s(x, y), \tag{12.82}$$

where $\mathscr{F}^{-1}$ denotes the inverse Fourier transform, $\sim$ denotes the similarity, and K is an arbitrary constant.

Now if the complex nonlinear filter is inserted in the frequency plane of the coherent optical processing system of fig. 12.4, the output signal may be approximated by

$$g(\alpha, \beta) \simeq \sum_{n=0}^{\infty} \int\!\!\int_{-\infty}^{\infty} B_n(p, q) F(p, q)$$
$$\times \cos\{n[x_0 p + \phi(p, q)] \exp[-i(p\alpha + q\beta)]\} \, dp \, dq \tag{12.83}$$

or

$$g(\alpha, \beta) \simeq f(x, y) * b_0(x, y) + \sum_{n=1}^{\infty} f(x, y) * b_n(x + nx_0, y)$$
$$+ \sum_{n=1}^{\infty} f(x, y) * b_n(-x - nx_0, -y), \tag{12.84}$$

where the $b_n(x, y)$ are the corresponding inverse Fourier transforms of

$\frac{1}{2}B_n(p, q) \exp[\phi(p, q)]$. From eq. (12.84), it can be seen that the first summation covers the corresponding convolutions, and that the last summation covers the crosscorrelations.

If the input signal is assumed to be

$$f(x, y) = s(x, y) + n(x, y),   \tag{12.85}$$

where $s(x, y)$ is the detecting signal, and $n(x, y)$ is additive white Gaussian noise, then the crosscorrelation terms of eq. (12.85) may be written as

$$R_{12}(\alpha, \beta) = \sum_{n=1}^{\infty} [s(x, y) + n(x, y)] * b_n(-x - nx_0, y).   \tag{12.86}$$

The crosscorrelations with respect to $n(x, y)$ and to $b_n(x, y)$ are approximately zero; also, $b_1(x, y)$ has a great deal of similarity to $s(x, y)$. That is,

$$b_1(x, y) \sim Ks(x, y),   \tag{12.87}$$

but

$$b_n(x, y) \not\sim Ks(x, y), \quad \text{for} \quad n \neq 1,   \tag{12.88}$$

where K is an arbitrary constant. Therefore the crosscorrelation of eq. (12.86) may be approximated by

$$R_{12}(\alpha, \beta) \simeq s(x, y) * b_1(-x - x_0, -y) \simeq s(x, y) * Ks(-x - x_0, -y),   \tag{12.89}$$

that is, R_{12} is proportional to the autocorrelation of $s(x, y)$.

It may be emphasized that eq. (12.88) is true in practice but not necessarily required, since the higher order crosscorrelation between $s(x, y)$ and $b_n(x, y)$ could be spatially separated from the first order crosscorrelation. Moreover, from eq. (12.89) it is clear that an optimum detection requires an optimum value of $b_1(x, y)$, which may be obtained by the linear optimization technique proposed in the previous section.

Figures 12.5 and 12.6 demonstrate the separation of signal from random noise. The input transparency consists of a signal embedded in random noise (fig. 12.5). The results of the signal detection for various values of signal/reference beam ratios, are given in fig. 12.6. The recording medium used for the filter synthesis was a Kodak 649F (D−19, 5 min.) photographic plate. These filters were synthesized at the linear optimization conditions obtained from interpolations of fig. 12.2. From fig. 12.6, it can be seen that the optimum signal detection occurs at a one-to-one signal/reference beam ratio.

In summary, we see that, instead of being restricted to the linear region

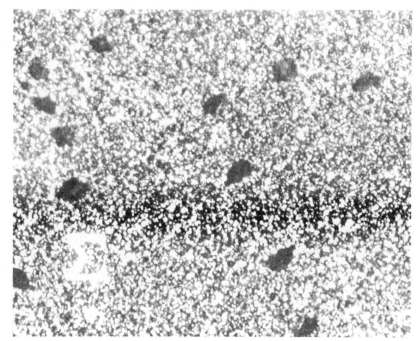

Fig. 12.5 Signal embedded in random noise.

of the transfer characteristic of the photographic emulsion, once again an optimum nonlinear spatial filter can be achieved from the system theory point of view. Although it may be difficult to apply this theory in the case of complicated signals, we may always obtain a generalized linear optimization. In the analysis of nonlinear spatial filtering, it has been shown that the signal detection (represented by the first order crosscorrelation) can be spatially separated from the unwanted higher order correlations. Therefore the ambiguity of identifying the first order from the higher order correlations can be resolved. In fact, it has also been shown that the higher order correlations are generally very weak.

To conclude this chapter, once again we have seen that the optimum wavefront recording exceeds the linear region of the H and D curve. Therefore it may be seen from sec. 8.3 that the signal-to-noise ratio, and the dynamic range that we have defined, are different from those of the conventional photographic process.

Problems

12.1 Using the T-E transfer characteristic of fig. 11.14:

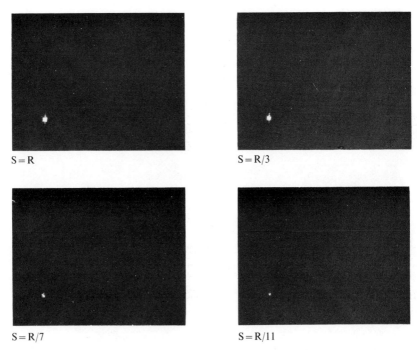

Fig. 12.6 Signal detection by means of complex spatial filtering, for various values of signal/reference beam ratios.

(a) By a trial-and-error curve-fitting process, approximate a third order polynomial to the corresponding $T\text{-}E$ curve.
(b) With the polynomial obtained in (a), compute the optimum exposures of a holographic process for the following object/reference beam ratios: $A/B=1$, $A/B=1/2$, $A/B=1/10$, and $A/B=1/30$.
(c) Determine the corresponding optimum first order linear transmittances for the reference/object beam ratios specified in (b).

12.2 With reference to the same $T\text{-}E$ transfer characteristic as described in prob. 12.1, for a certain holographic process:
(a) By means of a five-point analysis (sec. 11.1), make a rough plot of the first-order diffraction efficiency as a function of exposure, for the object/reference beam ratios $A/B=1$, $A/B=1/2$, $A/B=1/10$, and $A/B=1/30$.
(b) From the plotted curves of (a), determine the corresponding optimum exposures. Compare these results with the calculated values obtained in prob. 12.1.

12.3 Let us assume that the transmittance of a certain recording medium may be approximated by the nonlinear equation,

$T(E) = 1 + a_1 E + a_3 E^3$, for $E \geq 0$,

where E is the exposure. If the input exposure is a sinusoidal grating,

$E(x, y) = E_0 + E_1 \cos px$, for every y,

where E_0 and E_1 are arbitrary constants and p is the spatial frequency, then determine the first order and the higher order transmittance.

12.4 Repeat prob. 12.3, if the input exposure is a spatial random noise, which follows the one-sided Gaussian probability distribution

$p(E) = \dfrac{2}{\sqrt{2\pi}\, \rho^2} \exp\left(-\dfrac{E^2}{2\sigma^2} \right)$, for $E \geq 0$ and every (x, y),

where σ^2 is the variance of the Gaussian statistics. By means of the mean-square error criterion of eq. (12.14), determine the optimum linear transmittance.

12.5 In the synthesis of an optimum complex spatial filter, the Fourier transform of a detecting signal is $S(p, q)\exp[i\phi(p, q)]$. Take $S(p, q)$ as approximately uniform over the spatial bandwidth of interest, and use the T-E transfer characteristic of the filter emulsion as given in prob. 12.1 to (a) determine the approximate optimum linear transmittance of the complex filter; and (b) analyze the complex spatial filtering of fig. 12.4, to show that the output signal detection with respect to the filter synthesis in (a) is indeed optimum.

References
12.1 R. J. Kochenberger, A frequency Response Method for Analysis and Synthesizing Contactor Servomechanisms, *Trans. AIEE*, **69** (1950), 270.

12.2 J. Loeb, Un Criterium Général de Stabilité des Servomécanismes Siéges de Phénomènes Héréditaires, *Compt. Rend.* **233** (1951), 344.

12.3 A. Blaquière, Extension de la Théorie de Nyquist au cas de Caractéristiques Nonlinéaires, *Compt. Rend.* **233** (1951), 345.

12.4 A. Blaquière, *Nonlinear System Analysis*, Academic Press, New York, 1966, pp. 177 ff.

12.5 H. W. Rose, Effect of Carrier Frequency on Quality of Reconstructed Wavefronts, *J. Opt. Soc. Am.*, **55** (1965), 1565.

12.6 F. T. S. Yu, Optimal Linearization in Holography, *J. Opt. Soc. Am.*, **59** (1969), 490, and *Appl. Opt.* **8** (1969), 2483.

12.7 F. T. S. Yu, Method of Linear Optimization in the Wavefront Recording of Nonlinear Holograms, ICO Symposium on Application of Holography, Besançon, France, 1970.

12.8 F. T. S. Yu, Linear Optimization in the Synthesis of Nonlinear Spatial Filters, *IEEE Trans. Inform. Theory*, **IT-17** (1971), 524.

12.9 J. W. Goodman and G. R. Knight, Effects of Film Nonlinearities on Wavefront-Reconstruction Images of Diffuse Objects, *J. Opt. Soc. Am.*, **58** (1968), 1276.

12.10 O. Bryngdahl and A. Lohmann, Nonlinear Effects in Holography, *J. Opt. Soc. Am.*, **58** (1968), 1325.

Chapter 13 Applications of Holography

Leith and Upatnieks' revival and improvement of Gabor's holography have led to a great number of scientific applications. These have not been limited to optical wavelengths; applications are found in the microwave and radio-frequency regions (ref. 13.1), as well as in acoustics (refs. 13.2, 13.3). It would require an exhaustive effort to cover each of the many applications. This is not the main objective of this chapter. We will, however, devote some time to discussing a few of the most interesting applications in the optical region.

13.1 Microscopic Wavefront Reconstruction

The invention of holography was motivated by the hope for improvements in electron microscopy. Gabor's original work (refs. 13.4–13.6), of course, had this orientation, as did that of El-Sum (ref. 13.7).

Holographic microscopy may—at least in principle—have four advantages. First, *holographic magnifications* may be accomplished by constructing the wavefront with a very short wavelength and then reconstructing the hologram image by means of a longer wavelength. Second, the lateral resolution of the image reconstruction is limited by the size of the hologram aperture; the larger the hologram aperture, the better the resolution. Thus a high resolution image may be attainable for a large field of view. Third, wavefront aberrations in holography may be corrected to some degree by suitable recording and reconstructing procedures. Thus a nearly aberration-free hologram image may be achieved. Fourth, images of high longitudinal resolution may be achieved. Thus holographic microscopy may offer enormous depth of focus.

Several practical problems in the application of microscopic wavefront reconstruction prevent this new technique from being a serious threat to conventional microscopy at present. Figures 13.1 and 13.2 (from ref. 13.8) are examples of the results to be expected from the technique.

13.2 Multiexposure Holographic Interferometry

One of the most interesting and important of the applications of holography may be to interferometry. Holograms preserve the amplitude and phase distributions of the recording field. The same hologram can also be the record of two or more different sets of wavefronts (multiexposure holography). Therefore it is possible to produce a superposition of two or more hologram images, between which an interference image will be reconstructed. This powerful *holographic interferometry* is based on the

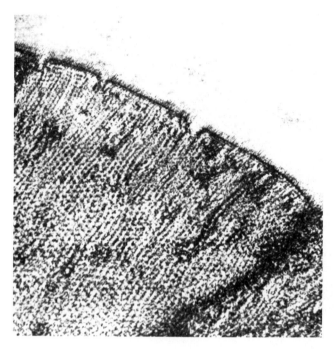

Fig. 13.1 Holographic microscopy of a fly's wing. The lateral magnification is about 60 ×. (Permission by E. N. Leith.)

principle of coherent additions of the complex wave fields (ref. 13.9).

To analyze the process, let us assume that a photographic plate has been holographically and sequentially exposed to N independent exposures. If we denote $I_n(x, y)$ as the nth irradiance of the recording, then the total exposure is

$$E(x, y) = \sum_{n=1}^{N} t_n I_n(x, y), \tag{13.1}$$

where t_n is the nth exposure time. The irradiance $I_n(x, y)$ may be written as

$$I_n(x, y) = |u_n(x, y)|^2 + |v|^2 + u_n(x, y)\, v^* + u_n^*(x, y)\, v, \tag{13.2}$$

where $u_n(x, y)$ and v are the complex light fields due to the object and reference waves, respectively.

If this multiexposure recording is properly biased in the linear region of the T-E curve of the photographic emulsion, then the transmittance of

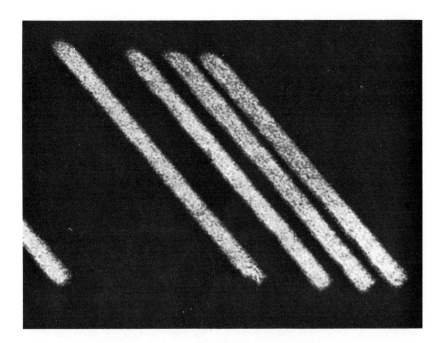

Fig. 13.2 Holographic microscopy of a test chart. The lateral magnification is about $120 \times$. The spacing between lines is about 10 microns. (Permission by E. N. Leith.)

the recorded hologram is given by

$$T(x, y) = K_1 \sum_{n=1}^{N} |u_n(x, y)|^2 + K_2 + K_3 \sum_{n=1}^{N} u_n(x, y) \, v^* + K_4 \sum_{n=1}^{N} u_n^*(x, y) \, v,$$
(13.3)

where the K's are the appropriate positive constants.

If the hologram of eq. (13.3) is illuminated by a monochromatic plane wave v, then a sequence of virtual images will be reconstructed. The resulting complex light field behind the hologram aperture may be written as

$$E_v(x, y) = |v|^2 \, K_3 \sum_{n=1}^{N} u_n(x, y),$$
(13.4)

where the subscript v denotes the virtual image diffraction. It can be seen from eq. (13.4) that the object light fields are superimposed on each other; thus they will mutually interfere.

It is obvious that, if the hologram is illuminated by a conjugate monochromatic plane wave v^*, then a sequence of real hologram images will be constructed. The complex light field due to these real images is

$$E_r(x, y) = |v^2| \, K_4 \sum_{n=1}^{N} u_n^*(x, y),$$ (13.5)

where the subscript r refers to the real image. These object light fields will also mutually interfere.

One of the most interesting applications of multiexposure holography is due to Heflinger, Wuerker, and Brooks (ref. 13.10). Examples of double-exposure holographic interferograms are given in figs. 13.3 and 13.4. The interferogram of fig. 13.3 was made by using a Q-switched pulsed ruby laser. The first pulse (i.e., the first exposure) records only the hologram of the diffuse background, and the second pulse records the hologram in

Fig. 13.3 A double exposure holographic interferogram of the shock wave of a 22-caliber bullet moving through argon gas at 1060 m/sec. The hologram was made with a Q-switched ruby laser. (Permission by L. O. Heflinger.)

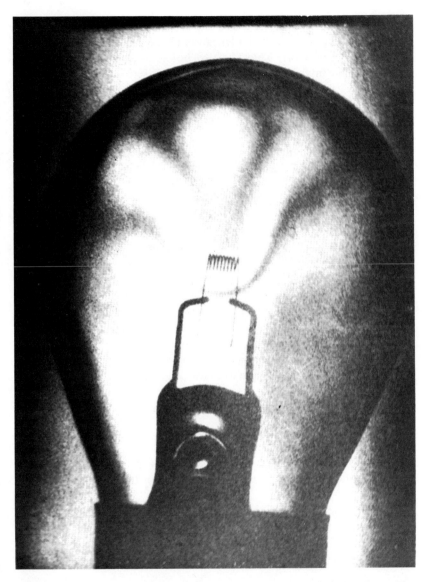

Fig. 13.4 A double exposure holographic interferogram of an incandescent lamp. The hologram was made with a Q-switched ruby laser. (Permission by L. O. Heflinger.)

which a bullet is traveling in argon gas in front of the background. The shock wave generated by the bullet causes changes in the refractive index of the surrounding gas. Thus the two hologram images will mutually interfere. This interference pattern in turn describes the shock wave behavior generated by the bullet. The resulting interference fringes form a three-dimensional pattern. The interferogram of fig. 13.4 was made in a similar fashion; one exposure was made with a cold filament and the other exposure was made with a heated filament. The distribution of the gas in the bulb is not the same for each case; thus the holographic image reconstructions will mutually interfere.

13.3 Time-Average Holographic Interferometry

The concept of multiexposure interferometry may be extended to continuous-exposure interferometry, which we will call *time-average holographic interferometry*. The principle of this technique is identical to that of multiexposure holography, except that the exposure is a continuous rather than discrete variable. That is, a large number of infinitesimal holographic recordings of the same object take place within a given exposure time. Each of these infinitesimal subholograms will produce its own holographic images, and an interference pattern will result. The earliest application of time-average holographic interferometry is the Powell and Stetson study of a vibrating object (ref. 13.11).

The analysis of this technique will now be given. Let a planar object be situated in the ξ coordinate system, in an arrangement similar to that shown in Fig. 10.7. If the object is vibrating in the $\xi\eta$ plane, the object function satisfies

$$O[\xi(t), \eta(t), \zeta'(t)] = O[\xi_0 + \xi'(t), \eta_0 + \eta'(t), \zeta'(t)], \qquad (13.6)$$

where ξ_0 and η_0 are some average coordinates of the object function, and $\xi'(t)$, $\eta'(t)$ and $\zeta'(t)$ are the time-dependent coordinate variables about $(\xi_0, \eta_0, 0)$.

By the Fresnel-Kirchhoff theory, the resulting complex light field on the recording medium may be written as

$$u(\rho; k, t) = \int\int_S O[\xi(t); k] \, E_i^+[\rho - \xi(t); k] \, d\xi(t) \, d\eta(t), \qquad (13.7)$$

where S denotes the surface integral, and where

$$E_i^+[\xi(t); k] = -\frac{i}{\lambda l} \exp\left\{ik\left[l - \zeta(t) + \frac{\xi^2(t) + \eta^2(t)}{2l}\right]\right\}$$

is the spatial impulse response.

The transmission function of the time-average hologram is

$$T(\boldsymbol{\rho}; k) = K \int_0^{\Delta t} \left[|u(\boldsymbol{\rho}; k, t)|^2 + |v|^2 + u(\boldsymbol{\rho}; k, t) v^* + u^*(\boldsymbol{\rho}; k, t) v \right] dt, \qquad (13.8)$$

where Δt is exposure time and K is an appropriate positive constant. It may be seen from eq. (13.8) that the reconstructed wavefront is indeed a summation of the wavefronts produced by the object at each of its positions. Since each wavefront and its associated subhologram is coherent with all the others, the summed holographic image will contain interference fringes.

Of special interest is the case of vibration along the optical axis only; i.e., $\xi(t) = \eta(t) = 0$, $\zeta(t) \neq 0$. The complex light field from the object of eq. (13.7) is then

$$u(\boldsymbol{\rho}; k, t) = -\frac{i}{\lambda l} \exp(ikl) \int\int_S O[\xi_0, \eta_0, \zeta'(t)] \exp[-ik\zeta'(t)]$$

$$\times \exp\left\{ \frac{k}{2l} [(x - \xi_0)^2 + (y - \eta_0)^2] \right\} d\xi\, d\eta. \qquad (13.9)$$

If we illuminate the hologram of eq. (13.8) by a monochromatic wavefront, the resultant virtual hologram image diffraction will be

$$E(\xi; k) = C \int_0^{\Delta t} O[\xi_0, \eta_0 \zeta'(t)] \exp[-ik\zeta'(t)] dt, \qquad (13.10)$$

where C is a complex constant.

If the displacement $\zeta'(t)$ is small, eq. (13.10) may be approximated by

$$E(\xi; k) = CO(\xi_0, \eta_0) \int_0^{\Delta t} \exp[-ik\zeta'(t)] dt. \qquad (13.11)$$

As an example, if the vertical motion is a sinusoidal vibration,

$$\zeta(t) = a(\xi_0, \eta_0) \cos \omega t, \quad \text{for } t \ll \Delta t, \qquad (13.12)$$

where $a(\xi_0, \eta_0)$ is the amplitude function over the planar object, then the solution of eq. (13.11) is

$$E(\xi; k) = CO(\xi_0, \eta_0) J_0[ka(\xi_0, \eta_0)], \qquad (13.13)$$

where J_0 is the zero order Bessel function.

That is to say, the resultant hologram image is the object function weighted by a zero order Bessel function, with a primary maximum value at $a(\xi_0, \eta_0) = 0$. The irradiance at each point of the resultant hologram image depends on the object amplitude function $a(\xi_0, \eta_0)$.

Examples of time-average holographic interferograms by Powell and Stetson are given in fig. 13.5. These interferograms are those of a vibrating diaphragm. The different order modes of vibration can be easily seen; by counting the number of fringes from the edge of the diaphragm, one may determine the vibration amplitude at any point.

13.4 Contour Generation

Contour generation by wavefront reconstruction processes was first reported by Hildebrand and Haines (refs. 13.12, 13.13). The two methods of contour generation are known as multisource generation and multiwavelength generation. Multisource generation is accomplished by two mutually coherent but separate point sources. Wavefront recording in this method can be obtained either by simultaneous illumination of the object (by these two coherent sources), or by double-exposure sequential illuminations of the object. For either type of illumination, interference fringes will appear on the surface of the object. For example, in the multi-source arrangement of fig. 13.6 point sources are arranged in such a way that the interference fringes follow hyperbolic curves of constant path-length difference. If this illuminated object is recorded on a plate oriented parallel to the fringe pattern of the object, surface variations of the object will give rise to contours in the hologram image.

However, the multisource technique has a practical disadvantage. It requires that the directions of the illuminations be precise. This constraint means that some parts of an irregularly shaped object may not be illuminated at all. This disadvantage can be easily removed by means of the multiwavelength technique. If the source emits only two wavelengths, then the wavefront construction will consist of two superposed patterns. If this two-wavelength hologram is illuminated by a monochromatic source, then two hologram images, with slightly different location and magnification, will be reconstructed in superposition. Thus interference takes place, and a bright and dark fringe pattern will be generated on the surface of the resultant image.

Figure 13.7 gives the reconstructed images of a U.S. quarter. The image reconstruction of fig. 13.7a was produced by a single-wavelength hologram, while the image shown in b was produced by a two-wavelength hologram. The wavelengths used for recording were two spectral lines

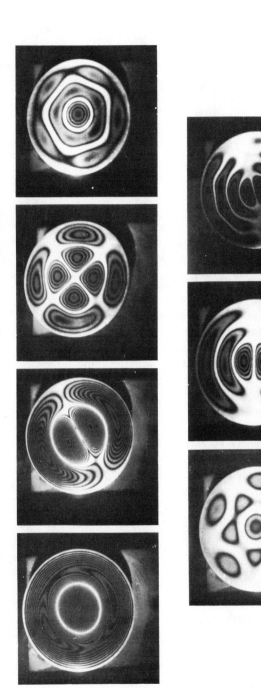

Fig. 13.5 Time-average holographic interferograms of a vibrating diaphragm in different modes of vibration. (Permission by R. L. Powell.)

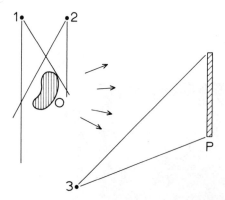

Fig. 13.6 Recording geometry for multisource contour generation. 1, 2, point source of illumination; 3, reference source; O, object; P, photographic plate.

produced by an argon laser. These two lines were separated by 65 Å and the elevations contoured in fig. 13.7*b* are spaced at about 0.02 mm.

Contour generation may also be accomplished by changing the refractive index of the region surrounding an object (fig. 13.8). A first exposure is made with only the object inside a transparent container. A second exposure takes place when the container, with the unmoved object, is filled with some appropriate liquid or gas. Thus, the optical path lengths are changed, and superposition gives rise to interference fringes on the reconstructed image.

13.5 Imaging through a Randomly Turbulent Medium

Coherent optical systems may be applied to the problem of imaging through randomly turbulent media. It is well known that the random refractive index of the turbulence interferes with light propagation. Thus conventional imaging techniques may suffer some major distortions. However, imaging by means of holographic techniques are often able to offset some of the degradation caused by the turbulence. Several holographic imaging techniques through a randomly aberrating medium were proposed by Leith and Upatnieks (ref. 13.14), Kogelnik (ref. 13.15), and Goodman et al, (ref. 13.16).

In this section we will briefly illustrate two of these techniques. First, let us consider the geometry shown in fig. 13.9. In the presence of the tur-

→

Fig. 13.7 Contour generation by the two-wavelength technique: *a*, single-wavelength image; *b*, two-wavelength image. (Permission by B. P. Hildebrand.)

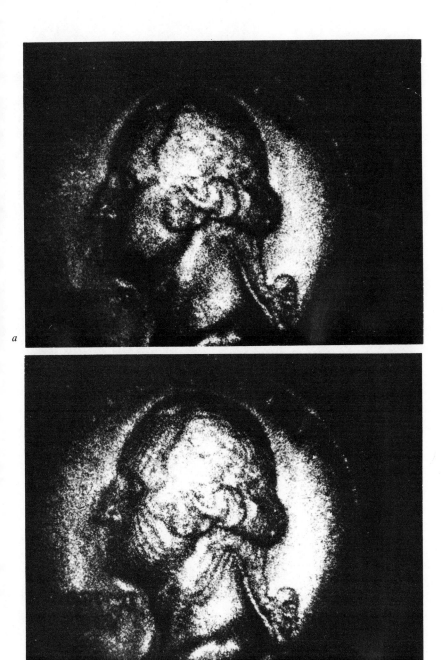

a

b

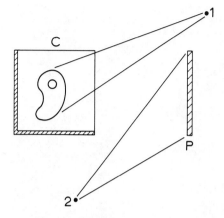

Fig. 13.8 Contour generation by means of changes in the refractive index of the fluid surrounding the object. C, transparent container; O, object; P, photographic plate; 1, illuminating source; 2, reference source.

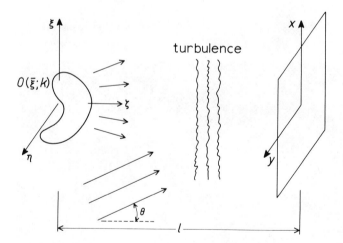

Fig. 13.9 Holographic imaging through a randomly turbulent medium.

bulent medium, the spatial impulse response is modified to give

$$E_{lN}^{+}(\xi, t, k) = E_{l}^{+}(\xi; k) N(\xi, t; k), \tag{13.14}$$

where $E_l^{+}(\xi; k)$ is the spatial impulse response in the absence of turbulence, $\xi(\xi, \eta, \zeta)$ is the coordinate system, and $N(\xi, t; k)$ is the time-dependent complex random function of the turbulence. $N(\xi, t; k)$ may be written as

$$N(\xi, t; k) = |N(\xi, t; k)| \exp[i\psi(\xi, t; k)], \tag{13.15}$$

where $\psi(\xi, t; k)$ is the random phase function. It is clear that $0 \le |N(\xi, t; k)| \le 1$, because the turbulence is assumed to be in a passive medium.

The complex light amplitude distribution on the photographic plate is determined as usual by means of the Fresnel-Kirchhoff theory,

$$u(\rho, t; k) = \iint_S O(\xi; k) E_{lN}^{+}(\rho - \xi, t; k) \, d\xi \, d\eta, \tag{13.16}$$

where $O(\xi; k)$ represents the object function, S denotes the surface integration, and $\rho(x, y)$ is the chosen coordinate system.

We would like to illustrate, however, a special case. Let there be a thin layer of turbulence located very close to the surface of the recording medium. The complex light field distributions on the recording surface due to the object and to the reference light field, respectively, may be approximated by

$$u(\rho, t; k) \simeq |N(\rho, t; k)| \exp[i\psi(\rho, t; k)] u_0(\rho; k) \tag{13.17}$$

and

$$v(\rho, t; k) \simeq |N(\rho, t; k)| \exp\{i[kx \sin\theta + \psi(\rho, t; k)]\}, \tag{13.18}$$

where

$$u_0(\rho; k) = |u_0(x, y)| \exp[i\phi(x, y)] = \iint_S O(\xi; k) E_l^{+}(\rho - \xi; k) \, d\xi \, d\eta$$

is the object light field distribution without the effect of the turbulence. The irradiance of the wavefront recording is therefore

$$
\begin{aligned}
I(\rho, t; k) &= [u(\rho, t; k) + v(\rho, t; k)] [u(\rho, t; k) + v(\rho, t; k)]^* \\
&= |N(\rho, t; k)|^2 \{1 + |u_0(x, y)|^2 + 2|u(x, y)| \cos[kx \sin\theta + \phi(x, y)]\}.
\end{aligned}
\tag{13.19}
$$

The transmittance of the recorded hologram may be shown to be

$$T(\boldsymbol{\rho}; k) = K \int^{\Delta t} I(\boldsymbol{\rho}, t; k) \, dt = K \{ 1 + |u_0(x, y)|^2$$
$$+ 2|u(x, y)| \cos[kx \sin\theta + \phi(x, y)] \int_0^{\Delta t} |N(\boldsymbol{\rho}, t; k)|^2 \, dt \}, \qquad (13.20)$$

where K is a proportionality constant and Δt is the exposure time.

If the time integral of eq. (13.20) is relatively *uniform* over the recording aperture, then the transmittance function may be written as

$$T(\boldsymbol{\rho}; k) = K \{ 1 + |u_0(x, y)|^2 + 2|u(x, y)| \cos[kx \sin\theta + \phi(x, y)] \} |N|^2 \, \Delta t.$$
$$(13.21)$$

Thus, for this special case, the random turbulence does not affect the image.

Typical results of holography through a randomly turbulent medium are shown in fig. 13.10. The hologram was made when a stationary phase-

Fig. 13.10 Holographic imaging through a randomly phase-distorting medium.

distorting medium (an ordinary shower glass) was introduced near the recording aperture. For comparison, fig. 13.11 shows the result obtained by replacing the recording medium by a conventional positive lens. We see that the image formed by the conventional technique suffers major distortion, while the holographic image has suffered only minor degradation.

If a layer of stationary phase-distorting medium is inserted between the object and the recording medium in the manner shown in fig. 13.12, then the resultant complex light field scattered from the distorting medium is

$$u(\boldsymbol{\rho}; k) = \iint\limits_{S_2} u_0(\xi', \eta') \exp\left[-i\psi(\xi', \eta')\right] E_{l_2}^+(\boldsymbol{\rho}-\boldsymbol{\xi}'; k) \, d\xi' \, d\eta', \qquad (13.22)$$

where

$$u_0(\xi', \eta') = \iint\limits_{S_1} O(\boldsymbol{\xi}; k) E_{l_1}^+(\boldsymbol{\xi}'-\boldsymbol{\xi}, k) \, d\xi \, d\eta$$

Fig. 13.11 Conventional imaging through a randomly phase-distorting medium.

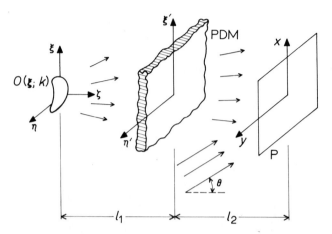

Fig. 13.12 Holographic imaging through a phase-distorting medium. O, object function; PDM, phase-distorting medium; P, photographic plate. Note the position of the plane reference wave.

is the object light field distribution on the ξ' coordinate plane, $\psi(\xi', \eta')$ is the stationary random phase function, S_1 and S_2 denote the surface integrations over the ξ and ξ' coordinate planes, and $E_{l_1}^{+}$ and $E_{l_2}^{+}$ are the respective spatial impulse responses from the ξ to ξ' coordinate systems and from the ξ' to ρ coordinate systems.

If a reference plane wave is applied *behind* the distorting medium, as shown in the same figure, then the transmittance of the recorded hologram will be

$$T(\rho; k) = K[|u|^2 + |v|^2 + uv^* + u^*v], \tag{13.23}$$

where K is the proportionality constant and $v = \exp(ikx \sin\theta)$ is the plane reference wave.

If this hologram described by eq. (13.23) is illuminated by an oblique monochromatic plane wave of negative θ, as shown in fig. 13.13, then the real image diffraction at a distance l_2 behind the hologram aperture can be shown to be proportional to the conjugate wave field $u^*(\alpha', \beta')$. If the same random phase-distorting medium is inserted at the σ' coordinate system, then at a distance l_1 behind the phase-distorting plate the complex light field can be shown to be

$$E(\sigma; k) = K_1 O^*(\sigma; k), \tag{13.24}$$

which is the conjugate of the object function. The phase distortion has been completely eliminated, and a distortion-free image is formed. How-

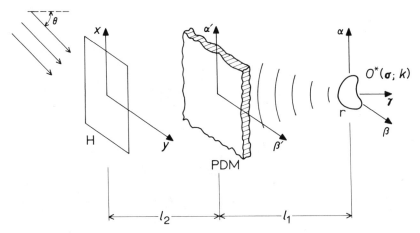

Fig. 13.13 Image reconstruction by means of identical phase compensation. H, hologram; r, real image. Plane monochromatic illumination, negative θ. The phase-distorting medium PDM is the same used in the construction shown in the previous figure.

ever, extremely accurate positioning of the phase-distorting medium is required during the holographic reconstruction process; otherwise the reconstructed image may not be completely free from distortion.

13.6 Coherent Optical Target Recognition through a Randomly Turbulent Medium

It is well recognized that conventional imaging is unsuccessful when a randomly phase-distorting medium closely overlays the observation system. However, in the last section, we have seen that holographic imaging is a possible solution, since identical phase distortions occurring in the signal and reference fields will be cancelled during wavefront recording. In practice, this means that a bright area or point must be positioned near the object being imaged. However, in this section (based upon refs. 13.17, 13.18), we will consider what degree of recognition can be attained even though it may not be possible to generate such a reference field.

The complex light field scattered from the target through a turbulent medium under the "far-field" geometry of fig. 13.14 may be written as the convolution integral

$$u(\rho, t; k) = \iint_S O(\xi; k)\, E_{IN}^+(\rho - \xi, t; k)\, d\xi\, d\eta, \tag{13.25}$$

where $O(\xi; k)$ is the target function, E_{IN}^+ is defined by eq. (13.14), and S denotes surface integration.

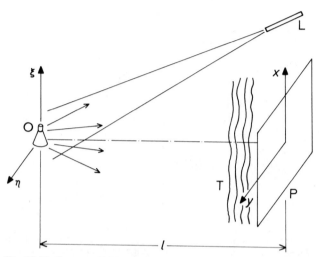

Fig. 13.14 Geometry of far-field object-spectrum recording. O, object; T, turbulence; L, laser; P, photographic plate.

If it is assumed that the separation l is very large, then eq. (13.25) may be approximated by

$$u(\boldsymbol{\rho}, t; k) \simeq -\frac{i}{\lambda l} \exp\left[ik\left(l+\frac{\rho^2}{2l}\right)\right] \iint_S O(\boldsymbol{\xi}; k)\, N(\boldsymbol{\rho}-\boldsymbol{\xi}, t; k)$$

$$\times \exp\left[-i\frac{k}{l}(x\xi+y\eta)\right] d\xi\, d\eta, \tag{13.26}$$

which is the Fourier transform of the convolution of the object function and the complex random function N.

If the turbulence is close to the surface of the recording aperture, then eq. (13.26) may be further approximated by

$$u(\boldsymbol{\rho}, t; k) \simeq -\frac{i}{\lambda l} \exp\left[ik\left(l+\frac{\rho^2}{2l}\right)\right] N(\boldsymbol{\rho}, t; k)\, \mathscr{F}[O(\boldsymbol{\xi}; k)]. \tag{13.27}$$

Thus the transmittance of the recorded photographic plate may be written as

$$T(\boldsymbol{\rho}; k) = K\, |\mathscr{F}[O(\boldsymbol{\xi}; k)]|^2 \int_0^{\Delta t} |N(\boldsymbol{\rho}, t; k)|^2\, dt, \tag{13.28}$$

where K is a proportionality constant, and Δt is the exposure time.

If the ensemble time average of eq. (13.28) is uniform over the recording aperture, then the above equation may be reduced to

$$T(\mathbf{\rho}; k) = K |N|^2 \, \Delta t \, |\mathscr{F}[O(\xi; k)]|^2, \tag{13.29}$$

which is proportional to the power spectrum of the target function. If the turbulence is a phase-only distorting medium, then eq. (13.29) may be written as

$$T(\mathbf{\rho}; k) = K \, \Delta t \, |\mathscr{F}[O(\xi; k)]|^2, \tag{13.30}$$

which, it can be seen, is completely free of phase disturbance.

Since the target (object) is far field, the power spectrum of the object function spreads all over the recording surface. In general, it is difficult to recognize the shape of the object by means of the recorded power spectrum. However, it may be possible to discover the shape of the object by the corresponding autocorrelation. To do this, we illuminate the recorded photographic plate by a monochromatic plane wave, as shown in fig. 13.15. Then the autocorrelation function of the recorded object may be obtained at the back focal plane of the transform lens, such that

$$R(\mathbf{\sigma}; k) = \iint T(\mathbf{\rho}; k) \, e^{-i(x\alpha + y\beta)} \, dx \, dy, \tag{13.31}$$

where $\mathbf{\sigma}(\alpha, \beta)$ is the coordinate system at the back focal plane.

To illustrate the effect of a close-lying phase distortion on an object's autocorrelation function relative to its effect on the conventional image, the two techniques were compared using a 6328 Å He-Ne laser for illumination. The far field condition of fig. 13.14 was simulated by means of a transform lens. A shower glass near the front of the recording aperture

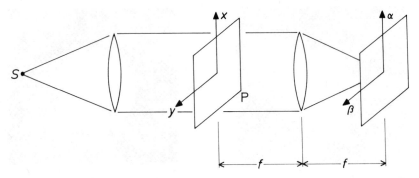

Fig. 13.15 Geometry for obtaining the autocorrelation function. S, monochromatic point source; P, recorded film.

produced the close-lying phase distortion. Conventional imaging was obtained by placing a positive lens behind the shower glass. The images with and without the intervening phase distortion are shown in figs. 13.16a and 13.16b, respectively.

For the autocorrelation technique, the power spectrum was recorded directly on a photographic plate. The developed transparency was then Fourier transformed by a lens, and the autocorrelation function was obtained at the back focal plane. The autocorrelation photographs with and without the intervening phase distortion are shown in figs. 13.17a and 13.17b. Comparison of figs. 13.16a and 13.17b shows that the autocorrelation method is significantly less sensitive to phase distortion than is the conventional imaging technique. Similar successful tests have been made on several other object shapes, and no difficulty was encountered in

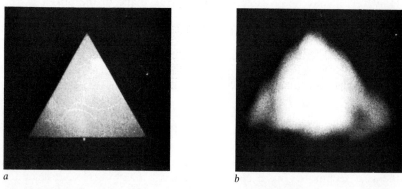

a   b

Fig. 13.16 Conventional imaging: a, without phase-distorting medium; b, with phase-distorting medium.

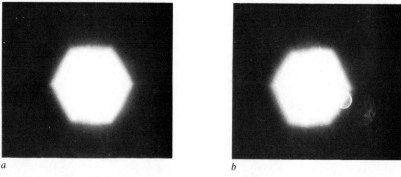

a   b

Fig. 13.17 Autocorrelation technique: a, without phase-distorting medium; b, with phase-distorting medium.

associating any of the distorted autocorrelation functions with their respective undistorted autocorrelation and original object functions.

We see that successful elimination of phase distortion may be accomplished by detecting the corresponding power spectrum of the object rather than the object function. Although there exists no unique relation between the object function and the corresponding power spectrum, for a small number of distinguishable objects it may be possible to recognize the individual objects by means of their corresponding autocorrelation functions. Of course this scheme eliminates the original phase information. However, for strong turbulence, the phase distortion hinders far more than the phase information helps.

References

13.1 G. L. Tyler, The Bistatic, Continuous-Wave Radar Method for the Study of Planetary Surfaces, *J. Geophys. Res.*, **71** (1966), 1559.

13.2 R. K. Mueller and N. K. Sheridan, Sound Holograms and Optical Reconstruction, *Appl. Phys. Letters*, **9** (1966), 328.

13.3 A. F. Metherell et al., Introduction to Acoustical Holography, *J. Acoust. Soc. Am.*, **42** (1967), 733.

13.4 D. Gabor, A New Microscope Principle, *Nature*, **161** (1948), 777.

13.5 D. Gabor, Microscopy by Reconstructed Wavefronts, *Proc. Roy. Soc.*, ser. A, **197** (1949), 454.

13.6 D. Gabor, Microscopy by Reconstructed Wavefronts, II, *Proc. Phys. Soc.*, ser. B, **64** (1951), 449.

13.7 H. M. A. El-Sum, Reconstructed Wavefront Microscopy, doctoral dissertation, Stanford University, 1952 (available from University Microfilms, Ann Arbor, Michigan).

13.8 E. M. Leith and J. Upatnieks, Microscopy by Wavefront Reconstruction, *J. Opt. Soc. Am.*, **55** (1965), 569.

13.9 D. Gabor et al., Optical Image Synthesis (Complex Amplitude Addition and Subtraction) by Holographic Fourier Transformation, *Phys. Letters*, **18** (1965), 116.

13.10 L. O. Heflinger, R. F. Wuerker, and R. E. Brooks, Holographic Interferometry, *J. Appl. Phys.*, **37** (1966), 642.

13.11 R. L. Powell and K. A. Stetson, Interferometric Vibration Analysis by Wavefront Reconstruction, *J. Opt. Soc. Am.*, **55** (1965), 1593.

13.12 B. P. Hildebrand and K. A. Haines, The Generation of Three-Dimensional Contour Maps by Wavefront Reconstruction, *Phy. Letters*, **21** (1966), 422.

13.13 B. P. Hildebrand and K. A. Haines, Multiple-Wavelength and

Multiple-Source Holography Applied to Contour Generation, *J. Opt. Soc. Am.*, **57** (1967), 155.

13.14 E. N. Leith and J. Upatnieks, Holographic Imagery through Diffusing Media, *J. Opt. Soc. Am.*, **56** (1966), 523.

13.15 H. Kogelnik, Holographic Image Projection through Inhomogeneous Media, *Bell System Tech. J.*, **44** (1965), 2451.

13.16 J. W. Goodman et al., Wavefront-Reconstruction Imaging through Random Media, *Appl. Phys. Letters*, **8** (1966), 311.

13.17 F. T. S. Yu and H. W. Rose, Coherent Optical Target Recognition through a Randomly Turbulent Medium, *J. Opt. Soc. Am.*, **59** (1969), 474.

13.18 H. W. Rose, T. L. Williamson, and F. T. S. Yu, Coherent Optical Target Recognition through a Phase Distorting Medium, *Appl. Opt.*, **10** (1971), 515.

Appendix A — Solution of a Set of Nonhomogeneous Kolmogorov Differential Equations

In order to solve the Kolmogorov differential equations

$$\frac{dP_M(t)}{dt} = -Mq(t)\,P_M(t), \quad \text{for} \quad m=M, \tag{A.1}$$

$$\frac{dP_m(t)}{dt} = -mq(t)\,P_m(t) + (m+1)\,q(t)\,P_{m+1}(t), \quad \text{for} \quad m<M, \tag{A.2}$$

with the initial conditions

$$P_M(0)=1, \tag{A.3}$$

$$P_m(0)=0, \quad \text{for} \quad m<M, \tag{A.4}$$

it is convenient to obtain a partial differential equation for the transition probability generating function

$$\psi_m(z,t) = \sum_{m=0}^{M} z^m P_m(t), \quad \text{for} \quad |z|\le 1. \tag{A.5}$$

The solution of eq. (A.1) is

$$P_M(t)=\exp\!\left[-M\int_0^t q(t')\,dt'\right], \tag{A.6}$$

and, if we apply the initial condition (A.3), then

$$\lim_{t\to 0}\int_0^t q(t')\,dt'=0. \tag{A.7}$$

Since the probability generating function $\psi_m(z,t)$ is a function of z and t, the partial derivatives of $\psi_m(z,t)$ with respect to z and to t are

$$\frac{\partial \psi_m(z,t)}{\partial z} = \sum_{m=0}^{M} P_m(t)\,mz^{m-1}, \tag{A.8}$$

$$\frac{\partial \psi_m(z,t)}{\partial t} = \sum_{m=0}^{M} \frac{\partial P_m(t)}{\partial t}\,z^m. \tag{A.9}$$

If we substitute eqs. (A.8) and (A.9) in eq. (A.2), we obtain the partial differential equation

$$\frac{\partial \psi_m(z,t)}{\partial t} = (1-z)\,q(t)\,\frac{\partial \psi_m(z,t)}{\partial z}, \tag{A.10}$$

with the boundary condition

$$\psi_M(z, 0) = z^M. \tag{A.11}$$

In order to solve eq. (A.10) we can use the following theorem.

Given the partial differential equation

$$\frac{\partial \psi_m(z, t)}{\partial t} = a(z, t) \frac{\partial \psi_m(z, t)}{\partial z}, \tag{A.12}$$

subject to the boundary condition (A.11), let us define a new function $u(z, t)$ such that the solution $z(t)$ of the differential equation

$$\frac{dz}{dt} + a(z, t) = 0 \tag{A.13}$$

satisfies

$$u(z, t) = \text{constant}. \tag{A.14}$$

Define a function g such that

$$g(z) = u(z, S), \tag{A.15}$$

and let $g^{-1}(x)$ be the inverse function of $g(z)$, so that

$$g^{-1}(x) = z \text{ if } x = g(z). \tag{A.16}$$

Then

$$\psi_m(z, t) = \{g^{-1}[u(z, t)]\}^M. \tag{A.17}$$

The proof of this theorem can be found in Syski's book.* From eq. (A.10) the form of $a(z, t)$ is

$$a(z, t) = q(t)(1 - z). \tag{A.18}$$

The ordinary differential equation is

$$\frac{dz}{dt} + q(t)(1 - z) = 0. \tag{A.19}$$

This may be written as

$$\frac{dz}{1-z} + q(t) \, dt = 0. \tag{A.20}$$

---

* R. Syski, *Introduction to Congestion Theory in Telephone Systems*, Oliver and Boyd Publishing Company, London, 1960, pp. 696ff.

By integrating the above equation, one obtains

$$-\ln(1-z)+\rho(t)=\text{constant}, \tag{A.21}$$

$$\rho(t)=\int_0^t q(t')\,dt'.$$

Any solution of (A.19) is therefore satisfied by

$$u(z,t)=\text{constant}, \tag{A.22}$$

if we define

$$u(z,t)=-\ln(1-z)+\rho(t). \tag{A.23}$$

To solve for $g^{-1}(x)$, we write

$$x=u(z,S)=-\ln(1-z)+\rho(S), \tag{A.24}$$

which implies

$$1-z=e^{[\rho(S)-x]} \tag{A.25}$$

from eq. (A.16), which implies

$$g^{-1}(x)=z=1-e^{[\rho(S)-x]}. \tag{A.26}$$

Therefore

$$g^{-1}[u(z,t)]=1-e^{[\rho(S)+\ln(1-z)-\rho(t)]}. \tag{A.27}$$

This can be written

$$g^{-1}[u(z,t)]=1-(1-z)\,e^{\rho(S)-\rho(t)}. \tag{A.28}$$

From eq. (A.17), the transition probability generating function is therefore

$$\psi_m(z,t)=[1-(1-z)\,e^{\rho(S)-\rho(t)}]^M. \tag{A.29}$$

Using the binomial expansion, eq. (A.29) can be written as

$$\psi_m(x,t)=\sum_{m=0}^M \frac{M!}{(M-m)!\,m!}\,[1-e^{\rho(S)-\rho(t)}]^{M-m}\,e^{[\rho(S)-\rho(t)]m}z^m. \tag{A.30}$$

Therefore, the conditional probability $P_m(t)$ can be identified from eq. (A.5) as

$$P_m(t)=\frac{M!}{(M-m)!\,m!}\,[1-e^{\rho(S)-\rho(t)}]^{M-m}\,e^{[\rho(S)-\rho(t)]m}. \tag{A.31}$$

If we let $S = 0$, and from eq. (A.7),

$$\rho(0) = 0, \tag{A.32}$$

then substitution of (A.32) in eq. (A.31) yields

$$P_m(t) = \frac{M!}{(M-m)! \, m!} [1 - e^{-\rho(t)}]^{M-m} \, e^{-m\rho(t)}, \quad \text{for} \quad m \le M. \tag{A.33}$$

The mean of m, $\bar{m} = \sum_{m=0}^{M} m P_m(t)$, can be obtained from the differentiation of the transition probability generating function of eq. (A.5), i.e.,

$$\bar{m} = \frac{\partial \psi_m(z, t)}{\partial z}\bigg|_{z=1} = \sum_{m=0}^{M} m P_m(t) \, z^{m-1}\bigg|_{z=1}. \tag{A.34}$$

By substituting eq. (A-29) in eq. (A.34), we obtain

$$\bar{m} = M e^{-\rho(t)} \tag{A.35}$$

Similarly, the second moment of $P_m(t)$ can be obtained from the second differentiation of the transition probability generating function by z,

$$\frac{\partial^2 \psi_m(z, t)}{\partial z^2}\bigg|_{z=1} = \sum_{m=0}^{M} m(m-1) \, P_m(t) \, z^{m-2}\bigg|_{z=1} = \overline{m^2} - \bar{m}. \tag{A.36}$$

Accordingly, from eq. (A-29) we have

$$\frac{\partial^2 \psi_m(z, t)}{\partial z^2}\bigg|_{z=1} = M(M-1) \, e^{-2\rho(t)}.$$

Then the second moment is

$$\overline{m^2} = \frac{\partial^2 \psi_m(z, t)}{\partial z^2} + m = [(M-1) \, e^{-\rho(t)} + 1] \, M e^{-\rho(t)}. \tag{A.37}$$

Therefore, the variance is

$$\sigma^2 = \overline{m^2} - \bar{m}^2 = M e^{-\rho(t)} (1 - e^{-\rho(t)}). \tag{A.38}$$

Appendix B The Fresnel-Kirchhoff Theory or Huygens' Principle

According to Huygens' principle, the amplitude observed at a point p' of a coordinate system $\sigma(\alpha, \beta, \gamma)$, due to a light source located in another coordinate system $\rho(x, y, z)$, as shown in fig. B.1, may be calculated by assuming that each point of the light source is an infinitesimal spherical radiator. Thus, the complex light amplitude $E_l^+(\rho; k)$ contributed by a point p in the ρ coordinate system can be considered to be that from an unpolarized monochromatic point source, such that

$$E_l^+ = -\frac{1}{\lambda r} \exp[i(kr - \omega t)], \tag{B.1}$$

where λ, k, ω, are the wavelength, wave number, and angular frequency, respectively, of the point source, and r is the distance between the point source and the point of observation:

$$r = [(l + \gamma - z)^2 + (\alpha - x)^2 + (\beta - y)^2]^{1/2}. \tag{B.2}$$

If the separation l of the two coordinate systems is assumed to be large compared to the magnitudes of ρ and σ, then r may be approximated by l in the denominator of eq. (B.1), and by

$$r = \left[l + \gamma - z + \frac{(\alpha - x)^2}{2l} + \frac{(\beta - y)^2}{2l} \right]^{1/2} \tag{B.3}$$

in the exponent. Therefore, eq. (B.1) may be written as

$$E_l^+(\sigma - \rho; k) \simeq -\frac{i}{\lambda l} \exp\left\{ ik\left[l + \gamma - z + \frac{(\alpha - x)^2}{2l} + \frac{(\beta - y)^2}{2l} \right] \right\}, \tag{B.4}$$

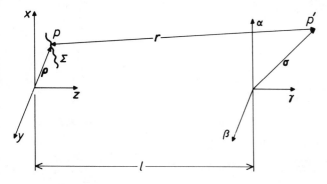

Fig. B.1 The coordinate systems.

where the time-dependent exponential has been dropped for convenience.

Furthermore, if the point of radiation and the point of observation are intercharged, then the complex light amplitude observed at $\rho(x, y, z)$ is

$$E_l^-(\rho - \sigma; k) \simeq -\frac{i}{\lambda l} \exp\left\{ ik\left[l + z - \gamma + \frac{(x - \alpha)^2}{2l} + \frac{(y - \beta)^2}{2l} \right] \right\}. \tag{B.5}$$

It is clear that eqs. (B.4) and (B.5) represent the free-space radiation from a monochromatic point source. They are also called *free-space impulse responses*.

Therefore, the complex amplitude produced at the σ coordinate system by a monochromatic radiating surface located in the ρ coordinate system can be written as

$$E(\sigma - \rho; k) = \int \int_\Sigma u(\rho; k) \, E_l^+(\sigma - \rho; k) \, d\Sigma, \tag{B.6}$$

where $u(\rho; k)$ is the complex field of the monochromatic radiating surface, and Σ denotes the surface integral.

Index